THIS BOOK ON LOAN
PROPERTY OF
DUNBARTON HIGH SCHOOL

DATE MADE	STUDENT'S NAME	HOME ROOM NO.	ISSUED BY:	CONDITION
	Sarah Robb			

053-00

THIS BOOK ON LOAN
PROPERTY OF
DUNBARTON HIGH SCHOOL

DATE SEP'L	STUDENT'S NAME	HOME ROOM NO.	ISSUED BY:	CONDITION

9

Scott

ONTARIO

ADDISON-WESLEY

Foundations of Mathematics 10

Addison-Wesley Secondary Mathematics Authors
Robert Alexander
Elizabeth Ainslie
Paul Atkinson
Maurice Barry
Cam Bennet
Barbara J. Canton
Ron Coleborn
Fred Crouse
Garry Davis
Jane Forbes
George Gadanidis
Liliane Gauthier
Florence Glanfield
Katie Pallos-Haden
Peter J. Harrison
Carol Besteck Hope
Terry Kaminski
Brendan Kelly
Stephen Khan
Ron Lancaster
Duncan LeBlanc
Kevin Maguire
Rob McLeish
Jim Nakamoto
Nick Nielsen
Linda Rajotte
Brent Richards
Margaret Sinclair
Kevin Spry
David Sufrin
Paul Williams
Elizabeth Wood
Rick Wunderlich
Leanne Zorn

Robert Alexander
Mathematics Teacher and Consultant
Richmond Hill

Barbara J. Canton
Head of Mathematics
Loyalist Collegiate and
Vocational Institute
Kingston
Limestone District School Board

Peter J. Harrison
Curriculum Coordinator: Mathematics
Toronto District School Board

Nick Nielsen
Mathematics Teacher
Alternative Scarborough
Education 2 (ASE 2)
Toronto District School Board

Margaret Sinclair
Vice Principal
Dante Alighieri Academy
Toronto

Kevin Spry
Head of Mathematics
Centre Dufferin District High School
Shelburne

Addison
Wesley

Toronto

Publisher
Claire Burnett

Managing Editor
Enid Haley

Product Manager
Reid McAlpine

Senior Consulting Editor
Lesley Haynes

Coordinating Editor
Mei Lin Cheung

Marketing Manager
Dawna Day Harris

Editorial Contributors
Rosina Daillie
Annette Darby
Gay McKellar
Tony Rodrigues
Christina Yu

Design/Production
Pronk&Associates

Art Direction
Pronk&Associates

Electronic Assembly/Technical Art
Pronk&Associates

Copyright © 2000 Pearson Education Canada Inc., Toronto, Ontario

ISBN: 0–201–68484–5
ISBN: 0–201–71066–8

This book contains recycled product and is acid free.

Printed and bound in Canada

1 2 3 4 5   BP   04 03 02 01 00

REVIEWERS

Contents

Contents

5 Similar Triangles and Trigonometry

6 Introduction to Quadratic Functions

Contents

7 From Algebra to Quadratic Equations

8 Analysing Quadratic Functions

The Ontario 9 – 12 Mathematics curriculum reflects important content changes at each grade level, with a new emphasis on technology. Here is how *Addison-Wesley Foundations of Mathematics 10* structures the grade 10 Applied course for greatest success with students.

Program Overview

1   Introduction to Functions

This informal introduction to functions reviews linear relationships from grade 9. It introduces graphing calculators and motion sensors as tools for exploring relationships.

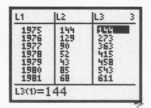

2   Linear Functions
AND
3   Linear Systems

These chapters cover linear functions, piecewise linear functions, and the solutions of linear systems. The work with piecewise linear functions is new for students. The chapter presents two methods of displaying piecewise linear functions on a graphing calculator; this responds to the curriculum requirement for the use of technology.

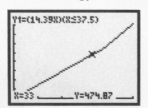

4   Proportional Reasoning

Research has shown that proportional reasoning skills are difficult for many people to acquire, and that some people never acquire them at all.

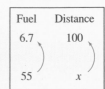

For this reason, this chapter is based on the recommendation in *Research Ideas for the Classroom: Middle Grades Mathematics*, published by the National Council of Teachers of Mathematics. Emphasis is placed on identifying proportional relationships and their properties, interpreting the properties in real situations, and using the properties to solve problems.

5   Similar Triangles and Trigonometry

Students start by examining similar triangles, and investigating connections among the measures of sides of similar right triangles. Right triangle trigonometry

is introduced with and without dynamic geometry software.

6 Introduction to Quadratic Functions

This introduction to quadratic functions emphasizes data collection activities. Students use a graphing calculator:

- to determine equations of parabolas of best fit using technology and estimation
- to solve quadratic equations
- to determine maximum/minimum values of quadratic functions

The next two chapters establish other methods for solving quadratic equations and determining maximum/minimum values.

7 From Algebra to Quadratic Equations

This chapter develops the algebraic skills for solving quadratic equations by factoring. Topics include:

- multiplying binomials
- factoring trinomials
- difference of squares
- binomial squares

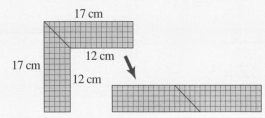

8 Analysing Quadratic Functions

This chapter develops quadratic functions with equations in the form $y = a(x - p)^2 + q$, and compares this equation with $y = ax^2 + bx + c$. Problems involving quadratic functions are presented with the equations or graphs given, as required by the curriculum.

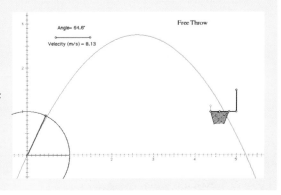

To the Student

Welcome to *Addison-Wesley Foundations of Mathematics 10*! This introduction explains how each chapter of this book is organized.

Mathematical Modelling

At the start of each chapter, a short **Mathematical Modelling** section introduces a problem. You may not have the skills you need to solve the problem when you first see it.

The chapter presents the problem again, in a later section, after developing related skills and concepts.

These questions encourage you to reflect on the problem.

This tells you when the problem reappears in the chapter.

FYI Visit refers you to our web site, where you can connect to other sites with information related to the chapter problem.

Preparing for the Chapter

Before you start new content, or explore the **Mathematical Modelling** problem in depth, you have an opportunity to review.

In this Preparing for the Chapter, you review graphing by considering the area and perimeter of rectangles.

This list tells you what skills or knowledge you need for the chapter.

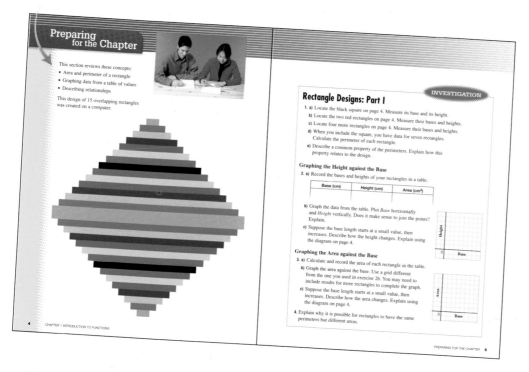

If you can complete all the exercises successfully, you're ready to proceed. If you have any difficulties, your teacher may select some activities or exercises from the Teacher's Resource Book to help you review.

Short review exercises also appear later in the chapter, for specific skills needed in a section.

Numbered Sections

The numbered sections in the chapter develop new mathematical concepts
and skills. These teaching sections offer a variety of approaches.

Investigations lead you to discover new concepts and the thinking
behind them. Investigations might involve:

- concrete materials
- an application
- graphing technology
- geometry software

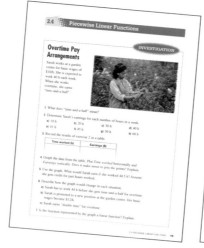

Examples with full solutions demonstrate new methods. Use the Examples
as a reference when you're completing homework.

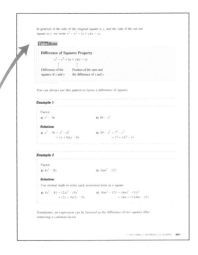

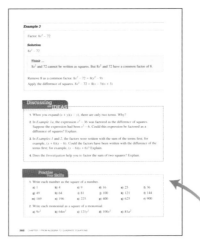

Look for the **Take Note** boxes. They highlight important results.

Practise Your Skills helps you review prerequisite skills that are
essential for this section.

Discussing the Ideas lets you talk about the Investigations and Examples with your teacher and other students. This can help to clarify your understanding.

Exercises help reinforce your understanding.

A exercises involve the simplest skills of the section.

B exercises usually require several steps, and they may involve applications.

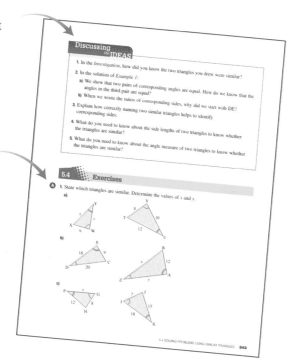

C exercises are more thought-provoking. They may call on previous knowledge or foreshadow upcoming work.

Communicating the Ideas helps you confirm that you understand important concepts. It may involve writing or drawing a diagram.

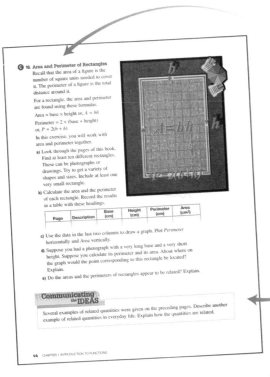

Focus Exercises

Some Exercise sections contain **Focus** exercises. They highlight connections to other subject areas, or to specific applications.

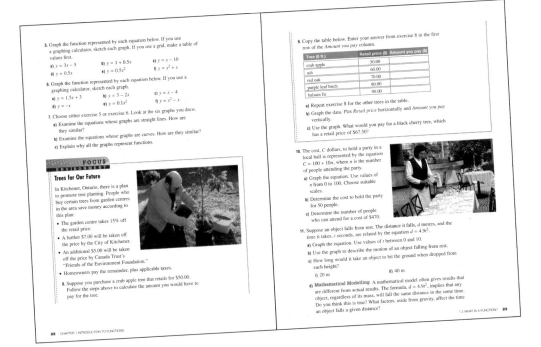

5. Graph the function represented by each equation below. If you use a graphing calculator, sketch each graph. If you use a grid, make a table of values first.

a) $y = 3x - 5$ b) $y = 1 + 0.5x$ c) $y = x - 10$

d) $y = 0.5x$ e) $y = 0.5x^2$ f) $y = x^2 + x$

6. Graph the function represented by each equation below. If you use a graphing calculator, sketch each graph.

a) $y = 1.5x + 3$ b) $y = 3 - 2x$ c) $y = x - 4$

d) $y = -x$ e) $y = 0.1x^2$ f) $y = x^2 - x$

7. Choose either exercise 5 or exercise 6. Look at the six graphs you drew.

a) Examine the equations whose graphs are straight lines. How are they similar?

b) Examine the equations whose graphs are curves. How are they similar?

c) Explain why all the graphs represent functions.

FOCUS
ENVIRONMENT

Trees for Our Future

In Kitchener, Ontario, there is a plan to promote tree planting. People who buy certain trees from garden centres in the area save money according to this plan:

• The garden centre takes 15% off the retail price.

• A further $7.00 will be taken off the price by the City of Kitchener.

• An additional $5.00 will be taken off the price by Canada Trust's "Friends of the Environment Foundation."

• Homeowners pay the remainder, plus applicable taxes.

8. Suppose you purchase a crab apple tree that retails for $50.00. Follow the steps above to calculate the amount you would have to pay for the tree.

9. Copy the table below. Enter your answer from exercise 8 in the first row of the *Amount you pay* column.

Tree (6 ft.)	Retail price ($)	Amount you pay ($)
crab apple	50.00	
ash	60.00	
red oak	70.00	
purple leaf birch	80.00	
balsam fir	90.00	

a) Repeat exercise 8 for the other trees in the table.

b) Graph the data. Plot *Retail price* horizontally and *Amount you pay* vertically.

c) Use the graph. What would you pay for a black cherry tree, which has a retail price of $67.50?

10. The cost, C dollars, to hold a party in a local hall is represented by the equation $C = 100 + 10n$, where n is the number of people attending the party.

a) Graph the equation. Use values of n from 0 to 100. Choose suitable scales.

b) Determine the cost to hold the party for 50 people.

c) Determine the number of people who can attend for a cost of $470.

11. Suppose an object falls from rest. The distance it falls, d metres, and the time it takes, t seconds, are related by the equation $d = 4.9t^2$.

a) Graph the equation. Use values of t between 0 and 10.

b) Use the graph to describe the motion of an object falling from rest.

c) How long would it take an object to hit the ground when dropped from each height?

i) 20 m   ii) 40 m

d) **Mathematical Modelling** A mathematical model often gives results that are different from actual results. The formula, $d = 4.9t^2$, implies that any object, regardless of its mass, will fall the same distance in the same time. Do you think this is true? What factors, aside from gravity, affect the time an object falls a given distance?

Complete all the exercises within the border for an in-depth look at this situation.

Other Focus exercises include:

• Sports
• Design
• Consumer
• Technology
• Science
• Employment
• Geography

Mathematical Modelling

In each chapter, a numbered section returns to the **Mathematical Modelling** problem introduced at the start of the chapter.

In Chapter 4, we consider the fat content of foods.

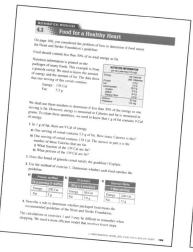

This section helps you develop a mathematical model to solve the problem.

By looking at many possible models, you learn to identify your assumptions, recognize how accurate a model might be, and revise a model if you think it is necessary.

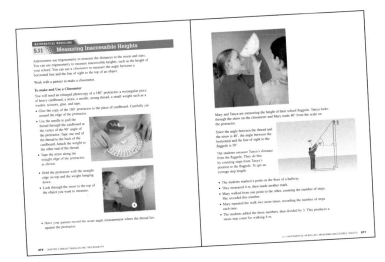

In Chapter 5, we measure the height of an inaccessible object.

Example 2

Graph each function.

a) $y = 2 - x$ **b)** $y = x^2 - 1$

Solution

Method 1 Using grid paper Create a
table of values and draw each graph.

a) $y = 2 - x$

x	y
−3	5
−2	4
−1	3
0	2
1	1
2	0
3	−1

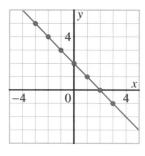

b) $y = x^2 - 1$

x	y
−3	8
−2	3
−1	0
0	−1
1	0
2	3
3	8

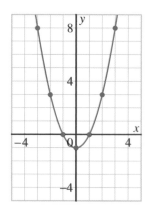

In the graph of $y = 2 - x$, the points lie along a line.

In the graph of $y = x^2 - 1$, the points lie along a curve.

Method 2 Using a graphing calculator Clear any Y= equations and turn
off any statplots.

Dynamic Geometry Software

Dynamic geometry software, such as *The Geometer's Sketchpad*, is another powerful technology tool. This software lets you create geometric figures that can be moved or altered as you observe results.

This book provides explicit instruction in the use of *The Geometer's Sketchpad*.

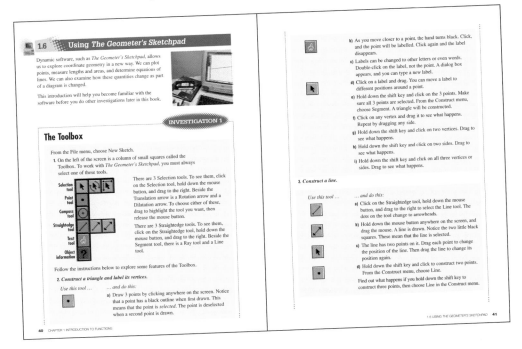

A computer logo indicates when a computer is required to complete an Investigation or exercise.

Testing Your Knowledge

At the end of each chapter, this section gives you suggestions for studying key concepts, and exercises to test your knowledge.

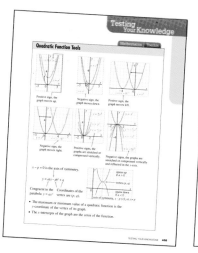

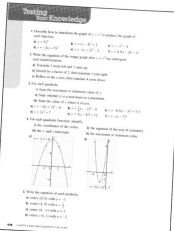

The **Mathematics Toolkit** summarizes key points for the chapter, and connects concepts when relevant. Use the Toolkit as a handy reference as you progress through the review exercises.

Exercises let you practise key mathematical concepts and problem-solving skills.

Mathematical Modelling exercises give you a chance to solve a problem related to the Mathematical Modelling problem for the chapter, or to solve an open-ended problem in which you develop new strategies.

Cumulative Review

A Cumulative Review appears after Chapter 3, Chapter 5, and Chapter 8. As you complete this section, you review the concepts and skills for all chapters that precede the Cumulative Review.

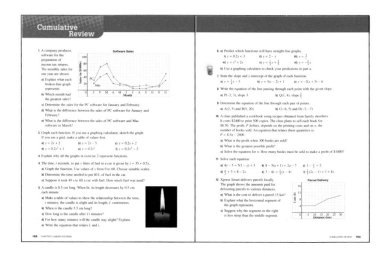

When people solve problems in the world of work, they are dealing with "big" problems that require analysis and a clear strategy. They use whatever mathematical concepts fit the problem: a formula, a graph, a number pattern, and so on.

This book presents a Project after Chapter 3, Chapter 5, and Chapter 8. The Projects enable you to work with an applied problem in a structured way, over an extended period of time.

This project appears after Chapter 3. You may need to use any of the skills or concepts you developed in Chapters 1 to 3.

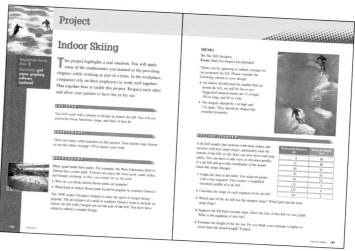

The Project doesn't have one "right" answer. Your teacher may choose to score your Project using a rubric. A rubric is a table with descriptions of four levels of performance in different categories.

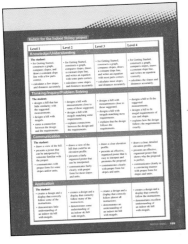

This is the rubric for the Project after Chapter 3.

Read the descriptions for each category in the rubric before you start the Project. This helps you understand what your teacher might expect in a top-rating performance.

1 Introduction to Functions

Are Capacities on Can Labels Correct?

Cans of food have different shapes and sizes. On many cans, the capacity is marked on the label.

How can we be sure that the capacities on the labels are correct? The way we solve this problem is an example of a mathematical model.

Take Note

A *mathematical model* is a way to solve an applied problem.

Until you develop a model to solve this problem, here are some questions to consider.

1. What is capacity?

2. How are capacity and volume related?

3. What does the symbol "mL" mean?

4. How is capacity marked on the label of a can?

5. Does every can have its capacity marked on the label? Explain.

You will return to this problem in Section 1.4.

 FYI Visit www.awl.com/canada/school/connections

For information related to the above problem, click on <u>MATHLINKS</u>, followed by Foundations of Mathematics 10. Then select a topic under Are Capacities on Can Labels Correct?

Preparing
for the Chapter

This section reviews these concepts:

- Area and perimeter of a rectangle
- Graphing data from a table of values
- Describing relationships

This design of 15 overlapping rectangles was created on a computer.

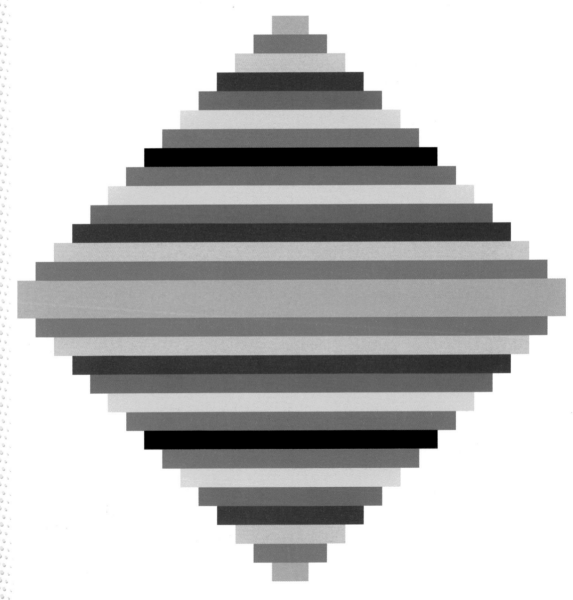

Rectangle Designs: Part I

1. a) Locate the black square on page 4. Measure its base and its height.

  b) Locate the two red rectangles on page 4. Measure their bases and heights.

  c) Locate four more rectangles on page 4. Measure their bases and heights.

  d) When you include the square, you have data for seven rectangles. Calculate the perimeter of each rectangle.

  e) Describe a common property of the perimeters. Explain how this property relates to the design.

Graphing the Height against the Base

2. a) Record the bases and heights of your rectangles in a table.

Base (cm)	Height (cm)	Area (cm²)

  b) Graph the data from the table. Plot *Base* horizontally and *Height* vertically. Does it make sense to join the points? Explain.

  c) Suppose the base length starts at a small value, then increases. Describe how the height changes. Explain using the diagram on page 4.

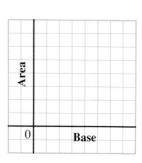

Graphing the Area against the Base

3. a) Calculate and record the area of each rectangle in the table.

  b) Graph the area against the base. Use a grid different from the one you used in exercise 2b. You may need to include results for more rectangles to complete the graph.

  c) Suppose the base length starts at a small value, then increases. Describe how the area changes. Explain using the diagram on page 4.

4. Explain why it is possible for rectangles to have the same perimeters but different areas.

This design of 15 overlapping rectangles was created on a computer.

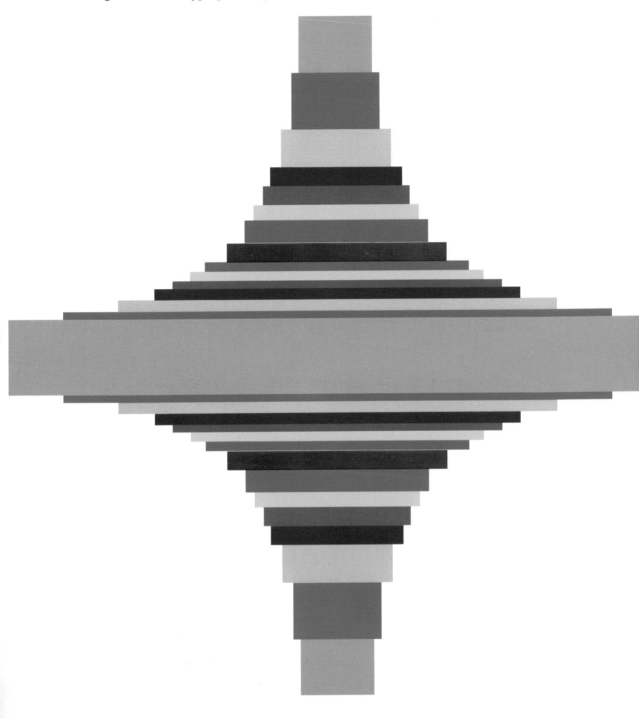

Rectangle Designs: Part II

1. a) Locate the black square on page 6. Measure its base and its height.

b) Locate the two red rectangles on page 6. Measure their bases and heights.

c) Locate four more rectangles on page 6. Measure their bases and heights.

d) You have data for seven rectangles. Calculate the area of each rectangle.

e) Describe a common property of the areas. Explain how this property relates to the design.

Graphing the Height against the Base

2. a) Record the bases and heights of your rectangles in a table.

Base (cm)	Height (cm)	Perimeter (cm)

b) Graph the data from the table. Plot *Base* horizontally and *Height* vertically. You may need to include results for more rectangles to complete the graph.

c) Suppose the base length starts at a small value, then increases. Describe how the height changes. Explain using the diagram on page 6.

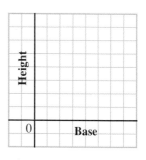

Graphing the Perimeter against the Base

3. a) Calculate and record the perimeter of each rectangle in the table.

b) Graph the perimeter against the base. Use a grid different from the one you used in exercise 2b. You may need to include results for more rectangles to complete the graph.

c) Suppose the base length starts at a small value, then increases. Describe how the perimeter changes. Explain using the diagram on page 6.

4. Explain why it is possible for rectangles to have the same areas but different perimeters.

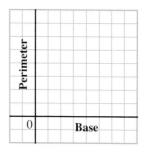

In the *Investigation*, you collected data, made tables, and drew graphs to show how quantities are connected. In mathematics, a connection between two quantities is called a *relation*.

You can collect data from graphs.

Example

A graph similar to this one appeared in a 1996 newspaper article.

a) Who are public sector employees?

b) What does the graph show?

c) Explain the vertical scale.

d) Determine the number of men and the number of women working in the public sector in 1976, 1981, 1986, 1991, and 1996.

e) Determine the total number of public sector employees in the years from 1976 to 1996.

f) Draw a graph to show the total number of employees.

g) What does the graph tell you about how the total number of employees changed between 1976 and 1996?

h) Use the graph. Estimate the total number of employees in 1994.

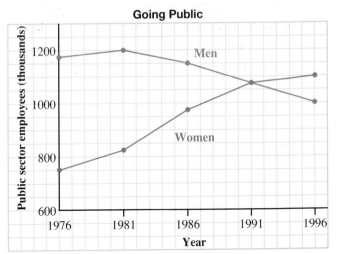

Solution

a) Public sector employees are people who work for the government.

b) The graph is two broken lines. One broken line shows the number of women employees. The other broken line shows the number of men employees.

c) Each number on the vertical scale represents thousands. For example, 600 represents 600 000; 1200 represents 1 200 000.

d) The data collected from the graphs are in the first three columns of this table.

e) To determine the total number of employees, add the numbers in the second and third columns.

f) Use the data in the first and last columns of the table to draw a graph. Join adjacent points with line segments.

Year	Women (thousand)	Men (thousand)	Total (thousand)
1976	750	1175	1925
1981	825	1200	2025
1986	975	1150	2125
1991	1075	1075	2150
1996	1100	1000	2100

g) From 1976 to 1991, the total number of employees increased. From 1991 to 1996, the total number of employees decreased.

h) Start at 1994 on the horizontal axis. Draw a vertical line up to the graph. Draw a horizontal line to meet the vertical axis. This line meets the axis at approximately 2125. In 1994, the total number of employees was about 2125 thousand, or 2 125 000.

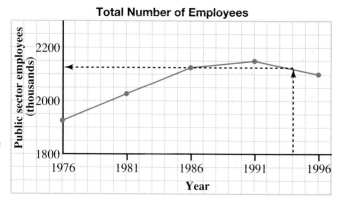

Total Number of Employees

Discussing the IDEAS

1. Refer to the *Investigation*, page 7. Visualize all the rectangles that have an area of 36 cm².

   a) Explain why the perimeters of these rectangles are not the same.

   b) Describe the rectangle with the minimum perimeter.

2. In the solution of the *Example*, part d, explain how the numbers of women were determined from the graph.

3. Refer to the graphs in the *Example*. Give a possible explanation for each statement.

   a) The number of female employees increased from 1976 to 1996.

   b) The number of male employees decreased from 1981 to 1996.

   c) The total number of employees began decreasing after 1991.

Practise Your Skills

1. Determine each percent.

   a) 50% of 30　　　　**b)** 25% of 400　　　　**c)** 12% of 60

   d) 43% of 200　　　　**e)** 68% of 250　　　　**f)** 79% of 188

2. Express as a percent.

   a) 0.27　　　　**b)** 0.82　　　　**c)** $\frac{3}{4}$

   d) $\frac{1}{3}$　　　　**e)** $\frac{2}{7}$　　　　**f)** $\frac{7}{12}$

A 1. The interest rate the Bank of Canada charges when it loans money to the banks is called "Canada's bank rate." The graph shows how this rate changed from January to December in one year.

a) What does the zigzag line on the vertical axis represent?

b) What was the bank rate in each month?
   i) March   ii) June   iii) July

c) What was the bank rate in August?

d) Explain why the graph consists of horizontal line segments.

e) Did any of the graphs you drew in the *Investigation* on page 7 or in the *Example* contain horizontal line segments? Explain.

B 2. This graph was drawn using Statistics Canada data for women, 25 years and over.

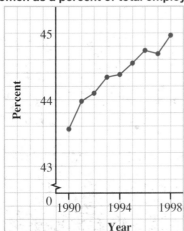

Working Women
Women as a percent of total employed

a) What is the scale on the vertical axis?

b) Use the graph. Estimate the percent of working women for each year from 1990 to 1998.

c) Determine the percent of working men for each year from 1990 to 1998.

d) Draw a similar graph with the title "Working Men".

e) Write to explain how the two graphs are related.

3. A graph similar to this appeared in a 1999 newspaper article. The source of the data is Groupe Léger et Léger.

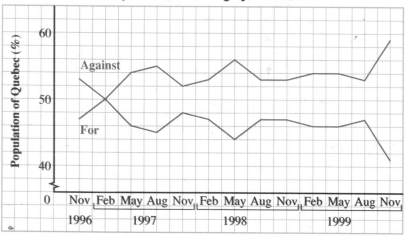

Opinions on sovereignty in Quebec

a) What is the scale on the vertical axis?

b) What is the scale on the horizontal axis?

c) Explain what each broken-line graph represents.

d) How are the two graphs related? Explain.

e) Suggest a reason why the relationship in part d might not be exact.

4. When you take a roll of film to a store for processing, the store develops the film and prints the negatives. A store advertises these prices.

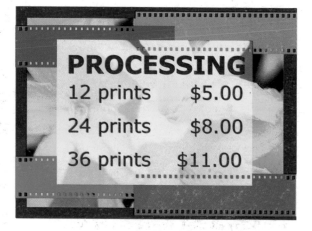

PROCESSING

12 prints $5.00

24 prints $8.00

36 prints $11.00

a) Graph the data. Plot *Number of prints* horizontally and *Cost ($)* vertically. Does it make sense to join the points? Explain.

b) Use the graph to determine the cost for 18 prints, and for 27 prints.

c) Explain why the cost for 24 prints is not double the cost for 12 prints.

5. When some people take film for processing, they order two copies of each picture. These are called doubles. The store charges $2.00 extra for doubles. Assume that doubles are chosen.

a) Use the data in exercise 4. List the prices for 12, 24, and 36 prints.

b) Graph the data from part a on the grid for exercise 4.

c) Use the new graph to determine the cost for 18 prints, and for 27 prints.

6. The stopping distance for a car depends on the driver's reaction time and the time to slow down the car after the brakes have been applied. The graph below shows typical reaction-time distances and braking distances for speeds up to 120 km/h.

Stopping Distance

a) What is meant by a "driver's reaction time"?

b) Copy and complete a table with the headings shown. Use the graph to complete the first three columns. Use the second and third columns to complete the fourth column.

Speed (km/h)	Reaction-time distance (m)	Braking distance (m)	Total stopping distance (m)

c) Draw a graph to show the total stopping distances for speeds up to 120 km/h.

d) What does the graph tell you about how the total stopping distance changes as the speed increases?

e) Use your graph. Estimate the total stopping distance when the speed is 90 km/h, and when it is 110 km/h.

f) **Mathematical Modelling** These stopping distances are only typical. Actual stopping distances may vary greatly from those shown here. List factors that can affect the stopping distance of a car.

The Growth of Internet Shopping

Each year, more people are using the Internet to make purchases. A newspaper article published in 1999 contained graphs similar to these. The data for 1999 to 2003 are predictions.

Internet Access

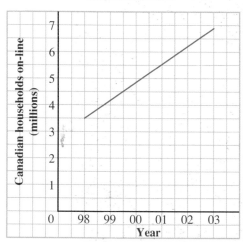

Internet Shopping

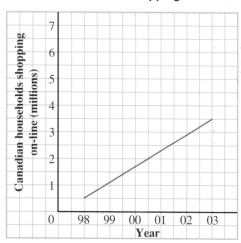

Use the graphs to complete these exercises.

7. Consider the data for 1998 on each graph.

   a) How many households were on-line that year?

   b) How many households were shopping on-line?

   c) Write the fraction $\frac{\text{number of households shopping on-line}}{\text{number of households on-line}}$.

   d) Write the fraction in part c as a percent.

   e) What percent of the households were shopping on-line in 1998?

8. Consider the predictions for 2003 on each graph.

   a) How many households were predicted to be on-line that year?

   b) How many households were predicted to be shopping on-line?

   c) What percent of the households were predicted to be shopping on-line in 2003?

9. What do your results in exercises 7 and 8 tell you about how the number of households shopping on-line is changing?

10. Area and Perimeter of Rectangles

Recall that the area of a figure is the number of square units needed to cover it. The perimeter of a figure is the total distance around it.

For a rectangle, the area and perimeter are found using these formulas.

Area = base × height or, $A = bh$

Perimeter = 2 × (base + height) or, $P = 2(b + h)$

In this exercise, you will work with area and perimeter together.

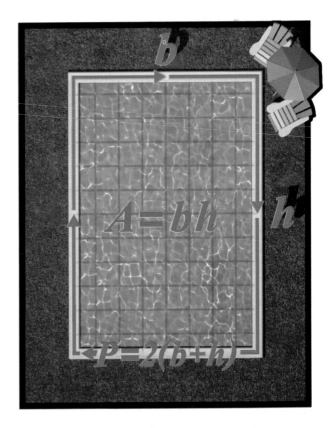

a) Look through the pages of this book. Find at least ten different rectangles. These can be photographs or drawings. Try to get a variety of shapes and sizes. Include at least one very small rectangle.

b) Calculate the area and the perimeter of each rectangle. Record the results in a table with these headings.

Page	Description	Base (cm)	Height (cm)	Perimeter (cm)	Area (cm^2)

c) Use the data in the last two columns to draw a graph. Plot *Perimeter* horizontally and *Area* vertically.

d) Suppose you had a photograph with a very long base and a very short height. Suppose you calculate its perimeter and its area. About where on the graph would the point corresponding to this rectangle be located? Explain.

e) Do the areas and the perimeters of rectangles appear to be related? Explain.

Communicating the IDEAS

Several examples of related quantities were given on the preceding pages. Describe another example of related quantities in everyday life. Explain how the quantities are related.

1.2 What Is a Function?

Many calculators have an $\boxed{x^2}$ key.

You enter a number, press $\boxed{x^2}$, and the calculator displays the result.

| Input 7 | → | Press $\boxed{x^2}$ | → | Output 49 |

The number you enter is the *input number*.

The number the calculator displays is the *output number*.

The $\boxed{x^2}$ key illustrates the idea of a function.

Take Note

A *function* is a rule that gives a single output number for every valid input number.

The rule that defines a function can be expressed in different ways. For the function above, we can express the rule as follows.

In words

Multiply the number by itself.

As an equation

$y = x^2$

We often use x to represent an input number, and y to represent the corresponding output number. We say that the equation *expresses y as a function of x.*

Any other letters could be used.

As a table of values

Input number x	Output number y
−4	16
−3	9
−2	4
−1	1
0	0
1	1
2	4
3	9
4	16

As a graph

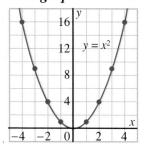

In the graph, the plotted points appear to lie on a curve. If we included more points corresponding to input numbers between those listed in the table of values, we would see that the points do lie on a curve.

Examples of Functions

1. Four rules for determining output numbers are given. For each rule:

a) Make a table of values.

b) Draw a graph.

c) Determine if the rule is a function. Explain.

d) If the rule is a function, write its equation.

Rule 1:
Add 3 to the number.

Rule 2:
Subtract the number from 3.

Rule 3:
Multiply the number by 3.

Rule 4:
Double the number, then add 3.

If the points on the graph of a function lie on a straight line, the function is called a *linear function*. If the points do not lie on a straight line, the function is *non-linear*. You should have found examples of linear functions in the *Investigation*. The function $y = x^2$ on page 15 is an example of a non-linear function. The next example involves a linear function.

Example 1

Canada started using the Celsius scale for temperatures in the 1970s. The Fahrenheit scale is still used in the United States. Suppose a tourist wants to convert a Celsius temperature to Fahrenheit. An approximate rule for converting Celsius to Fahrenheit is to "double and add 30."

a) Use the rule to convert each Celsius temperature to Fahrenheit:
0°C, 11°C, 25°C

b) Write an equation for the approximate rule.

c) Graph the equation in part b.

d) Use the graph to convert each Celsius temperature to Fahrenheit:
17°C, 34°C

Solution

a) Multiply each Celsius temperature by 2, then add 30.

$$2(0) + 30 = 0 + 30 \qquad 2(11) + 30 = 22 + 30 \qquad 2(25) + 30 = 50 + 30$$
$$= 30 \qquad\qquad\qquad = 52 \qquad\qquad\qquad\qquad = 80$$
$$0°C \doteq 30°F \qquad\qquad 11°C \doteq 52°F \qquad\qquad 25°C \doteq 80°F$$

b) Let F and C represent the Fahrenheit and Celsius temperatures, respectively. The equation is $F = 2C + 30$.

c) **Method 1 Using grid paper** Make a table of values. Use the results of part a to plot the points. Use a scale of 1 square to 10°C and 10°F. Use a ruler to draw a straight line through the points.

Degrees Celsius C	Degrees Fahrenheit F
0	30
11	52
25	80

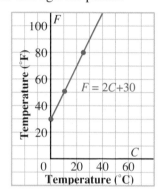

Method 2 Using a graphing calculator Rewrite the formula using x and y: $y = 2x + 30$

Set-Up

- Press MODE to get the Mode menu. Make sure that the items are highlighted as shown in the first screen below. If any of these items are not highlighted, use the arrow keys and ENTER to highlight them.

- Press 2nd ZOOM to get the Format menu. Make sure that the items are highlighted as shown in the second screen below. If any of these items are not highlighted, use the arrow keys and ENTER to highlight them.

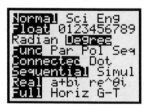
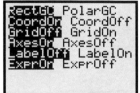

- Press Y=. Use CLEAR and the arrow keys to clear all equations.

- If any of Plot 1, Plot 2, or Plot 3 at the top of the screen is highlighted, use the arrow keys and ENTER to remove the highlighting.

- Make sure the cursor is beside Y1=. Press 2 $\boxed{\text{x,т,θ,n}}$ $\boxed{\ +\ }$ 30. You have entered the equation $y = 2x + 30$. Note that $\boxed{\text{x,т,θ,n}}$ is a variable key. You always use this key to enter the variable in an equation. Your screen should look like the first one below.
- Press $\boxed{\text{WINDOW}}$, and change the settings to those shown in the second screen below.
- Press $\boxed{\text{GRAPH}}$. A graph of the function $y = 2x + 30$ appears as shown in the third screen below.

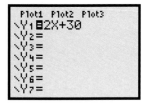

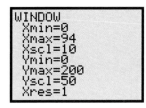

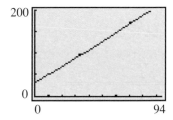

d) **Method 1 Using grid paper** Use the graph in part c. To convert 17°C, start at 17°C on the horizontal axis. Draw a vertical line up to the graph, then a horizontal line that meets the vertical axis at 64°F.
Hence, 17°C ≐ 64°F

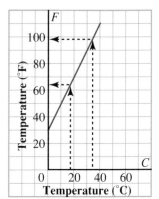

To convert 34°C, start at 34°C on the horizontal axis. Draw a vertical line up to the graph, then a horizontal line that meets the vertical axis at 98°F.
Hence, 34°C ≐ 98°F

Method 2 Using a graphing calculator
Press $\boxed{\text{TRACE}}$. Values of x and y appear. Press $\boxed{\blacktriangleright}$ or $\boxed{\blacktriangleleft}$ to move a cursor along the line. The values of x and y show the coordinates of the point on the line where the cursor is located. Move the cursor until $x = 17$. The value of y is 64. Hence, 17°C ≐ 64°F (below left)

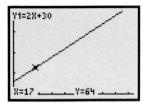

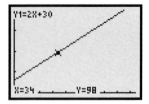

Press 34 $\boxed{\text{ENTER}}$. The cursor moves to the point where $x = 34$. The value of y is 98. Hence, 34°C ≐ 98°F (above right)

Example 2

Graph each function.

a) $y = 2 - x$ **b)** $y = x^2 - 1$

Solution

Method 1 Using grid paper Create a table of values and draw each graph.

a) $y = 2 - x$

x	y
−3	5
−2	4
−1	3
0	2
1	1
2	0
3	−1

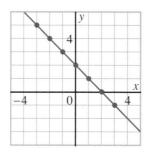

b) $y = x^2 - 1$

x	y
−3	8
−2	3
−1	0
0	−1
1	0
2	3
3	8

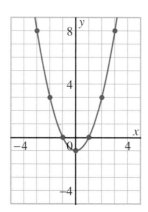

In the graph of $y = 2 - x$, the points lie along a line.

In the graph of $y = x^2 - 1$, the points lie along a curve.

Method 2 Using a graphing calculator Clear any Y= equations and turn off any statplots.

a) Press $\boxed{\text{WINDOW}}$, and change the settings to those shown below left.
Press $\boxed{\text{Y=}}$ 2 $\boxed{-}$ $\boxed{\text{X,T,}\theta,n}$ $\boxed{\text{ENTER}}$ $\boxed{\text{GRAPH}}$ to enter and display the graph
below right.

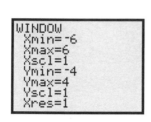

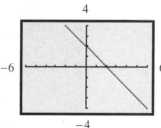

b) Use the same window as in part a.
Press $\boxed{\text{Y=}}$ $\boxed{\text{CLEAR}}$ to clear the first function.
Press $\boxed{\text{X,T,}\theta,n}$ $\boxed{x^2}$ $\boxed{-}$ 1 $\boxed{\text{ENTER}}$ $\boxed{\text{GRAPH}}$ to enter and display
the graph below right.

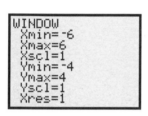

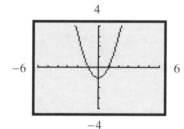

Discussing the IDEAS

1. What is a function?

2. **a)** What ways are there to express a function?

   b) How are these ways similar? How are they different?

3. For the equation in *Example 1*, is it possible to have input values of C that are not integers, such as 12.5 or 23.6? Describe the types of output values of F you might expect from such input values. Explain how these examples support the statement that the points on the graph can be joined with a straight line.

4. *Example 2* demonstrated two ways to graph a function from its equation. What are some advantages of using a graphing calculator to graph a function? What might be some advantages of using grid paper?

1. Simplify.

   a) $2(-3) + 5$ **b)** $4(-1) - 2$ **c)** $3 - 4(-1)$

   d) $6 + 5(-2)$ **e)** $-2(-3) + 4$ **f)** $-5(4) + 8$

2. Evaluate each expression for $x = 4$.

   a) $x + 5$ **b)** $2x - 6$ **c)** $x^2 + 2$

   d) $0.5x + 1$ **e)** $4 - x$ **f)** $4 - 3x$

3. Evaluate each expression in exercise 2 for $x = -1$.

4. Evaluate the expression $3x - 2$ for each value of x.

   a) 5   **b)** 2   **c)** -1 **d)** -4 **e)** -5

5. Evaluate the expression $6 - 2x$ for each value of x.

   a) -2 **b)** -5 **c)** 1   **d)** 3   **e)** 8

1.2   Exercises

A   **1. a)** Graph each table of values.

i)

x	y
1	4
2	5
3	6
4	7

ii)

x	y
-2	-6
0	0
1	3
2	6

iii)

x	y
1	4
2	3
3	2
4	1

  b) State whether each table in part a represents a linear or non-linear function. Explain.

B   **2.** Write an equation for each graph in exercise 1.

3. For each function, choose 4 input values and calculate the corresponding output values. Use positive and negative input values. Graph each function on a separate grid.

  a) *Rule 1*: Triple the number, then subtract 3.

  b) *Rule 2*: Add 1 to the number, then double the result.

  c) *Rule 3*: Subtract the number from 12.

4. Write an equation for each graph in exercise 3.

5. Graph the function represented by each equation below. If you use a graphing calculator, sketch each graph. If you use a grid, make a table of values first.

a) $y = 3x - 5$ **b)** $y = 1 + 0.5x$ **c)** $y = x - 10$

d) $y = 0.5x$ **e)** $y = 0.5x^2$ **f)** $y = x^2 + x$

6. Graph the function represented by each equation below. If you use a graphing calculator, sketch each graph.

a) $y = 1.5x + 3$ **b)** $y = 3 - 2x$ **c)** $y = x - 4$

d) $y = -x$ **e)** $y = 0.1x^2$ **f)** $y = x^2 - x$

7. Choose either exercise 5 or exercise 6. Look at the six graphs you drew.

a) Examine the equations whose graphs are straight lines. How are they similar?

b) Examine the equations whose graphs are curves. How are they similar?

c) Explain why all the graphs represent functions.

Trees for Our Future

In Kitchener, Ontario, there is a plan to promote tree planting. People who buy certain trees from garden centres in the area save money according to this plan:

- The garden centre takes 15% off the retail price.

- A further $7.00 will be taken off the price by the City of Kitchener.

- An additional $5.00 will be taken off the price by Canada Trust's "Friends of the Environment Foundation."

- Homeowners pay the remainder, plus applicable taxes.

8. Suppose you purchase a crab apple tree that retails for $50.00. Follow the steps above to calculate the amount you would have to pay for the tree.

9. Copy the table below. Enter your answer from exercise 8 in the first row of the *Amount you pay* column.

Tree (6 ft.)	Retail price ($)	Amount you pay ($)
crab apple	50.00	
ash	60.00	
red oak	70.00	
purple leaf birch	80.00	
balsam fir	90.00	

a) Repeat exercise 8 for the other trees in the table.

b) Graph the data. Plot *Retail price* horizontally and *Amount you pay* vertically.

c) Use the graph. What would you pay for a black cherry tree, which has a retail price of $67.50?

10. The cost, C dollars, to hold a party in a local hall is represented by the equation $C = 100 + 10n$, where n is the number of people attending the party.

a) Graph the equation. Use values of n from 0 to 100. Choose suitable scales.

b) Determine the cost to hold the party for 50 people.

c) Determine the number of people who can attend for a cost of $470.

11. Suppose an object falls from rest. The distance it falls, d metres, and the time it takes, t seconds, are related by the equation $d = 4.9t^2$.

a) Graph the equation. Use values of t between 0 and 10.

b) Use the graph to describe the motion of an object falling from rest.

c) How long would it take an object to hit the ground when dropped from each height?

   i) 20 m   ii) 40 m

d) **Mathematical Modelling** A mathematical model often gives results that are different from actual results. The formula, $d = 4.9t^2$, implies that any object, regardless of its mass, will fall the same distance in the same time. Do you think this is true? What factors, aside from gravity, affect the time an object falls a given distance?

12. Graph each function. What do the graphs have in common?

a) $y = x + 5$ b) $y = 0.5x + 5$ c) $y = 5 - x$

d) $y = x^2 + 5$ e) $y = (x - 1)(x - 5)$ f) $y = 5x^2 + 5$

13. Here are three statements about functions.

Statement 1:
When you graph a function, you should join the plotted points.

Statement 2:
A table of values shows all possible input numbers.

Statement 3:
For every input number, there is only one output number.

a) Only one statement is always true. Which statement is it? Explain.

b) Choose one of the other two statements. Use examples to explain why this statement is sometimes true and sometimes false.

14. A computer salesperson earns a monthly salary of $2083.33, plus 3% commission on all sales. Here is an equation for her monthly income, I dollars, when she has sales totalling s dollars for the month.
$I = 2083.33 + 0.03s$

a) Graph the equation. Use these window settings:

```
WINDOW
 Xmin=0
 Xmax=47000
 Xscl=500
 Ymin=2000
 Ymax=3500
 Yscl=500
 Xres=1
```

b) Use the graph to describe as much as you can about the salesperson's earnings.

c) How much does she earn when her sales are $28 500?

d) What must her sales be for her to earn $2500 in a month?

e) Sketch the graph. Show the results of parts c and d on the sketch.

f) Suppose the salesperson has a goal to earn $40 000 in a year. How much is this each month? What would her sales have to be? Do you think it is realistic for a computer salesperson to sell that much? Explain.

g) Did you use the graph, the equation, or another approach to complete part f? Write to explain your work.

15. Mathematical Modelling In *Example 1*, this rule was given to convert Celsius temperatures to Fahrenheit temperatures: $F = 2C + 30$.
This is only an approximate rule.
The exact rule is $F = 1.8C + 32$.

When the approximate rule is used, the results may not be accurate. The difference between the result given by the approximate rule and the result given by the exact rule is called the *approximation error*. Convert each temperature to Fahrenheit. Determine the approximation error for each temperature.

a) On a hot summer day, the temperature can be 30°C or more.

b) On a pleasant spring day, the temperature is about 20°C.

c) When it is freezing, the temperature is 0°C.

16. **Mathematical Modelling**
Psychologists have experimented to measure how much a person remembers of material that was learned. The results of one experiment are shown in the table. The variable p represents the percent of the material remembered after t days.

Time in days (t)	1	5	15	30	60
Percent remembered (p)	84%	71%	61%	56%	54%

a) Draw a graph of p against t on grid paper. Draw a smooth curve through the plotted points.

b) Suppose q represents the percent forgotten instead of the percent remembered. Draw a similar graph of q against t.

c) Write to explain how the two graphs are related.

C 17. **a)** Each rule represents a function. Graph each function on the same grid.
   i) *Rule 1*: Add 5 to the number, then double the result.
   ii) *Rule 2*: Double the number, then add 5 to the result.

b) For each rule in part a, is it possible for two or more input numbers to have the same output number? Explain.

Communicating *the* IDEAS

Describe how it is possible to obtain a graph from an equation. Explain the significance of the points that lie on the graph of a function.

Sports of all kinds carry a risk of serious eye injury or blindness. But with proper protection, virtually all eye injuries are preventable. Statistics for hockey-related eye injuries show that wearing protective equipment works. Since 1972, no player has been blinded while wearing a CSA-certified full-face protector.

Dr. Tom Pashby, an opthalmologist, has worked to protect amateur athletes from eye injuries since the 1970s. He has kept detailed statistics showing the number of eye injuries in all sports. Data for hockey are shown in the table.

Eye Injuries in Canadian Hockey

1975	1976	1977	1978	1979	1980	1981	1982	1983	1984	1985	1986
144	129	90	52	43	85	68	119	115	124	121	123

1987	1988	1989	1990	1991	1992	1993	1994	1995	1996	1997
93	62	37	33	21	28	32	16	19	22	12

You can use a graphing calculator to display graphs to help you analyse these data. To complete the investigations, you will need a TI-83 graphing calculator. Since other students may have used the calculator previously, you may need to reset it. Ask your teacher or consult the manual for instructions about how to do this.

Graphing Eye Injuries in Hockey

We will investigate these questions:

- What does a graph of the number of eye injuries against years look like?

- Are there any patterns in the data?

Step 1. Enter the data.

- Press [MODE] to get the Mode menu. Make sure that the items are highlighted as shown in the first screen below. If any of these items are not highlighted, use the arrow keys and [ENTER] to highlight them.

- Press [2nd] [ZOOM] to get the Format menu. Make sure that the items are highlighted as shown in the second screen below. If any of these items are not highlighted, use the arrow keys and [ENTER] to highlight them.

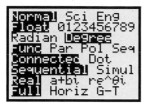 

- Press [STAT] 1. A screen appears showing lists L1, L2, and L3. This screen is the *list editor*.

- If there are any numbers in the columns, they must be cleared. Move the cursor into each column heading, then press [CLEAR] [ENTER].

- Move the cursor to the space below L1. In list L1, enter the years from the table on page 26.

- Move the cursor to the space below L2. In list L2, enter the numbers of eye injuries from the table.

- Check that the data entered are correct. Use the arrow keys to go back to fix any errors you may have made entering the data.

When you have finished, your screen should look like this.

L1	L2	L3	2
1992	28		
1993	32		
1994	16		
1995	19		
1996	22		
1997	13		

L2(24) =

Step 2. Set up the graph.

You must tell the calculator the lists to plot along the axes and the type of graph you want.

- Press ⬚2nd⬚ ⬚Y=⬚ to get the Stat Plot menu. This menu shows the following information for three plots:
  - whether it is turned on or off
  - the type of graph
  - the lists used to create the graph
  - the plotting symbol used

- Press **1** to select Plot 1.
- Press ⬚ENTER⬚ to turn on Plot 1.
- Press ⬚▼⬚ ⬚►⬚ ⬚ENTER⬚ to select the second plot type.
- You want the numbers in L1 to be graphed horizontally. Press ⬚▼⬚ ⬚2nd⬚ **1** to make sure that L1 is beside Xlist.
- You want the numbers in L2 to be graphed vertically. Press ⬚▼⬚ ⬚2nd⬚ **2** to make sure that L2 is beside Ylist.
- Press ⬚▼⬚ ⬚ENTER⬚ to select the square as the first plotting symbol.

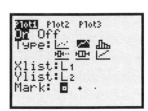

The screen should look exactly like that at the right.

You must clear any equations that may have been left by the previous user.

- Press ⬚Y=⬚. Use ⬚CLEAR⬚ and the arrow keys to clear any equations that appear on the screen.
- Plot 1 should be highlighted. If Plot 2 or Plot 3 is highlighted, move the cursor to this plot then press ⬚ENTER⬚ to remove the highlighting.

The screen should look exactly like this.

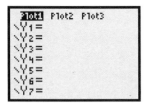

Step 3. Set up the window.

- Press ⬚WINDOW⬚.
- The horizontal axis must show years from 1975 to 2000. Enter Xmin = 1975, Xmax = 2000, Xscl = 5.

- The vertical axis must show numbers of injuries up to 150.
  Enter Ymin = 0, Ymax = 150, Yscl = 50.

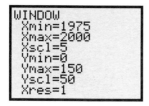

Step 4. Graph the data.

- Press GRAPH.

1. If you have not already done so, follow the steps above to display the graph of the number of eye injuries against years. Your graph will appear on a screen like this.

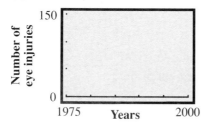

2. Press TRACE. A cursor will appear on the first data point. Its coordinates will also appear on the screen. Press ▶ to move the cursor among the data points.

3. To draw the graph on grid paper:

   a) Draw and label the axes, showing the scales.

   b) Use the table on page 26 to plot the data.

4. Give possible reasons for each statement.

   a) the decrease in the number of eye injuries from 1975 to 1979

   b) the increase in the number of eye injuries from 1979 to 1986

   c) the decrease in the number of eye injuries since 1986

5. Try some of the other settings in the Stat Plot menu:

   a) Try using the first plot type.   b) Try using a different plotting symbol.

Comparing Cumulative Eye Injuries

The graph you made in *Investigation 1* shows the number of eye injuries each year. We can also make a graph to show the total number of injuries since 1975 for each year. This is called the *cumulative* number of injuries. For example:

Year	Number of eye injuries	Cumulative injuries	
1975	144	144	
1976	129	273	← 144 + 129 = 273
1977	90	363	← 144 + 129 + 90 = 363

We can use the calculator to do these calculations and graph the results.

Step 1. Enter the data.

- Press [STAT] **1**. The data you entered in *Investigation 1* should be in lists L1 and L2. If not, follow *Step 1* on page 27 to enter the data from the table on page 26.

- Move the cursor into the column heading for list L3. The cursor must be on L3, and not in the space below L3.

- Press [2nd] [STAT] [▶] to get the Ops menu.

- Press **6** to choose cumSum(.

- Press [2nd] **2** to choose list L2.

- Press [ )] [ENTER]. The calculator calculates all the cumulative sums for list L2, then enters them in list L3.

When you have finished, your screen should look like this.

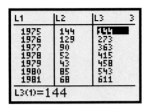

Step 2. Set up the graph.

You must tell the calculator the lists to plot along the axes and the type of graph you want.

- Press [2nd] [Y=] to get the Stat Plot menu.

- Press **2** to select Plot 2.

- Press [ENTER] to turn on Plot 2.
- Press [▼] [ENTER] to select the first plot type.
- You want the numbers in L1 to be graphed horizontally.
  Press [▼] [2nd] **1** to make sure that L1 is beside Xlist.
- You want the numbers in L3 to be graphed vertically.
  Press [▼] [2nd] **3** to make sure that L3 is beside Ylist.
- Press [▼] [►] [►] [ENTER] to select the dot as the first plotting symbol.

The screen should look exactly like this.

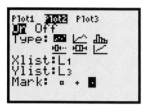

- Press [Y=]. Plot 1 will probably still be selected. To clear it,
  press [▲] [ENTER].
- If necessary, clear Plot 3 and any equations that may be present.

The screen should look exactly like this.

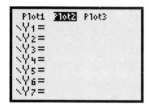

Step 3. Set up the window.

- Press [WINDOW].
- The horizontal axis should show years from 1975 to 2000.
  Enter Xmin = 1975, Xmax = 2000, Xscl = 5.
- The vertical axis should show cumulative numbers of injuries up to
  2000. Enter Ymin = 0, Ymax = 2000, Yscl = 500.

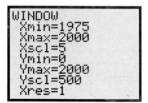

Step 4. Graph the data.

- Press GRAPH.

1. If you have not already done so, follow the steps above to display the graph of the cumulative numbers of eye injuries against years. Your graph will appear on a screen like this one.

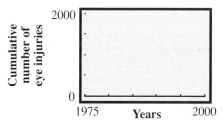

2. Press TRACE. A cursor appears on the first data point. Its coordinates will also appear on the screen. Press ► to move the cursor among the data points.

3. To draw the graph on grid paper:

   a) Draw and label the axes, showing the scales.

   b) Use the table on page 26 to calculate cumulative injuries for each year.

4. Compare your graph with the graph from *Investigation 1*.

   a) Suppose the number of eye injuries keeps decreasing. How will this be shown on each graph?

   b) Explain why the graph in *Investigation 2* will never start going down like the graph in *Investigation 1*.

Communicating the IDEAS

Write a report of your findings that someone who did not do the investigations could understand. Describe the situations you investigated, the conclusions you reached, and how you arrived at your conclusions. Use written explanations, tables, graphs, equations, or calculations.

1.4 Are Capacities on Can Labels Correct?

On page 2, you considered a problem about the labels on cans. How can we be sure that the capacities marked on the labels are correct?

Recall that volume, *V*, can be measured in two kinds of units.

Cubic units refer to the amount of space occupied by an object. For the cube shown, $V = 1 \text{ cm}^3$

Capacity units refer to the amount a container holds. For the same cube, $V = 1 \text{ mL}$

You can use the relationship between millilitres and cubic centimetres to solve the problem above. You will:

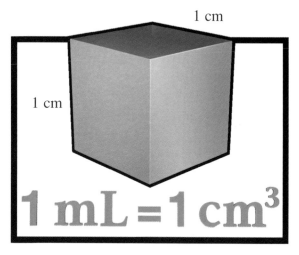

1 cm

1 cm

$$1 \text{ mL} = 1 \text{ cm}^3$$

- Record the capacity in millilitres of several different cans.

- Measure the diameter and the height of each can.

- Use a graphing calculator to calculate each volume in cubic centimetres.

- Compare the volume in cubic centimetres with the capacity in millilitres marked on the label, both numerically and graphically.

You will need several cans of different sizes, with capacities marked in millilitres, a ruler, and a graphing calculator.

1. Copy the table below, with a row for each can. Measure and record the diameter and the height of each can. Record the contents and the capacity in millilitres.

Contents of can	Diameter (cm)	Height (cm)	Capacity (mL)

2. Follow these steps to enter the data into a graphing calculator.

   a) Press STAT 1.

   b) If there are any numbers in columns L1, L2, or L3, they must be cleared. Move the cursor into each column heading, then press CLEAR ENTER.

   c) Enter the diameters in list L1, the heights in list L2, and the capacities in list L3.

   d) Check that the data entered are correct. Use the arrow keys to go back to fix any errors you may have made entering the data.

3. You will need to calculate the volumes of the cans. Since the formula for volume involves the radius, the radius of each can is required instead of its diameter.

a) Move the cursor into the heading of the L1 column.

b) Press $\boxed{\text{2nd}}$ $\boxed{1}$ $\boxed{\div}$ $\boxed{2}$ $\boxed{\text{ENTER}}$. The calculator divides every number in L1 by 2 and inserts the results in L1. List L1 now contains radii instead of diameters.

4. You are now ready to calculate the volumes of the cans.

a) Up to now, you have used lists L1, L2, and L3. The list editor contains three more lists that you can access using the arrow keys. Move the cursor to the heading of the L4 column, and clear L4 if necessary.

b) Move the cursor into the heading of the L4 column again. Press $\boxed{\text{2nd}}$ $\boxed{\wedge}$ $\boxed{\text{2nd}}$ $\boxed{1}$ $\boxed{x^2}$ $\boxed{\text{2nd}}$ $\boxed{2}$ $\boxed{\text{ENTER}}$. The calculator will multiply π by the square of each number in L1 then by the corresponding number in L2 and place the product in L4. Hence, L4 contains the volumes of the cans calculated using the formula $V = \pi r^2 h$.

5. You can now compare the calculated volumes with the capacities that were marked on the labels. Move the cursor to the top of the L3 column. Compare the numbers in columns L3 and L4. Does there appear to be a relationship between the numbers in these columns? Explain.

Do either Part A, Part B, or both.

Part A Numerical Model

In this model, you will divide the volume by the capacity and determine the mean of the results.

6. Move the cursor into the heading of the L5 column.

a) Press $\boxed{\text{2nd}}$ $\boxed{4}$ $\boxed{\div}$ $\boxed{\text{2nd}}$ $\boxed{3}$ $\boxed{\text{ENTER}}$. The calculator divides every number in L4 by the corresponding number in L3, then inserts the results in L5.

b) Look at the numbers in the L5 column. These numbers should be close to 1. Explain.

7. You can calculate the mean of the numbers in the L5 column.

a) Press $\boxed{\text{2nd}}$ $\boxed{\text{MODE}}$ to leave the list editor. Press $\boxed{\text{2nd}}$ $\boxed{\text{STAT}}$ for LIST. Press $\boxed{\blacktriangleright}$ $\boxed{\blacktriangleright}$ to get to the Math menu. Press $\boxed{3}$ to choose mean(. Press $\boxed{\text{2nd}}$ $\boxed{5}$ $\boxed{)}$ $\boxed{\text{ENTER}}$. The calculator will calculate the mean of the numbers in list L5.

b) What does the mean tell you about how the volumes compare with the capacities?

Part B Graphical Model

In this model, the calculator displays a graph of the volumes against the capacities. When the graph appears on the screen, the axes will be the ones shown in the blank screen at the right.

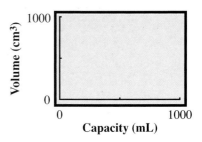

8. Press ⬚2nd⬚ ⬚Y=⬚ for the Stat Plot menu. Choose Plot 1. Make sure that L3 appears beside Xlist and L4 appears beside Ylist.

9. a) Press ⬚WINDOW⬚. Look at the screen above. Enter appropriate values for Xmin, Xmax, Xscl, Ymin, Ymax, and Yscl.

   b) Press ⬚Y=⬚, and clear any equations. Check that only Plot 1 is highlighted. If Plot 2 or Plot 3 is highlighted, use the arrow keys and ⬚ENTER⬚ to turn it off. Then press ⬚GRAPH⬚. You should see a scatter plot of the data.

10. a) Press ⬚Y=⬚. Make sure the cursor is beside Y1=. Press ⬚X,T,θ,n⬚ ⬚GRAPH⬚. The calculator graphs the line $y = x$. On this line, the x- and y-coordinates of the points are equal.

   If the points on the scatter plot appear to be above the line, complete part b. If they appear to be below the line, complete part c.

   b) Press ⬚Y=⬚ ⬚☐⬚ ⬚X,T,θ,n⬚ ⬚ENTER⬚ ⬚GRAPH⬚, where ⬚☐⬚ represents a number slightly greater than 1. Repeat using other numbers until your line passes as close as possible through the plotted points.

   c) Press ⬚Y=⬚ ⬚☐⬚ ⬚X,T,θ,n⬚ ⬚ENTER⬚ ⬚GRAPH⬚, where ⬚☐⬚ represents a number slightly less than 1. Repeat using other numbers until your line passes as close as possible through the plotted points.

11. Write the equation of the line you found in exercise 10. What does this equation tell you about how the volumes compare with the capacities?

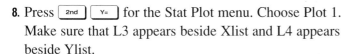

Write a report of your findings that someone who did not do the investigation could understand. Describe the problem you investigated, the conclusions you reached, and how you arrived at your conclusions. Use written explanations, tables, graphs, equations, or calculations.

1.5 Using a Motion Detector to Collect Data

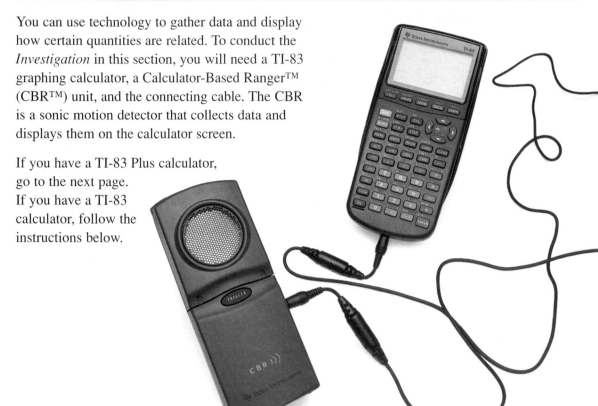

You can use technology to gather data and display how certain quantities are related. To conduct the *Investigation* in this section, you will need a TI-83 graphing calculator, a Calculator-Based Ranger™ (CBR™) unit, and the connecting cable. The CBR is a sonic motion detector that collects data and displays them on the calculator screen.

If you have a TI-83 Plus calculator, go to the next page. If you have a TI-83 calculator, follow the instructions below.

Checking for the RANGER program

Your calculator must have a program called RANGER. Follow these steps to check this:

- Turn on the calculator, and press [PRGM]. A list of programs stored in memory will appear. If RANGER appears in the list, your calculator has the program, and you can go on to the next page. If RANGER does not appear, continue with the following steps.

Transferring the RANGER program to the TI-83 calculator

- Connect the CBR to the calculator with the connecting cable.
- On the calculator, press [2nd] [X,T,θ,n] [▶] [ENTER]. The calculator indicates that it is waiting to receive the program.
- On the CBR, open the pivoting head and press (82/83). The CBR will send the program to the calculator.

How Fast Are You?

A CBR can measure and record your distance as you walk away from the CBR. The calculator displays a graph to show how this distance changes as you move. It also records the data in a table. You will use the data collected to calculate your speed in metres per second.

Step 1. Set-Up

- Work with a partner.

- Work in a space where one person can walk some distance away from the CBR, and the other can hold the CBR and the calculator. The CBR will record distances up to about 6 m. Make sure there are no objects in the clear zone.

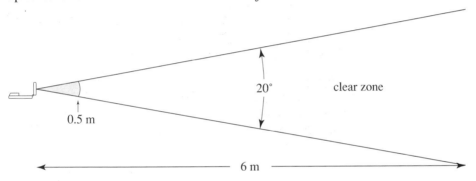

20°   clear zone

0.5 m

6 m

- Connect the CBR to the calculator with the connecting cable. Make sure the connections are secure.

For the TI-83 Plus

- Turn on the calculator, and press [APPS].
- Press **2** to choose CBL/CBR.
- Press any key.
- Press **3** to choose RANGER.
- Press [ENTER] to display the main menu.
- Press **1** to choose SETUP/SAMPLE.

For the TI-83

- Turn on the calculator, and press [PRGM].
- Choose RANGER.
- Press [ENTER] [ENTER] to display the main menu.
- Press **1** to choose SETUP/SAMPLE.

- Make sure the cursor is beside START NOW. If it is not, use the arrow keys to move it there.

Step 2. Collect and graph the data.

- When you are ready to start, press [ENTER]. Follow the instructions on the screen. As one person walks away from the CBR, the other will see the calculator display a distance-time graph recording the distance to the first person.

- To make the calculator redraw the graph in a better position on the screen, press [ENTER] **1**. The graph should look similar to this screen. It shows that the person started about 0.5 m from the CBR, walked at a steady rate for about 8 s, then stood still about 4.5 m from the CBR.

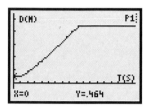

- To repeat, press [ENTER] **3**. Try to get the points plotted during walking as close to a straight line as possible.

- Leave the graph on the screen while you complete the following exercises. Since the data for this graph are now in your calculator, you can disconnect the cable so that others can use the CBR.

1. What quantity is represented along each axis?

   a) the horizontal axis   **b)** the vertical axis

2. Describe what would happen to the plotted points in each case.

   a) The person moves faster.   **b)** The person moves slower.
   c) The person moves toward the CBR instead of away from it.

3. a) Sketch the graph you obtained with the CBR, or use a computer linkup to print it. Label the axes, including units.

   b) Describe what your screen represents in terms of how the person moved toward or away from the wall.

Viewing the coordinates and calculating the average speed

You can view the coordinates of some points on the graph. These correspond to the data the CBR gathered.

- Press [ENTER] and use the arrow keys to move a flashing cursor along the graph. Its coordinates will appear on the screen. The *x*-coordinate shows the time in seconds, and the *y*-coordinate shows the distance from the CBR in metres.

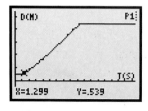   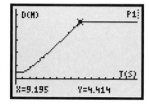

4. The first screen shows the cursor where the person started walking. The second screen shows the cursor where the person stopped. Use the coordinates on the screens to answer these questions.

a) On the first screen, how far was the person from the CBR when the calculator plotted the point under the flashing cursor? How many seconds after the plotting did this occur?

b) Repeat part a using the second screen.

c) How far did the person walk? How many seconds did it take?

d) What was the person's average speed?

5. a) On your screen, use the arrow keys to move the cursor to the points where the person started and stopped walking. Write the coordinates of these points on the sketch of the screen you made in exercise 3.

b) Repeat exercise 4, parts c and d.

Viewing the data in lists

You can also view the data in lists, but you have to exit the program first.

- Press ENTER to get to the main menu, then choose **6** to quit.
- Press STAT **1** to get to the list editor.

List L1 contains the times in seconds. List L2 contains the corresponding distances from the person to the CBR in metres. The calculator used these numbers to plot the points on the graph. The results are shown to 5 decimal places, but the CBR cannot measure distances and times this accurately.

L1	L2	L3	1
0	.46447	0	
.19988	.47133	0	
.29983	.47806	0	
.49971	.48863	0	
.69959	.49728	0	
.79954	.50112	0	
.99942	.5062	0	

L1(1)=0

6. Use the arrow keys to move to the rows in the table that correspond to the points you used to calculate the average speed in exercise 5. Verify that the coordinates in the table are those you used to calculate the average speed.

- After viewing the data, press 2nd MODE to quit.

Communicating *the* IDEAS

In the *Investigation*, the graphs generated by the CBR showed how distance travelled and elapsed time are related for a person walking away from the CBR. Suppose the person walks at a constant speed. Write to explain why the graph should be a straight line.

Using *The Geometer's Sketchpad*

Dynamic software, such as *The Geometer's Sketchpad*, allows us to explore coordinate geometry in a new way. We can plot points, measure lengths and areas, and determine equations of lines. We can also examine how these quantities change as part of a diagram is changed.

This introduction will help you become familiar with the software before you do other investigations later in this book.

INVESTIGATION 1

The Toolbox

From the File menu, choose New Sketch.

1. On the left of the screen is a column of small squares called the Toolbox. To work with *The Geometer's Sketchpad*, you must always select one of these tools.

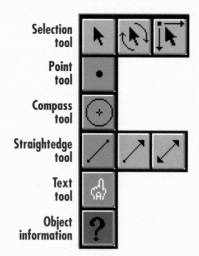

Selection tool

Point tool

Compass tool

Straightedge tool

Text tool

Object information

There are 3 Selection tools. To see them, click on the Selection tool, hold down the mouse button, and drag to the right. Beside the Translation arrow is a Rotation arrow and a Dilatation arrow. To choose either of these, drag to highlight the tool you want, then release the mouse button.

There are 3 Straightedge tools. To see them, click on the Straightedge tool, hold down the mouse button, and drag to the right. Beside the Segment tool, there is a Ray tool and a Line tool.

Follow the instructions below to explore some features of the Toolbox.

2. *Construct a triangle and label its vertices.*

Use this tool ...   *... and do this:*

a) Draw 3 points by clicking anywhere on the screen. Notice that a point has a black outline when first drawn. This means that the point is *selected*. The point is deselected when a second point is drawn.

b) As you move closer to a point, the hand turns black. Click, and the point will be labelled. Click again and the label disappears.

c) Labels can be changed to other letters or even words. Double-click on the label, not the point. A dialog box appears, and you can type a new label.

d) Click on a label and drag. You can move a label to different positions around a point.

e) Hold down the shift key and click on the 3 points. Make sure all 3 points are selected. From the Construct menu, choose Segment. A triangle will be constructed.

f) Click on any vertex and drag it to see what happens. Repeat by dragging any side.

g) Hold down the shift key and click on two vertices. Drag to see what happens.

h) Hold down the shift key and click on two sides. Drag to see what happens.

i) Hold down the shift key and click on all three vertices or sides. Drag to see what happens.

3. *Construct a line.*

Use this tool ...   *... and do this:*

a) Click on the Straightedge tool, hold down the mouse button, and drag to the right to select the Line tool. The dots on the tool change to arrowheads.

b) Hold down the mouse button anywhere on the screen, and drag the mouse. A line is drawn. Notice the two little black squares. These mean that the line is selected.

c) The line has two points on it. Drag each point to change the position of the line. Then drag the line to change its position again.

d) Hold down the shift key and click to construct two points. From the Construct menu, choose Line.

Find out what happens if you hold down the shift key to construct three points, then choose Line in the Construct menu.

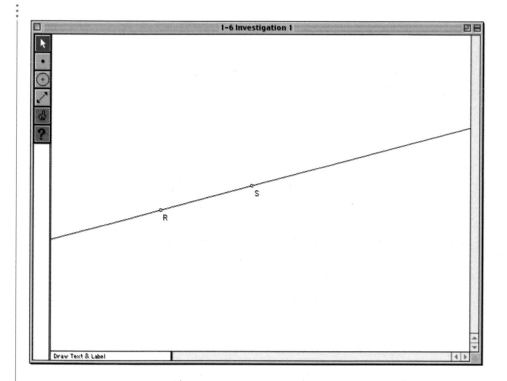

Sketchpad Tips

When an object has black squares on it, or when a point is black, the object is *selected*. To do something with an object, it must be selected. To *deselect* an object, click on the Selection tool, then click anywhere on a blank part of the screen.

To select two or more objects at the same time, hold down the shift key and click on the objects with the Selection tool. This is called *shift-clicking*.

INVESTIGATION 2

Graphing Points and Lines

From the File menu, choose New Sketch.

There are nine menus at the top of the screen. Click on each menu to display its contents.

Many of the options are grey. These cannot be chosen unless certain objects are selected.

In this investigation, you will explore the Graph menu and learn how to use *The Geometer's Sketchpad* to work with coordinates of points and equations of lines.

Set-Up

- From the Display menu, choose Preferences.
  In the window that appears, click Autoshow Labels for Points. This will make sure that every point is automatically labelled with a letter when it is constructed.
  The distance unit should be "cm." If it is not, click the down arrow beside Distance Unit, and choose "cm." Also, make sure the three Precision boxes show "hundredths." Then click OK.
- Make sure the Line tool is in the toolbox, and the Selection tool is chosen.

Construct a coordinate system.

1. From the Graph menu, choose Create Axes. The axes will appear.
   To show a grid, from the Graph menu choose Show Grid.
   You can use the same method to hide the grid if you want.

2. Two small circles show the points A(0, 0) and B(1, 0). You can double-click the labels A and B with the text tool and change them to other letters if you want.

   a) Predict what will happen if you drag point A. Check your prediction.

   b) Predict what will happen if you drag point B. Check your prediction.

   c) Predict what will happen if you drag either axis. Check your prediction.

Plot points and lines.

3. Follow these steps to plot the line through C(0, 3) and D(1, 5).

   a) From the Graph menu, choose Plot Points. Type 0 to enter 0 under "*x*." Press the tab key and type 3 to enter 3 under "*y*." Press the return (or enter) key. Type 1, press the tab key, type 5, and press return (or enter). Click Plot.

   b) From the Construct menu, choose Line. The computer graphs the line through C and D.

Calculate the equation of a line.

4. From the Measure menu, choose Slope. The equation of the line through C and D will be displayed.

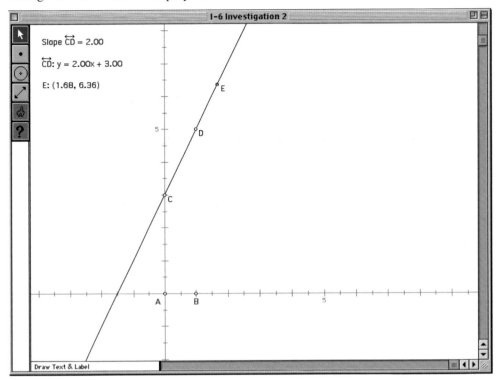

Verify properties of coordinates.

5. a) Make sure line CD is still selected. From the Construct menu, choose Point On Object. A new point E will be constructed on the line. To check that it is on the line, try dragging the point. It will only move along the line.

b) Make sure point E is selected. From the Measure menu, choose Coordinates. The coordinates of E will be displayed. Drag point E along the line and observe what happens to its coordinates.

6. In exercise 4, the equation of the line CD was displayed as $y = 2.00x + 3.00$, or $y = 2x + 3$. This means that for any point (x, y) on the line, its coordinates satisfy this equation. That is, if you double the x-coordinate and add 3, the result should be the y-coordinate. You can demonstrate this is true for many different points on the line.

a) Click on the coordinates of E to select them. From the Measure menu, choose Calculate. A calculator appears.

b) Click 2 $\boxed{*}$. Click on Values, click Point E, and click x. Click $\boxed{+}$ 3. The expression $2 \cdot x_E + 3$ appears on the calculator screen. Click OK. The computer displays the value of this expression.

c) Compare the value of this expression with the y-coordinate of E. What do you notice?

d) Predict what will happen to the value of the expression if you drag E along the line. Drag to check your prediction.

7. You can also demonstrate that the coordinates of points that are not on the line do not satisfy its equation.

a) Use the Point tool to construct a point F anywhere on the screen.

b) From the Measure menu, choose Coordinates. The coordinates of F appear.

c) Click on the coordinates of F to select them. From the Measure menu, choose Calculate.

d) Click 2 $\boxed{*}$. Click on Values, click Point F, and click x. Click $\boxed{+}$ 3. Click OK. The computer displays the value of $2x + 3$, where x is the x-coordinate of F.

e) Compare the value of this expression with the y-coordinate of F. What do you notice?

f) Predict what will happen to the value of the expression if you drag point F around the screen. Drag the point to check your prediction.

Communicating the IDEAS

List some advantages and disadvantages of using computer software, such as *The Geometer's Sketchpad*, to explore coordinate geometry.

Testing Your Knowledge

The concepts in this chapter form a toolkit of important ideas. You can use these concepts to solve problems.

Mathematics Toolkit

Function Tools

- A function is a rule that gives a single output number for every valid input number.
- Points on the graph of a linear function lie on a straight line.
- Points on the graph of a non-linear function do not lie on a straight line.

1. A 1999 newspaper article about the Canadian National Exhibition contained a graph similar to this.

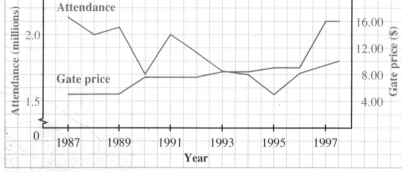

a) What does "gate price" mean?

b) Copy this table. Show each year from 1987 to 1998. Use the graph to complete the second and third columns.

Year	Gate price ($)	Attendance (millions)	Revenue from ticket sales (millions of $)
1987			
.			
.			
.			
1998			

c) Multiply the second and third columns to complete the fourth column.

d) Use the first and fourth columns. Draw a graph to show the revenue from ticket sales each year from 1987 to 1998.

e) What does the graph tell you about how the revenue from ticket sales has changed since 1987?

2. Graph the function represented by each equation below. If you use a graphing calculator, sketch each graph.

a) $y = 2x - 4$ **b)** $y = 3 + 0.25x$ **c)** $y = x - 5$

d) $y = 1.5x$ **e)** $y = 2x^2$ **f)** $y = x^2 - x$

3. The approximate temperatures of Earth's atmosphere at different altitudes up to 10 km are shown.

a) Draw a graph to show temperature as a function of altitude.

b) Use the graph.
  i) Determine the temperature at an altitude of 7 km.
  ii) Determine the altitude where the temperature is 0°C.

c) Above 11 km, the temperature remains fairly constant at −56°C. Show this on the graph.

Altitude (km)	Temperature (°C)
0	15
2	2
4	−11
6	−24
8	−37
10	−50

4. The sign at the right was seen on the back of a tractor trailer.

a) Draw a graph to show how stopping distance is related to the truck's speed.

b) Use the graph. Predict the stopping distance for a truck travelling at 100 km/h.

c) Use the graph. Suppose it took a truck 145 m to stop. Estimate how fast the truck was travelling before the driver applied the brakes.

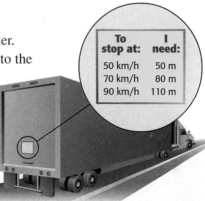

To stop at:	I need:
50 km/h	50 m
70 km/h	80 m
90 km/h	110 m

5. When you exercise, your heartbeat increases. There is a recommended safe maximum rate that a person's heartbeat should not exceed. This maximum rate depends on age. The chart shows the maximum recommended heart rates for different ages.

Age (in years)	Maximum heart rate (beats/min)
30	194
40	182
50	170

a) Draw a graph to show the recommended maximum heart rate as a function of age.

b) Use the graph. Determine the recommended safe maximum heart rates for a 20-year-old, a 70-year-old, and a 35-year-old person.

c) What is your recommended maximum rate?

d) Is it reasonable to use the graph for people older than 70 years or younger than 20 years? Explain.

6. The boiling point of water, T degrees Celsius, depends upon the height, H metres, above sea level. An equation that relates these two variables is $T = -0.0034H + 100$.

a) Graph the equation. Sketch the graph.

b) The highest mountain in the world is Mt. Everest at 8850 m. Use the graph. Determine the temperature at which water boils at the top of Mt. Everest.

7. The Dutch painter, Piet Mondrian, is known for his modern paintings containing only rectangles. *Tableau 1* on page 49 is one example.

a) Measure the base and the height of each rectangle in the painting. Summarize the results in a table.

b) Use a graphing calculator. Go to the list editor, and clear lists L1 and L2. Enter the bases in list L1 and the heights in list L2.

c) Graph *Height* against *Base* using an appropriate window.

d) On the same screen, graph the line $y = x$.

e) Sketch the graph on grid paper. Show the plotted points and the line $y = x$.

f) Some plotted points should be above the line, and others should be below the line. How do the rectangles for the points above the line differ from the rectangles for the points below the line? Explain.

g) Would it be possible for a rectangle to correspond to a point on the line? Explain.

2 Linear Functions

MATHEMATICAL MODELLING

When Will the Competitors Appear?

The Ironman Canada Triathlon is a race consisting of a 3.8-km swim, a 180-km bike race, and a 42.2-km marathon run. The best athletes in the world take just over 8 h to complete it. Some competitors take as long as 17 h.

Suppose you are a newspaper reporter writing an article about this year's race. You want to include maps to show the times when the competitors are expected to appear at key locations along the route. How can you predict when the competitors are likely to appear at these locations?

Here are some questions to consider.

1. Which part of the race do you think would take the most time? the least time?

2. What are some factors that might affect a competitor's performance in the race?

3. Will you be able to predict exactly when the competitors will appear at key locations along the route? Explain.

You will return to this problem in Section 2.6.

 FYI Visit **www.awl.com/canada/school/connections**

For information related to the above problem, click on <u>MATHLINKS</u>, followed by Foundations of Mathematics 10. Then select a topic under When Will the Competitors Appear?

Preparing for the Chapter

This section reviews these concepts:

- The slope of a line
- The distributive law
- Solving an equation
- Substituting

INVESTIGATION

The Speed of an Airplane

The table shows the time taken and the distance travelled by an airplane after it has reached its cruising altitude.

Time (h)	0	0.1	0.2	0.3	0.4	0.5
Distance (km)	0	60	120	180	240	300

We know that average speed = $\frac{\text{distance travelled}}{\text{time taken}}$.

1. Use the table and the equation above. Determine the average speed of the plane.

2. On grid paper, plot the data from the table above. Graph distance against time, with *Time* along the horizontal axis. Choose suitable scales.

3. Choose two points on the graph. Calculate the slope of the line. What are the units of the slope?

4. How does the slope of the line compare with the average speed of the plane?

5. On grid paper, draw a line with each slope.

   a) $\frac{2}{3}$ b) $-\frac{4}{3}$ c) -2

6. Expand.

   a) $4(3x + 5)$ b) $5(2x - 3)$ c) $-4(3x + 6)$ d) $-2(7x - 3)$

7. Solve each equation and check.

   a) $x + 9 = 5$ b) $x - 3 = 5$ c) $x - 4 = 2x$ d) $8x = 24$

8. Substitute $x = 2$ in each equation to determine y.

   a) $y = x + 1$ b) $y = 3(x + 1)$ c) $y = 3(x + 1) + 2$

   d) $y = x - 1$ e) $y = 3(x - 1)$ f) $y = 3(x - 1) - 2$

Relationships between two quantities can be described in many ways: in words, by a graph, a table, or an equation.

During the summer, Ryan works in a bicycle store. He is paid $6.50 an hour. He is paid for part hours worked.

If he knows the number of hours he has worked, he can calculate the amount he has earned.

This calculation can be thought of as the working of a machine:

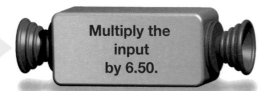

Input:
Number of hours worked

Multiply the input by 6.50.

Output:
Pay in dollars

INVESTIGATION 1

Relations and Equations

1. Determine Ryan's earnings for each number of hours worked.

    a) 1   **b)** 2   **c)** 3   **d)** 4   **e)** 5   **f)** 6   **g)** 7

2. Copy and complete this table.

Hours worked	Earnings ($)	Difference ($)
1	6.50	
2	13.00	$13.00 - 6.50 = 6.50$
3		
4		
5		
6		
7		

3. Describe as many patterns in the table as you can.

4. a) Graph the data in the table. Plot *Hours worked* horizontally and *Earnings* vertically.

b) Decide whether it is appropriate to join the points on the graph. Explain your decision.

c) Explain why the points lie on a straight line.

5. a) One day, Ryan felt ill and went home after working for only 3.5 h. Use the graph to estimate his earnings that day.

b) If Ryan did not go to work the next day, he would earn nothing. Explain how this is represented on the graph.

6. a) Select two points on the graph. Label them A and B. Determine the slope of the segment AB.

b) Select a different point on the graph and label it C. Determine the slope of the segment AC.

c) Select a fourth point on the graph and label it D. Determine the slope of the segment AD.

d) Compare the slopes you calculated in parts a, b, and c with the numbers in the *Difference* column in the table. Describe what you notice. How does the slope relate to Ryan's rate of pay?

7. Suppose h represents the number of hours worked and e represents Ryan's earnings in dollars.

a) Write an equation to express e in terms of h.

b) Check that your equation is correct by testing some numbers from the table.

In *Investigation 1*, you wrote an equation that expressed the earnings, e, in terms of the number of hours worked, h. If you know a value for h, the corresponding value for e can be calculated. This means that e is a *function* of h.

The equation showed that e is a constant multiple of h. We say that e *varies directly* as h.

Take Note

Direct Variation

- Direct variation is a relationship in which one quantity is a constant multiple of the other.
- The graph of a direct variation situation is a straight line through the origin.
- The equation of a direct variation situation has the form $y = mx$, where m is the slope of the graph.

Linear Functions

Set-Up

- Press [Y=]. Use [CLEAR] and the arrow keys to clear all equations.

- If any of Plot 1, Plot 2, or Plot 3 at the top of the screen is highlighted, use the arrow keys and [ENTER] to remove the highlighting.

```
WINDOW
 Xmin=-4.7
 Xmax=4.7
 Xscl=1
 Ymin=-6.2
 Ymax=6.2
 Yscl=1
 Xres=
```

- Press [WINDOW], then change the settings to those shown at the right.

1. Press [Y=]. Make sure the cursor is beside Y1. Press 2 [X,T,θ,n] [+] 3 to enter the first equation in the list below. Press [GRAPH]. Sketch the graph and write the equation beside it.

   a) $y = 2x + 3$ **b)** $y = x^2 - 2$ **c)** $y = 3x$ **d)** $y = \frac{1}{x}$

   e) $y = -x^2 + 3$ **f)** $y = -2x + 3$ **g)** $y = x + \frac{1}{2}$ **h)** $y = 0.4x + 0.2$

   i) $y = \sqrt{x - 2}$ **j)** $y = x^3 + 1$

2. Press [Y=]. Move the cursor to the equals sign beside Y1. Press [ENTER] to remove the highlighting. Press [▼] [►] to move the cursor to the line below. Enter the equation in exercise 1b. Press [GRAPH]. The calculator graphs only the equation with the equals sign highlighted. Sketch the graph and label it with its equation.

3. Repeat exercise 2 for each function in exercise 1. All ten equations remain in the Y= list.

4. Use your results from exercises 1, 2, and 3. List the functions that have straight-line graphs. Identify any equation that represents a direct variation.

5. Describe the similarities among the equations you listed in exercise 4.

6. Examine the differences in the tables of values for the functions. Press [Y=]. Use the arrow keys and [ENTER] to highlight the equals signs of all the equations. Press [2nd] [WINDOW] for [TBLSET]. Create the settings below left.

```
TABLE SETUP
 TblStart=0
 ΔTbl=1
 Indpnt: Auto Ask
 Depend: Auto Ask
```

X	Y1	Y2
0	3	-2
1	5	-1
2	7	2
3	9	7
4	11	14
5	13	23
6	15	34

X=0

Press 2nd GRAPH for TABLE to see tables of values for the first two functions (page 55, right).

Press ► to examine tables of values for all ten functions. Describe any patterns you see in the differences in *y*-values for each function.

7. Use the tables of values to determine the slope, *m*, of each straight line graph. Write the slope on the sketch of the graph of the function.

8. Recall that the *y*-intercept is the *y*-coordinate of the point of intersection of a graph with the *y*-axis.

   a) What is the *x*-coordinate of every point on the *y*-axis?

   b) Use the tables of values to determine the *y*-intercept of each straight line graph. Write these on the sketches of the graphs.

9. Use the results of exercises 7 and 8. Explain how you can find the slope and *y*-intercept of a line from its equation.

The functions you studied in *Investigations 1* and *2* that had straight line graphs also had constant differences in their tables of values. These functions are called *linear functions*.

Take Note

Linear Function

- The graph of a linear function is always a straight line.
- A linear function can be represented by an equation of the form $y = mx + b$, where m and b are numbers. The number, m, is the slope of the line and the number, b, is its *y*-intercept.
- The differences of successive *y*-values are constant for consecutive *x*-values.

Example

The cost, C dollars, to rent a room for a party is given by the equation $C = 100 + 5n$, where n is the number of people who plan to attend. The room can accommodate a maximum of 120 people.

a) Graph the equation.

b) Determine the slope of the line. Explain what it means.

Solution

a) Write $C = 100 + 5n$ as $y = 100 + 5x$.

Set-Up

- Press WINDOW. Change the settings to Xmin = 0, Xmax = 120, Xscl = 10, Ymin = 0, Ymax = 1000, Yscl = 100.
- Press Y= . Use CLEAR and the arrow keys to clear all equations.
- If any of Plot1, Plot 2, or Plot 3 at the top of the screen is highlighted, use the arrow keys and ENTER to remove the highlighting.

Place the cursor beside Y1= and press 1 0 0 + 5 X,T,θ,n to enter the equation $y = 100 + 5x$.

Press GRAPH to display the graph below left.

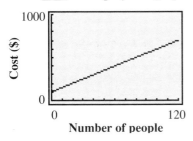

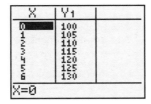

b) To determine the slope of the line, display a table of values.

Press 2nd WINDOW for TBLSET. Make sure that 0 is beside Tblstart and 1 is beside △Tbl. This sets the table so the values of x begin at 0 and increase by 1. Make sure that Auto is highlighted in the last two lines.

Press 2nd GRAPH for TABLE. A table of values appears (above right). The constant difference in the y-values is 5.

This difference, 5, represents the increase in cost for an increase of 1 in the number of people attending.

Recall that the slope of a line is the constant difference for successive x-values. The slope of the line is 5. It represents the cost in dollars for each person who attends the party.

In the *Example*, the graph is a straight line that does not pass through the origin. We say that the cost C *varies partially* as n, the number of people attending the party.

Partial Variation

- Partial variation is a relationship in which one quantity is a multiple of another, plus a constant.

- The graph of a partial variation situation is a straight line that does not pass through the origin.

- The equation of a partial variation situation has the form $y = mx + b$, where m is the slope and b is the y-intercept.

Discussing the IDEAS

1. In *Investigation 1*, suppose Ryan's pay was increased to $7 per hour. Describe how each item would be affected.

   a) the table of values (hours worked, earnings, difference)

   b) the graph   **c)** the equation

2. In the *Example*, what does 100 represent in the equation $C = 100 + 5n$?

3. Explain how you can tell from the table of values of a function whether its graph is a straight line.

4. Explain how to determine the slope of a line from its table of values.

5. How can you tell from the equation of a function whether its graph is a straight line?

6. Suppose the graph of a relationship is a non-vertical straight line. How can you tell whether it represents a direct variation relationship or a partial variation relationship?

2.1 Exercises

  1. Complete a table of values for each function.

   a) $y = 3x - 4$ **b)** $y = x + 2$ **c)** $y = -5x + 3$

Input x	Output y	Differences
0		
1		
2		
3		

B **2. a)** Predict which functions have straight-line graphs.

   i) $y = x^2 + 5$ **ii)** $y = x + 2$ **iii)** $y = -5x$

   iv) $y = -2x - 1$ **v)** $y = 2x^3$ **vi)** $y = 0.1x - 1.3$

   vii) $y = \frac{2}{5}x - \frac{1}{5}$ **viii)** $y = \sqrt{2x - 1}$ **ix)** $y = 5 - 2x^2$

   b) Check your predictions.

3. An input/output machine does this:

   Multiplies the input by 4, then subtracts 5.

   Write the equation of the function represented by this machine.

4. Describe each equation in terms of an input/output machine.

   a) $y = -2x + 5$ **b)** $y = 3x + 3$ **c)** $y = 2 - 5x$

5. Melissa works in a grocery store. She earns $6.80 per hour. She is paid for part hours worked.

   a) Create a table with a difference column for Melissa's earnings. Include the earnings for up to 20 h of work.

   b) Graph the data in the table. Plot *Hours worked* horizontally and *Earnings* vertically.

   c) Use the graph to determine Melissa's earnings when she works 15.5 h.

   d) Explain how the graph would change if Melissa's rate of pay increased to $7.25/h.

6. a) This table shows some data for a function that is a direct variation. Copy and complete the table.

Input x	Output y	Differences
0		
1		
2	6	
3		
4		
5	15	

   b) Describe this function in terms of an input/output machine. Write the equation that describes the function.

7. a) This table shows some data for a function that is a partial variation. Copy and complete the table.

Input x	Output y	Differences
0	3	
		4
1	7	
		4
2		
		4
3		

b) Describe this function in terms of an input/output machine. Write the equation that describes the function.

8. The cost to rent a powerboat from a local marina is $35, plus $5 per hour.

a) Suppose you pay for part of an hour. What will it cost to rent the boat for 3.5 h?

b) Write an equation that relates the cost to rent the boat, C dollars, to the number of hours, h.

c) Suppose the graph of this function were sketched with C on the vertical axis and h on the horizontal axis. What would be the slope of the line?

9. A salesperson works in a cosmetics booth. The salesperson earns a basic salary of $325 per week plus 10% of the total weekly sales.

a) Copy and complete this table.

Weekly sales ($)	10% of weekly sales ($)	Earnings ($)
200	$0.1 \times 200 = 20$	$325 + 20 = 345$
400	$0.1 \times 400 = 40$	$325 + 40 = 365$
600		
800		
1000		
1200		

b) Write an equation that relates the earnings, E dollars, to the weekly sales, s dollars.

c) Use the table to explain why this function is linear.

10. Use this input/output machine.

a) Find the output when the input is 5.

b) Find the output when the input is –3.

c) The output is 10. What was the input?

d) The output is the same as the input. What was the input?

11. a) Copy and complete this table.

Input x	Output y	Differences
0		
		2
1		
		2
2	8	
		2
3		
		2
4		

b) Write the equation that describes the function.

C **12. a)** Describe the graph of the data in each table.

i)

x	y	Differences
−3	28	
−2	24	
−1	20	
0	16	
1	12	

ii)

x	y	Differences
0	4	
1	4	
2	4	
3	4	
4	4	

b) Write the equation that describes each function.

Communicating *the* IDEAS

A function can be described in different ways: by an equation, by a table of values, by a graph. For each method, write to describe how you can tell if the function is linear.

The Equation of a Line: $y = m(x - p) + q$

In an earlier grade, you learned that the graph of the equation $y = mx + b$ is a straight line with slope m and y-intercept b. This is called the *slope y-intercept* form of the equation of a line. If we know the slope and the coordinates of the point where the line intersects the y-axis, we can use this form to write the equation.

Often we have information about the slope of the graph of a linear function and we know the coordinates of one point on the line.

INVESTIGATION

The Point Slope Form of the Equation of a Line

1. **a)** Copy and complete this table for the equation $y = 3(x - 1) + 2$.

    b) Graph the equation.

    c) Describe as many patterns in the table and in the graph as you can. What are the relationships among the slope of the line, the differences in the table, and the numbers in the equation?

x	*y*	Differences
0		
1		
2		
3		
4		
5		
6		
7		
8		

   d) The numbers 3, 1, and 2 appear in the equation. What do these numbers represent? Explain.

2. Repeat exercise 1 for each equation.

    a) $y = 3(x - 5) + 14$ **b)** $y = 3(x - 8) + 23$

3. **a)** Look at the patterns in the equations and in the tables in exercises 1 and 2. How are they similar?

    b) Simplify each equation using the rules of algebra.
      i) $y = 3(x - 1) + 2$ **ii)** $y = 3(x - 5) + 14$ **iii)** $y = 3(x - 8) + 23$

    c) Compare the answers in part b. Explain why all the equations in part b generate the same table of values.

4. a) Use the patterns in exercises 1, 2, and 3 to write the equation of the line with slope 4, which passes through the point (5, 2).

b) Show that the line in part a has a slope of 4.

c) Show that the line in part a passes through the point (5, 2).

In the *Investigation*, you should have found that the equation of a line with slope 3, which passes through the point (1, 2), can be written as $y = 3(x - 1) + 2$. This relationship is true for all lines.

Take Note

The Point Slope Form of the Equation of a Line

The equation of a line with slope m, which passes through the point with coordinates (p, q) can be written in the form $y = m(x - p) + q$.

Example 1

a) Determine the equation of each line.
   i) the line through the point (4, –5), with slope 2
   ii) the line through the point (–2, 3), with slope –4

b) Determine the y-intercept of each line in part a.

Solution

a)

Think ...

The equation of a line with slope m through the point with coordinates (p, q) is $y = m(x - p) + q$. Identify the values of m, p, and q for each line. Substitute these numbers in the equation $y = m(x - p) + q$.

i) The line passes through (4, –5), so $p = 4$ and $q = -5$.
The slope is 2, so $m = 2$.
Substitute in the equation $y = m(x - p) + q$.
$$y = 2(x - 4) + (-5)$$
$$y = 2(x - 4) - 5$$

ii) The line passes through (–2, 3), so $p = -2$ and $q = 3$.
The slope is –4, so $m = -4$.
Substitute in the equation $y = m(x - p) + q$.
$$y = -4(x - (-2)) + 3$$
$$y = -4(x + 2) + 3$$

b)

> *Think ...*
>
> The y-intercept of the line with equation $y = mx + b$ is b. Simplify the point slope equations to this form to determine b.

i) $y = 2(x - 4) - 5$

Use the distributive law.

$y = 2x + 2(-4) - 5$

$y = 2x - 8 - 5$

$y = 2x - 13$

Compare this equation with $y = mx + b$: $b = -13$

The y-intercept is -13.

ii) $y = -4(x + 2) + 3$

Use the distributive law.

$y = -4x - 4(2) + 3$

$y = -4x - 5$

Compare this equation with $y = mx + b$: $b = -5$

The y-intercept is -5.

Example 2

Determine the equation of the line through the points A(−3, 6) and B(3, 4).

Solution

> *Think ...*
>
> To use the point slope form of the equation of a line, we need to know the slope of the line, and the coordinates of one point on the line. Use the coordinates of A and B to calculate the slope of the line.

Plot the points on a grid.

The slope of the line joining A and B $= \dfrac{\text{rise}}{\text{run}}$

$$= \frac{4 - 6}{3 - (-3)}$$

$$= \frac{-2}{6}$$

$$= -\frac{1}{3}$$

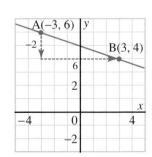

The line passes through B(3, 4), with slope $-\frac{1}{3}$.

Compare with the point slope form of the equation of a line through (p, q) with slope m: $\quad y = m(x - p) + q$

Substitute $p = 3$, $q = 4$, and $m = -\frac{1}{3}$.

The equation of the line through A and B is $y = -\frac{1}{3}(x - 3) + 4$.

Discussing *the* IDEAS

1. In the *Investigation*, you determined that all three equations represented the same line. Suppose you were asked to show that two linear equations had the same graph. How could you do this?

2. In *Example 1*, the point slope form of the equation of a line was used to determine the equations of lines. Describe another method you could use to determine the equations. Which method do you think is easier?

3. In *Example 2*, what is the equation of the line if we use point A rather than point B? Why was point B used?

Practise Your Skills

1. Simplify.
   a) $\frac{3+2}{3-2}$
   b) $\frac{3-2}{-3-2}$
   c) $\frac{-2+3}{-3+2}$
   d) $\frac{-2-3}{2-3}$
   e) $\frac{6+4}{5+1}$
   f) $\frac{6-4}{5-1}$
   g) $\frac{-6+4}{-5-1}$
   h) $\frac{-6-4}{-5+1}$

2. Determine the slope of the line through each pair of points.
   a) A(3, 2), B(4, 7)
   b) C(−2, 1), D(3, 6)
   c) E(−3, −7), F(5, 2)
   d) G(0, 4), H(−1, 6)
   e) J(5, 0), K(2, −3)
   f) L(4, 2), M(−3, −5)

2.2 Exercises

A 1. State the slope of each line and the coordinates of one point on the line.
   a) $y = 3(x - 4) + 1$
   b) $y = 4(x - 4) - 3$
   c) $y = 2(x + 1) + 3$
   d) $y = -2(x - 3) + 2$
   e) $y = -3(x + 2) + 6$
   f) $y = -0.5(x + 3) - 6$

B **2.** Write the equation of the line passing through each point with the given slope.

    a) (3, 2), slope 4   **b)** (–1, 4), slope 2   **c)** (3, –5), slope 3

    d) (5, –6), slope $\frac{1}{2}$ **e)** (2, 3), slope $\frac{2}{3}$ **f)** (3, 5), slope $-\frac{2}{5}$

3. Determine an equation of the line with these properties.

    a) slope 2.5, passing through the point (4, –2)

    b) slope –1, passing through the point (4, 0)

    c) slope 3, passing through the point (–6, 3)

    d) slope –2, passing through the point (–2, –3)

4. Determine the equation of the line joining each pair of points.

    a) (3, 4) and (8, 6)   **b)** (3, –2) and (6, 1)   **c)** (–4, 2) and (2, 5)

    d) (–4, 0) and (2, –6)   **e)** (–2, 0) and (0, 2)   **f)** (5, 0) and (8, 2)

5. Determine the y-intercept of each line.

    a) $y = 3x + 12$ **b)** $y = -2x - 3$ **c)** $y = -6x$

    d) $y = 3(x - 4) + 1$ **e)** $y = -2(x + 2) + 5$ **f)** $y = -3(x - 2) + 3$

    g) $y = -\frac{1}{2}(x + 4) + 5$ **h)** $y = 0.5(x - 3) - 1$ **i)** $y = -0.1(x - 2) - 1$

6. Determine the equation of each line on the grid, below left.

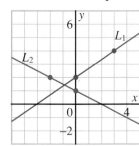 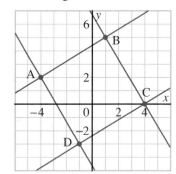

7. Determine the equation of each line on the grid, above right.

8. a) Draw a graph that describes each situation.

      i) y varies directly as x. When $x = 4$, $y = 8$

      ii) y increases by 5 for each consecutive value of x. When $x = 5$, $y = 8$

      iii) y is a multiple of x. When $x = 2$, $y = -8$

      iv) y varies partially as x. The y-intercept is 3, and the differences in y-values for consecutive x-values is 2.

    b) Determine an equation for each graph in part a.

9. Plot the points A(3, 2), B(7, 3), and C(4, 4) on a grid. Determine the equation of the line joining each pair of points.

Patterns in Lines

Some computer programs graph equations. This graph was produced
with *Graphmatica*.

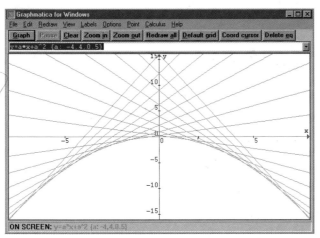

The graphing window shows values of x from -8 to 8 and values of y
from -16 to 16.

10. One line passes through the points $(0, 16)$ and $(4, 0)$. What is the
equation of this line? Rearrange the equation to the form $y = mx + b$.

11. Other lines pass through the following pairs of points. Determine the
equation of each line, then write it in the form $y = mx + b$.

a) $(0, 9)$ and $(3, 0)$ **b)** $(0, 4)$ and $(2, 0)$ **c)** $(0, 1)$ and $(1, 0)$

12. Examine the equations from exercises 10 and 11.

a) Identify any patterns in the equations.

b) Use the patterns to determine the equations of other lines on the
graph. Try to determine the equations of all the lines.

13. Suppose the patterns in the equations were extended to include more
equations. Describe how the graph would change.

Communicating *the* IDEAS

There is certain information you must have before you can determine the equation of a line.
Write to describe two possible ways that you could be given this information. Explain in
each case how you would determine the equation of the line.

2.3 Applications of Linear Equations

Many relationships in business, industry, science, and medicine can be represented by linear functions.

When we substitute a number for one variable in the equation of a linear function, we have a linear equation in one variable.

Example 1

The student council is organizing a dance. The profit from the dance is a function of the number of tickets sold. The function relating the profit, P dollars, and the number of tickets sold, T, is represented by the equation $P = 4T - 1200$, where $T \leq 500$.

a) What is the profit when 450 tickets are sold?

b) How many tickets have to be sold for the profit to be $500?

c) How many tickets have to be sold for the dance to break even?

Solution

a) Use the equation $P = 4T - 1200$.

Since 450 tickets are sold, substitute $T = 450$ then solve for P.
$$P = 4(450) - 1200$$
$$= 1800 - 1200$$
$$= 600$$
The profit is $600.

b) Use the equation $P = 4T - 1200$.
Since the profit is $500, substitute $P = 500$ then solve for T.
$$500 = 4T - 1200$$
Add 1200 to each side.
$$1700 = 4T$$
Divide each side by 4.
$$\frac{1700}{4} = T$$
$$T = 425$$
Four hundred twenty-five tickets must be sold.

c) "Break even" means the profit is zero.

Use the equation $P = 4T - 1200$.

Substitute $P = 0$, then solve for T.

$$0 = 4T - 1200$$

Add 1200 to each side.

$$1200 = 4T$$

Divide each side by 4.

$$300 = T$$

Three hundred tickets must be sold to break even.

In *Example 1b*, an alternative solution would be to solve the equation for T before substituting for P. This is called *isolating the variable*.

$$P = 4T - 1200$$

To solve for T, isolate $4T$ on the right side.

Add 1200 to each side.

$$P + 1200 = 4T$$

Divide each side by 4.

$$\frac{P + 1200}{4} = T$$

The variable T has been isolated.

Now substitute $P = 500$.

$$\frac{500 + 1200}{4} = T$$

$$\frac{1700}{4} = T$$

$$T = 425$$

Example 2

A computer salesperson earns a basic salary of $300 a week, plus a commission of 2% of her weekly sales. The function relating the salary, W dollars, to the sales, s dollars, is represented by the equation $W = 300 + 0.02s$.

a) What is the salary when sales are $15 000?

b) What are the sales when the salary is $550?

c) Solve the equation for s. Determine the sales when the salary is $725.

Solution

a) Use the equation $W = 300 + 0.02s$.

For sales of $15 000, substitute $s = 15\,000$, then solve for W.
$W = 300 + 0.02(15\,000)$
$W = 300 + 300$
$W = 600$
The salary is $600.

b) Use the equation $W = 300 + 0.02s$.
For a salary of $550, substitute $W = 550$, then solve for s.
$550 = 300 + 0.02s$
Subtract 300 from each side.
$250 = 0.02s$
Divide each side by 0.02.
$\frac{250}{0.02} = s$
$s = 12\,500$
The sales are $12 500.

c) Use the equation $W = 300 + 0.02s$.
To solve for s, isolate the variable s.
Subtract 300 from each side.
$W - 300 = 0.02s$
Divide each side by 0.02.
$\frac{W - 300}{0.02} = s$
Substitute $W = 725$.
$\frac{725 - 300}{0.02} = s$
$\frac{425}{0.02} = s$
$s = 21\,250$
The sales are $21 250.

Some linear equations contain fractions. Before we solve an equation of this type, we have to eliminate the denominator(s) of the fraction(s).

Example 3

Solve and check. $\frac{x}{2} + 5 = 3x - 7$

Solution

$\frac{x}{2} + 5 = 3x - 7$

Multiply each side by the denominator 2.

$2\left(\frac{x}{2} + 5\right) = 2(3x - 7)$

Use the distributive law.

$2\left(\frac{x}{2}\right) + 2(5) = 2(3x) - 2(7)$

$x + 10 = 6x - 14$

Add 14 to each side.

$x + 24 = 6x$

Subtract x from each side.

$24 = 5x$

Divide each side by 5.

$\frac{24}{5} = x$

$x = 4.8$

Check.

Substitute 4.8 for x in each side of the equation $\frac{x}{2} + 5 = 3x - 7$.

Left side $= \frac{x}{2} + 5$ Right side $= 3x - 7$

$\quad\quad\quad = \frac{4.8}{2} + 5$ $= 3(4.8) - 7$

$\quad\quad\quad = 2.4 + 5$ $= 14.4 - 7$

$\quad\quad\quad = 7.4$ $= 7.4$

Since both sides are equal, $x = 4.8$ is correct.

Discussing the IDEAS

1. In the solution of *Example 1b*

    a) Why did we add 1200 to each side?

    b) Why did we then divide each side by 4?

2. Following *Example 1*, there is an alternative method to complete part b. Look at the two methods. Which do you prefer? Explain.

3. In the solution of *Example 3*, suppose the first step had been to collect like terms. Then, we would multiply by the denominator 2. Complete the solution this way. Which solution is easier? Explain.

1. Solve each equation.

a) $2x = 4$
b) $2x = -4$
c) $-2x = 4$
d) $-2x = -4$

e) $\frac{x}{2} = 4$
f) $\frac{x}{2} = -4$
g) $-\frac{x}{2} = 4$
h) $-\frac{x}{2} = -4$

2.3 Exercises

A 1. Solve each equation.

a) $x + 3 = 2x - 5$
b) $2a - 4 = -3a - 6$

c) $7 - x = 3x - 13$
d) $-8 + 3x = 5x + 7$

e) $-4x - 1 = -2x + 10$
f) $3a - 7 = 7a - 3$

2. The cost to produce a school yearbook is given by the equation $C = 950 + 4n$, where C dollars is the cost to produce n yearbooks.

a) What is the cost to produce 350 yearbooks?

b) How many books can be produced for $2150?

B 3. A school's student council raises money by selling hot dogs at basketball games. The profit, P dollars, from selling n hot dogs is represented by the function $P = 2n - 60$.

a) What is the profit when 100 hot dogs are sold?

b) How many hot dogs must be sold to make a profit of $240?

c) Solve the equation for n. Calculate how many hot dogs must be sold for a profit of $350.

4. A biologist studied a certain tree for 10 years. She found that the diameter of the tree increased 2 cm each year. When the biologist began measuring, the diameter was 80 cm. This situation can be described by the equation $D = 80 + 2t$.

a) What do D and t represent in the equation?

b) What was the diameter of the tree after 7 years?

c) After how many years was the diameter 92 cm?

5. Solve each equation.

a) $2(x + 3) = 3x - 7$
b) $-9a + 4 = -2(a + 5)$

c) $4x + 2 = 3(x - 1) + 4$
d) $7 + 2n = 3 - 2(n + 4)$

e) $-2 - 3h = 2h - (1 - h)$
f) $-3(x + 4) = 2x + 8$

6. An object is thrown downward with a speed of 14 m/s. Its speed increases due to the force of gravity. Its speed, s metres per second, after t seconds is given by the equation $s = 14 + 9.8t$. The number 9.8 is the acceleration due to gravity.

a) Calculate the speed of the object after 5 s.

b) How long does it take for the speed to reach 121.8 m/s?

c) Solve the equation for t. Determine how long it takes for the speed to reach 82.6 m/s.

7. The circumference of a circle is given by the formula $C = \pi d$.

a) What do C and d represent in the equation?

b) What is the circumference of a circle with diameter 3 m?

c) What is the circumference of a circle with radius 10 cm?

d) Suppose you want to draw a circle with circumference approximately 15 cm. At what radius would you set the compasses?

8. Solve each equation, then check.

a) $\frac{x}{2} - 3 = 6$ b) $4 - \frac{x}{3} = 10$ c) $2 - \frac{x}{4} = -8$

d) $1 + \frac{x}{5} = 3$ e) $5 - \frac{x}{10} = 1$ f) $-1 - \frac{x}{4} = 3$

9. A car uses fuel at the rate of 8 L/100 km. The gas tank holds 60 L. The distance the car drives is d kilometres. At the start of a highway journey, the tank is full. The volume of fuel in the gas tank, V litres, is a function of d. The equation is $V = 60 - 0.08d$.

a) Explain the term $0.08d$ in the equation. Why is this term subtracted from 60?

b) What volume of fuel remains after the car has travelled 350 km?

c) How far has the car travelled when only 5 L of fuel remain?

d) The car is travelling on a highway with a gas station every 200 km. How many gas stations has the car passed before it must stop to fill up?

e) Solve the equation for d. Suppose the driver forgets to check the gas consumption and runs out of gas. How far has the car travelled?

10. Solve each equation, then check.

a) $\frac{x}{2} + 3 = x + 5$ b) $-\frac{x}{2} + 2 = 2x - 8$

c) $\frac{x}{2} - 3 = 3x - 8$ d) $1 - \frac{x}{2} = 4x + 1$

e) $3x - 3 = \frac{x}{2} + 7$ f) $-1 - \frac{x}{2} = 2 - 2x$

11. Solve each equation.

a) $\frac{1}{2}(x + 3) = 2x + 6$ b) $5 - 2x = \frac{1}{2}(1 - x)$

c) $x + 6 = \frac{1}{3}(x + 1)$ d) $3 + x = \frac{1}{4}(2 - x)$

e) $x + 4 = \frac{2}{3}(2x - 3)$ f) $-\frac{3}{4}(2x - 1) = 2 - 2x$

12. The formula for converting Fahrenheit to Celsius temperatures is
$C = \frac{5}{9}(F - 32)$.

a) Convert 88°F to Celsius.

b) Solve the equation for F. Convert 88°C to Fahrenheit.

c) Which is warmer: 88°F or 88°C? Explain.

13. A snowplow has a maximum speed of 50 km/h on a dry highway. Its maximum speed decreases by 1 km/h with every 2 cm of snow on the highway.

a) What is the speed of the snowplow with each depth of snow?

   i) 2 cm   **ii)** 4 cm   **iii)** 6 cm
   iv) 8 cm   **v)** 10 cm   **vi)** 12 cm

b) What is the speed of the snowplow with d centimetres of snow on the highway?

c) Write an equation that relates the speed of the plow, s kilometres per hour, to the depth of snow, d centimetres.

d) Use the equation in part c. Solve the equation for d. According to this model, at what snow depth will the plow be unable to move?

C 14. An object is thrown upward with a speed of 12 m/s. Its speed decreases due to the force of gravity. Its speed, s metres per second, after t seconds is given by $s = 12 - 9.8t$.

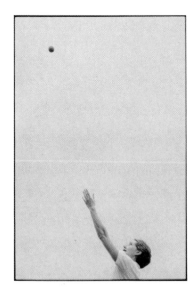

a) Compare this equation with the equation in exercise 6. How are the equations similar? How are they different?

b) What does −9.8 represent in the equation above? Explain.

c) Calculate the speed of the object after 1 s.

d) Calculate the speed of the object after 2 s. What do you notice? Explain.

e) Solve the equation for t. Determine the time the object takes to reach its greatest height.

Communicating the IDEAS

Write to explain why every example of direct variation is an example of a linear function, but not all linear functions are examples of direct variation. Use examples in your explanation.

Overtime Pay Arrangements

Sarah works at a garden centre for basic wages of $10/h. She is expected to work 40 h each week. When she works overtime, she earns "time-and-a-half".

1. What does "time-and-a-half" mean?

2. Determine Sarah's earnings for each number of hours in a week.

   a) 10 h   **b)** 20 h   **c)** 30 h   **d)** 40 h

   e) 41 h   **f)** 45 h   **g)** 50 h   **h)** 60 h

3. Record the results of exercise 2 in a table.

Time worked (h)	Earnings ($)

4. Graph the data from the table. Plot *Time worked* horizontally and *Earnings* vertically. Does it make sense to join the points? Explain.

5. Use the graph. What would Sarah earn if she worked 44.5 h? Assume she gets credit for part hours worked.

6. Describe how the graph would change in each situation.

   a) Sarah has to work 44 h before she gets time-and-a-half for overtime.

   b) Sarah is promoted to a new position at the garden centre. Her basic wages become $12/h.

   c) Sarah earns "double time" for overtime.

7. Is the function represented by the graph a linear function? Explain.

In the *Investigation*, you drew a graph similar to that below left. The second graph below appeared in Chapter 1, page 10. Although each graph contains line segments, they are not graphs of linear functions. The graphs are examples of *piecewise linear functions*. Each graph consists of line segments that do not form a single line.

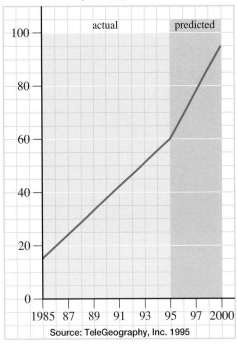

Global minutes of telephone traffic (billions) 1985 – 2000

actual   predicted

Source: TeleGeography, Inc. 1995

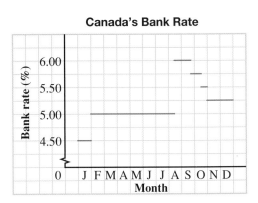

Canada's Bank Rate

To graph a piecewise linear function, we draw each line segment separately. As the above examples show, the line segments may or may not be connected.

Example 1

Maria works in construction. She earns a basic wage of $15/h for a 38-h work week. Union agreements require that she be paid time-and-a-half for any overtime hours worked.

a) Draw a graph to show Maria's earnings as a function of the number of hours worked, for up to 50 h in a week.

b) Use the graph. Estimate Maria's earnings if she works 43 h.

c) Explain why this situation involves a piecewise linear function.

Solution

a)

> **Think ...**
>
> Make a table to show Maria's earnings for 0 h, 38 h, and 50 h. Plot the points. Join them with line segments.

For the first 38 h, Maria earns 38 × $15 = $570.
50 h is 12 h more than 38 h.

For each hour of overtime, Maria earns 1.5 × $15 = $22.50.
For 12 h of overtime, she earns 12 × $22.50 = $270.
Maria's wages for 50 h are $570 + $270 = $840.

Time worked (h)	Earnings ($)
0	0
38	570
50	840

Use these results to make a table of values. Plot the data from the table.
Join the plotted points with line segments.

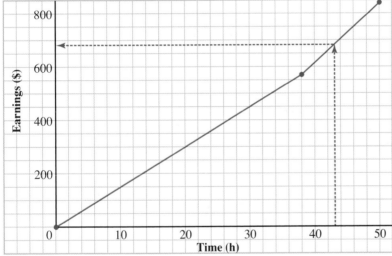

Earnings from Construction

b) Start at 43 h on the horizontal axis. Draw a vertical line up to the graph.
Then draw a horizontal line that meets the vertical axis at about 680. If
Maria works 43 h, her earnings will be about $680.

c) During the first 38 h, the earnings increase at a constant rate of $15/h.
Therefore, this part of the graph has slope 15. During the next 12 h, the
earnings increase at a constant rate of $22.50/h. Therefore, this part of the
graph has slope 22.5. The graph contains line segments, but it is not the graph
of a linear function. Therefore, it is the graph of a piecewise linear function.

In *Example 1b*, when we estimated the earnings for 43 h, we were estimating
between points on the graph. This is called *interpolating*.

Example 2

Bill works part time selling magazine subscriptions over the telephone. He earns $2 for each subscription he sells. If he reaches 100 sales in a week, he gets a $50 bonus. If he reaches 150 sales in a week, he gets an additional $50 bonus.

a) Draw a graph to show how Bill's earnings depend on the number of subscriptions he sells. Show sales up to 175 subscriptions.

b) Use the graph. Estimate Bill's earnings if he sells 190 subscriptions.

c) Calculate Bill's earnings if he sells 190 subscriptions.

d) Explain why this situation involves a piecewise linear function.

Solution

a)

> **Think ...**
>
> Approach the problem in stages — sales with no bonus, sales with a $50 bonus, and sales with a $100 bonus.

Sales with no bonus
Bill earns $2 per sale up to 99 sales.
His earnings for 99 sales are $99 \times \$2 = \$198$.

Sales with a $50 bonus
Bill's earnings for 100 sales are $100 \times \$2 + \$50 = \$250$.
His earnings for 149 sales are $149 \times \$2 + \$50 = \$348$.

Sales with a $100 bonus
Bill's earnings for 150 sales are $150 \times \$2 + \$100 = \$400$.
His earnings for 175 sales are $175 \times \$2 + \$100 = \$450$.

Use these results to make a table of values. Plot the data from the table. Join the plotted points from each section of the table with line segments.

Sales	Earnings ($)
0	0
99	198
100	250
149	348
150	400
175	450

b) Extend the line segment at the top of the graph beyond the last plotted point. Start at 190 on the horizontal axis. Draw a vertical line up to the graph. Then draw a horizontal line that meets the vertical axis at 480. If Bill sells 190 subscriptions, he will earn $480.

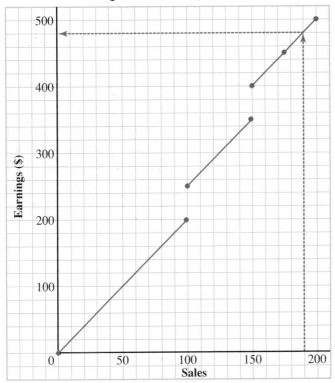

Earnings from Subscription Sales

c) Bill's earnings for 190 sales are 190 × $2 + $100 = $480.

d) During each stage, the earnings increase at a constant rate of $2 per sale. However, from 99 to 100 sales, the earnings increase by $52. From 149 to 150 sales, the earnings also increase by $52. The graph contains line segments, but it is not the graph of a linear function. Therefore, it is the graph of a piecewise linear function.

Since the sales and earnings are integers, each line segment on the graph in *Example 2* consists of many individual points. Since it is impractical to draw these points individually, we draw line segments instead.

In *Example 2b*, when we estimated the earnings for 190 subscriptions, we were estimating *beyond* the points originally plotted on the graph. This is called *extrapolating*.

1. In *Example 1*, why is the graph of the second line segment steeper than the first?

2. In *Example 2*, why do the three line segments have the same slope?

3. How would the graph in *Example 2* change in each situation?
   a) The bonuses are reduced to $30.
   b) Bill earns $2.50 for each subscription he sells.

4. Consider the graphs in *Examples 1* and *2*. Why are the line segments joined in *Example 1* but not in *Example 2*?

5. Is a piecewise linear function also a linear function? Explain.

Practise Your Skills

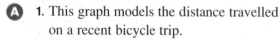

1. Determine each percent.
   a) 10% of $48.00   b) 20% of $160.00   c) 30% of $80.00

2. Determine the sale price.
   a) The regular price of a radio is $36.00. It is on sale at 10% off.
   b) The regular price of a VCR is $250.00. It is on sale at 20% off.
   c) The regular price of a TV is $360.00. It is on sale at 30% off.

2.4   Exercises

A 1. This graph models the distance travelled on a recent bicycle trip.

   a) During the trip, the cyclists stopped for lunch. When did they have lunch?

   b) When were the cyclists farthest from home?

   c) When were the cyclists travelling the fastest?

   d) Describe the cyclists' trip.

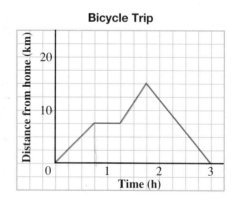

Bicycle Trip

2. This graph shows how the amount of gasoline in a car's fuel tank changed over a four-hour period.

a) What was the car doing during the first hour?

b) What happened at the end of the first hour?

c) What was the car doing during the second hour?

d) What was the car doing during the last two hours?

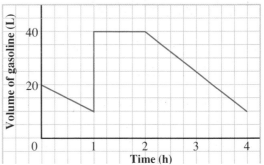

Gasoline in a Fuel Tank

e) Explain why the volume of gasoline is represented by a piecewise linear function.

f) Is it possible for any line segment on this graph to go up to the right? Explain.

B **3.** This graph shows a person's accumulated savings at the end of every two months, for one year.

a) During which 2-month period did the person not save any money?

b) During which 2-month period did the person save the most money?

c) During the following year, the person followed the same savings pattern. Draw a graph to show the accumulated savings for the 2-year period.

Accumulated Savings

4. A person walks for exercise. The graph shows her distance from home during one walk. Write to describe the person's walk.

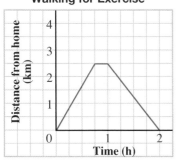

Walking for Exercise

5. A landscape architect designs ornamental pools. Three pools have rectangular surfaces. Their cross-sections are shown below. Water is pumped in at a constant rate on the left side. The depth of the water is measured on the vertical scale in the pool. Select the graph that best illustrates how the depth of the water in each pool changes. Explain your choice.

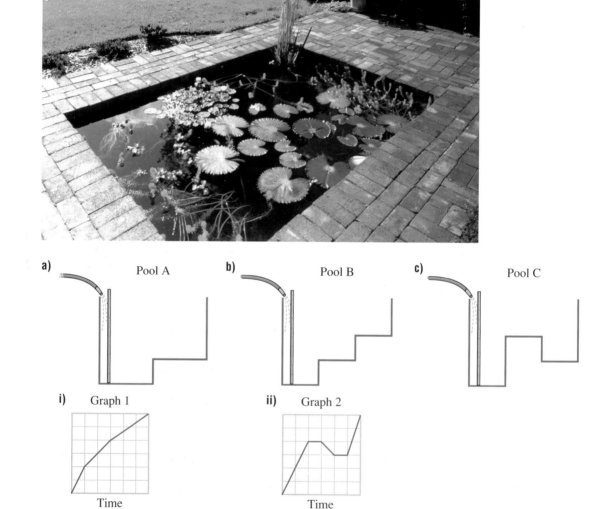

a) Pool A

b) Pool B

c) Pool C

i) Graph 1

Time

ii) Graph 2

Time

iii) Graph 3

Time

iv) Graph 4

Time

6. In exercise 5, would it make a difference if water is pumped in on the right side? Explain.

7. The local gas company charges the following rates.

From 0 m^3 to 49.9 m^3 of gas costs \$0.05/m^3.

From 50 m^3 to 99.9 m^3 of gas costs \$0.04/m^3.

Anything over 100 m^3 of gas costs \$0.03/m^3.

a) How much would a customer be charged for each volume of gas?
   i) 40 m^3 ii) 50 m^3 iii) 60 m^3 iv) 90 m^3
   v) 100 m^3 vi) 110 m^3 vii) 150 m^3 viii) 200 m^3

b) Draw a piecewise linear graph to show how the cost depends on the volume of gas. Plot *Volume* horizontally and *Cost* vertically.

c) Explain why the gas company has this billing scheme.

8. To promote a sale, a store mails this notice to customers.

   a) Determine the amount paid for each value of purchases:
      i) \$50   ii) \$100
      iii) \$150   iv) \$200
      v) \$250   vi) \$300

> **SAVE! SAVE! SAVE! SAVE! SAVE!**
>
> # UP TO 30% OFF!
>
> Next Saturday, you will receive a discount of up to 30% off the total value of all your purchases!
>
> Purchases up to \$100 .... Get 10% OFF!
> Purchases from \$100 to \$200 .... Get 20% OFF!
> Purchases of \$200 or more .... Get 30% OFF!
>
> **SAVE! SAVE! SAVE! SAVE! SAVE!**

b) Draw a graph with the *Value of purchases* plotted horizontally and the *Amount paid* plotted vertically. Join plotted points with line segments.

c) Use the graph. Estimate the amount paid for each value of purchases.
   i) \$83.50   ii) \$138.75   iii) \$266.25

d) Explain why this situation is represented by a piecewise linear function.

9. A Chinese restaurant gives a 10% discount on pick-up orders of \$20 or more.

a) Determine the cost of a pick-up order for each total price.
   i) \$15.00   ii) \$18.00   iii) \$19.00   iv) \$19.95
   v) \$20.00   vi) \$21.00   vii) \$25.00   viii) \$30.00

b) Record the results from part a in a table.

Total price ($)	Cost of order ($)

c) Graph the data from the table. Plot *Total price* horizontally and *Cost of order* vertically.

d) Use the graph. Estimate the cost of the order when the total price is \$26.75.

e) The graph in part c expresses the cost of the order as a function of the total price. Explain why this function is a piecewise linear function.

C **10.** In exercise 9, you should have found that if a customer chooses items with a total price of $19.00, the cost of the order is $19.00. However, if a customer chooses items with a total price of $20.00, the cost of the order is only $18.00.

a) Do you think this is fair? Explain.

b) Suppose the total price is close to $20.00, but less than $20.00. Describe a strategy a customer could use to obtain a lower cost for the order.

c) For what range of values for the total price could the strategy in part b be used?

d) The restaurant manager is concerned that if people use the strategy in part b, some food will be wasted. How could the manager modify the discount structure to prevent people wasting food, while giving them the benefits of using the strategy?

e) Redraw the graph in exercise 9 to represent the modified discount structure.

11. A donut shop has these prices.

Each	$ 0.70
1 dozen	$ 4.50
1/2 dozen	$ 3.15

A customer bought 9 donuts. The price was $5.25.

a) How did the clerk determine this price?

b) Do you think this was a fair price for 9 donuts? Explain.

c) Use the prices on the sign. Determine the price for each number of donuts up to 24 donuts. Record the results in a table.

d) Draw a graph. Plot the *Number of donuts* horizontally and the *Price* vertically.

12. a) In exercise 11, what could the customer do to get 9 donuts more cheaply? Do you think this strategy is reasonable? Explain.

b) Find the most economical way to buy each number of donuts up to 24 donuts. Record the results in a table.

c) Draw a graph. Plot the *Number of donuts* horizontally and the *Most economical price* vertically.

Communicating *the* IDEAS

Write to explain the difference between a linear function and a piecewise linear function. Use examples to illustrate your explanation.

Using a Graphing Calculator to Graph Piecewise Linear Functions

Suppose a hospital worker earns a basic wage of $14.39 an hour for a 37.5-h work week. The person is paid time-and-a-half for any overtime hours worked. This graph was created on a graphing calculator. It shows the earnings as a function of the number of hours worked, for up to 50 h in a week. The graph consists of two line segments joined together.

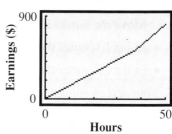

There are two different ways to create this graph on a graphing calculator. Each method requires some calculations before using the calculator.

Using a Statistical Plot to Graph a Piecewise Linear Function

We can use a statistical plot to graph any piecewise linear function. We enter the coordinates of the endpoints of the segments to be graphed in the list editor.

Example 1

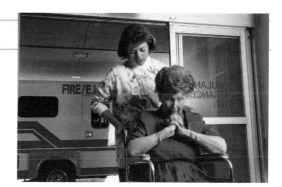

Use a statistical plot. Create the graph above to show the earnings for up to 50 h worked in a week.

Solution

Before using the calculator, calculate the coordinates of the endpoints of the segments OA and AB that form the graph.

For the first 37.5 h, the worker earns
$37.5 \times \$14.39 = \$539.63$.
The coordinates of A are (37.5, 539.63).

For each hour of overtime, the worker earns
$1.5 \times \$14.39 = \$21.59$.

50 h is 12.5 h more than 37.5 h.
For 12.5 h of overtime, the worker earns $12.5 \times \$21.59 = \$269.88$.
The total wages for 50 h are $\$539.63 + \$269.88 = \$809.51$.
The coordinates of B are (50, 809.51).

These are the data to be entered in the calculator.

Time worked (h)	Earnings ($)
0	0
37.5	539.63
50	809.51

10. a) Create a graph to show the federal income tax as a function of taxable income. Use taxable incomes up to $55 000.

   b) Trace to determine the federal income tax on each taxable income.
      i) $18 650   **ii)** $43 290   **iii)** $51 825

   c) Sketch the graph. Show the results of part b on your sketch.

11. A pizza restaurant distributes a flyer with these coupons.

$10.00 OFF	**$5.00 OFF**
ANY ORDER OF $60 OR MORE	ANY ORDER OF $30 OR MORE

$4.00 OFF	**$3.00 OFF**
ANY ORDER OF $25 OR MORE	ANY ORDER OF $20 OR MORE

$2.00 OFF	**$1.00 OFF**
ANY ORDER OF $15 OR MORE	ANY ORDER OF $10 OR MORE

a) Suppose the order is in each range of values below. Describe the range of values for the cost of the order.
   i) between $10 and $14.99
   ii) between $15.00 and $19.99
   iii) between $20.00 and $24.99
   iv) between $25.00 and $29.99
   v) between $30.00 and $59.99
   vi) $60.00 and over

b) Create a graph to show how the cost of the order is related to the order.

c) Sketch the graph. Label the axes and the scales.

d) Explain why this situation involves a piecewise linear function.

Communicating *the* IDEAS

Two methods of using a graphing calculator to graph a piecewise linear function were explained in this section. Choose one method. Without describing the keys to press, write to explain how to create the graph using this method.

On page 50, you imagined that you are a newspaper reporter writing an article about the Ironman Canada Triathlon. This event is held each summer in Penticton, British Columbia.

The race begins at 7 A.M. with a 3.8-km swim in Lake Okanagan. After the swim, there is a 180-km bike race through southern British Columbia. Then there is a 42.2-km marathon run from Penticton to Okanagan Falls and back. Maps of the bike and marathon courses are on pages 95 and 96. You plan to include these maps in your article, to show the estimated times when the competitors are expected to pass through the communities along the route. This will help people who want to watch the race.

From the Ironman Canada web site, these are the data for the 1999 race. The table shows the times for each part of the race for the winners of the women's and men's events.

Winner	Swim (h:min)	Bike (h:min)	Run (h:min)
Lori Bowden, Canada	0:58	5:03	3:09
Chuckie Veylupek, USA	0:51	4:53	3:01

Each person takes about 1 h to complete the swim, 5 h to complete the bike race, and 3 h to complete the run. Since the swim starts at 7 A.M., we make the following assumptions.

- The bike race starts at 8 A.M.
- The run starts at 1 P.M.
- The winner reaches the finish line at 4 P.M.

Part A A Graphical Model

1. Draw a large graph on grid paper like the one on the next page. Point T represents the transition from the bike race to the run. Point F represents the finish. Write the coordinates of T and F on your graph.

2. You have made some assumptions. Drawing the line segments OT and TF on the graph involves making other assumptions.

  a) What are these assumptions?

  b) Do you think these assumptions are reasonable? Explain.

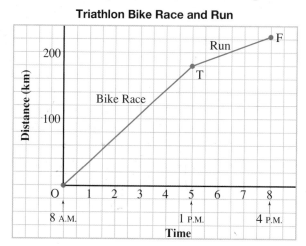

Triathlon Bike Race and Run

Recall that a mathematical model is a way to solve an applied problem. The graph in exercise 2 is a mathematical model that you can use to predict the times when the competitors will arrive at different locations along the route. In Part C, you will improve the model.

3. a) Copy this table. Use the map of the bike course on page 95 to complete the second column. Round each distance to the nearest kilometre. Use the graph to complete the third column.

  b) The times in part a are elapsed times after 8 A.M. Convert these times to actual times on the clock. Write these times in the fourth column.

Location	Distance from start (km)	Elapsed time (hours after 8 A.M.)	Arrival time (clock time)
Penticton	0	0	8:00 A.M.
Okanagan Falls			
Oliver			
Osoyoos			
Richter Mountain			
Keremeos			
Olalla			
Yellow Lake			
Kaleden			
Penticton	180	5	1:00 P.M.

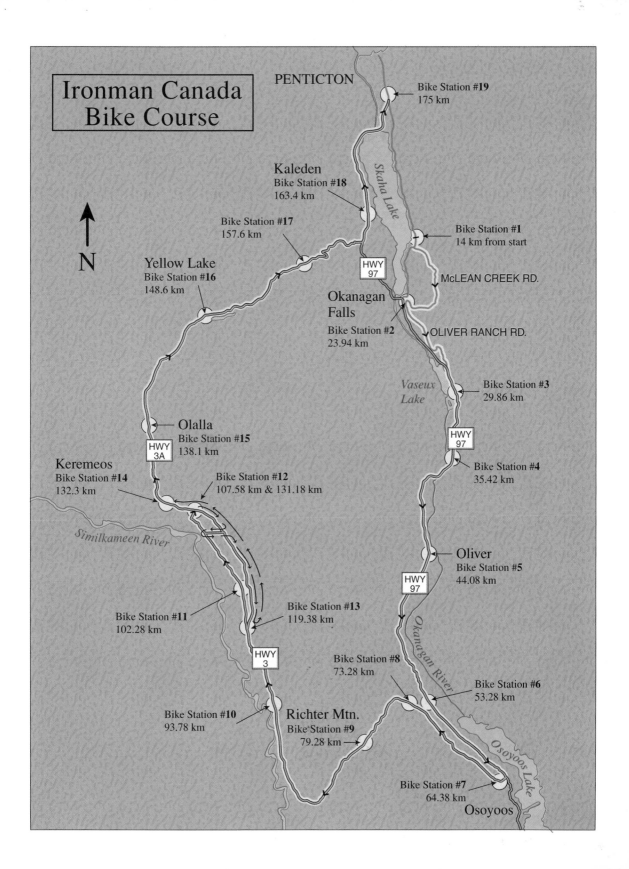

Ironman Canada Bike Course

PENTICTON

Bike Station #19
175 km

Kaleden
Bike Station #18
163.4 km

Bike Station #17
157.6 km

Bike Station #1
14 km from start

McLEAN CREEK RD.

Yellow Lake
Bike Station #16
148.6 km

HWY 97

Okanagan Falls
Bike Station #2
23.94 km

OLIVER RANCH RD.

Vaseux Lake

Bike Station #3
29.86 km

Olalla
Bike Station #15
138.1 km

HWY 3A

HWY 97

Bike Station #4
35.42 km

Keremeos
Bike Station #14
132.3 km

Bike Station #12
107.58 km & 131.18 km

Similkameen River

Oliver
Bike Station #5
44.08 km

HWY 97

Okanagan River

Bike Station #11
102.28 km

Bike Station #13
119.38 km

HWY 3

Bike Station #8
73.28 km

Bike Station #6
53.28 km

Bike Station #10
93.78 km

Richter Mtn.
Bike Station #9
79.28 km

Osoyoos Lake

Bike Station #7
64.38 km

Osoyoos

N

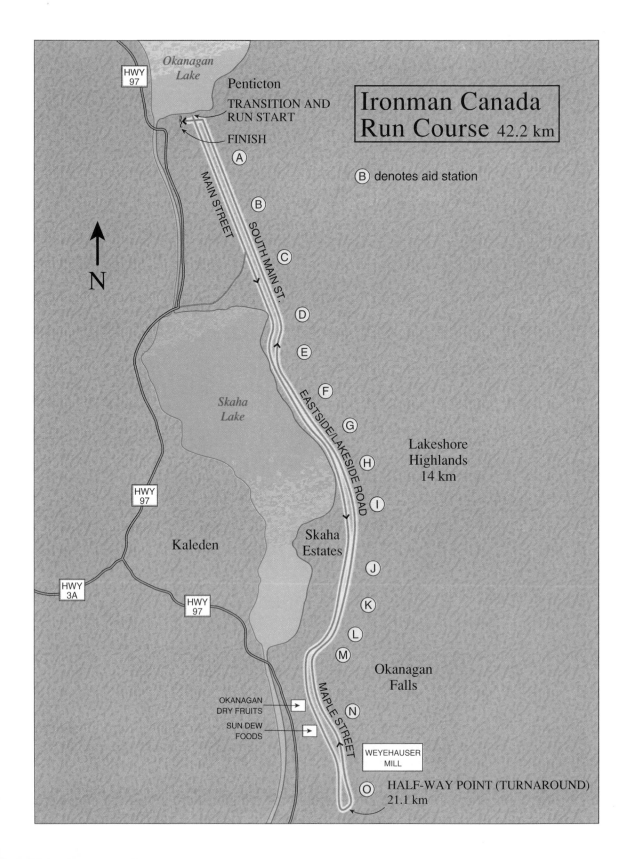

Ironman Canada
Run Course 42.2 km

Ⓑ denotes aid station

Okanagan Lake

HWY 97

Penticton

TRANSITION AND RUN START

FINISH

Ⓐ

MAIN STREET

Ⓑ

SOUTH MAIN ST.

Ⓒ

Ⓓ

Ⓔ

Skaha Lake

Ⓕ

EASTSIDE/LAKESIDE ROAD

Ⓖ

Lakeshore Highlands 14 km

Ⓗ

Ⓘ

HWY 97

Kaleden

Skaha Estates

Ⓙ

HWY 3A

HWY 97

Ⓚ

Ⓛ

Ⓜ

Okanagan Falls

OKANAGAN DRY FRUITS

MAPLE STREET

Ⓝ

SUN DEW FOODS

WEYEHAUSER MILL

Ⓞ HALF-WAY POINT (TURNAROUND) 21.1 km

4. a) Copy the table below. Use the map of the run course on page 96 to complete the second column. Use the graph to complete the third column.

b) Convert the times in part a to actual times on the clock. Write these times in the fourth column.

Location	Distance from start of bike race (km)	Elapsed time (hours after 8 A.M.)	Arrival time (clock time)
Penticton	180	5	1:00 P.M.
Lakeshore Highlands			
Half-way point			
Lakeshore Highlands			
Penticton	222	8	4:00 P.M.

Note: Save your graph for use in Part C.

Part B An Algebraic Model

For more accurate results, you will use an algebraic model. You will determine the equations of the line segments on the graph on page 94. Then you will use a graphing calculator to create the graph. You can trace along the graph to determine the times for different distances. This is an algebraic model because the calculator uses the equations to determine the coordinates of points on the graph.

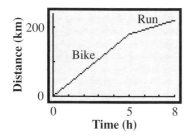

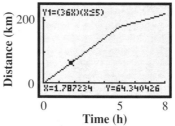

5. Determine the equation of each line segment on the graph you drew in exercise 1. Write the equations on the graph.

6. Use the Y= list. Create a graph to show distance against time for the bike race and the run. Your graph should look like the one above left.

On the map of the bike course, Osoyoos is approximately 64 km from the start. To predict the time to Osoyoos, press [TRACE]. Move the cursor until y is as close to 64 as possible. The screen on the preceding page shows a typical result. It takes about 1.79 h to reach Osoyoos. To convert to hours and minutes, multiply the decimal part by 60 to obtain $0.79 \times 60 = 47.4$. Hence, 1.79 h is about 1 h 47 min. This is the time it takes to reach Osoyoos. Since the bike race started at 8 A.M., we predict that the competitors will reach Osoyoos at about 9:47 A.M.

7. a) Copy the table below. Use the graphing calculator to complete the third column.

   b) Determine the predicted arrival times, as in the above example. Write these times in the fourth column.

Location	Distance from start of bike race (km)	Elapsed time (hours after 8 A.M.)	Arrival time (clock time)
Penticton	0	0:00	8:00 A.M.
Okanagan Falls	24		
Oliver	44		
Osoyoos	64	1:47	9:47 A.M.
Richter Mountain	79		
Keremeos	132		
Olalla	138		
Yellow Lake	149		
Kaleden	163		
Penticton	180	5:00	1:00 P.M.

8. a) Copy the table below. Use the graphing calculator to complete the third column.

   b) Convert the times in part a to actual times on the clock. Write these times in the fourth column.

Location	Distance from start of bike race (km)	Elapsed time (hours after 8 A.M.)	Arrival time (clock time)
Penticton	180	5:00	1:00 P.M.
Lakeshore Highlands			
Half-way point	201		
Lakeshore Highlands			
Penticton	222	8:00	4:00 P.M.

You now have the predicted arrival times to include on the maps of the courses in your article.

Part C Improving the Model

The distance-time graphs you have drawn show a piecewise linear function consisting of two line segments.

Improving the model for the bike course

When you used the first line segment on the graph to predict the arrival times, you were assuming that the cyclists were riding at a constant speed for 5 h.

9. Give some reasons why the cyclists could not ride at a constant speed for 5 h.

10. The bike course profile below shows the elevation above sea level for the bike course. Compare this profile with the map on page 95.

 a) On which parts of the course will the cyclists need more time?

 b) On which parts will they need less time?

 c) Describe how your answers in parts a and b affect your predicted arrival times along the bike route.

 d) Are the hills on the bike course as steep as they appear to be on the profile? Explain.

 e) Do you think the profile is misleading? Explain.

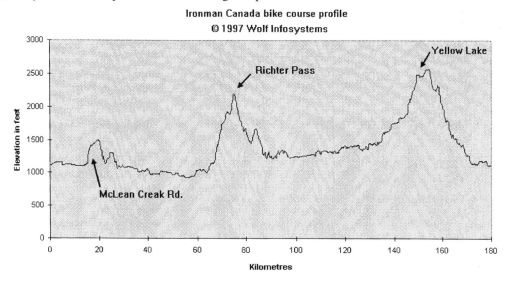

Ironman Canada bike course profile
© 1997 Wolf Infosystems

The table shows the times when the winners of the women's and men's events in 1999 reached Richter Pass, which is 75 km from the start of the bike race.

Winner	Richter Pass (h:min)
Lori Bowden, Canada	3:02
Chuckie Veylupek, USA	2:50

It takes about 3 h for the cyclists to reach Richter Pass.

11. Use the graph you drew in exercise 1.

    a) Plot point R(3, 75), which represents reaching Richter Pass 3 h after the start of the bike race. Join OR and RT. These segments represent the race more accurately than segment OT.

    b) Use the improved graph to change your predictions in exercise 3.

12. Determine the equations of line segments OR and RT. Write the equations on the graph.

13. a) Use the Y= list to create the improved graph.

    b) Trace to improve your predictions in exercise 7.

Improving the model for the run

When you used the second line segment on the graph to predict the arrival times, you were assuming that the runners were running at a constant speed for 3 h.

14. Do you think this assumption is reasonable? Explain.

The table shows the times (from the start of the bike race) when the winners of the women's and men's events in 1999 reached the half-way point of the run.

Winner	Half-way point (h:min)
Lori Bowden, Canada	6:37
Chuckie Veylupek, USA	6:25

15. It takes about 6.5 h for the competitors to reach the half-way point of the run. Compare this time with your results from exercise 8. Explain why it is not necessary to change the model for the run.

Communicating
the IDEAS

You are a newspaper reporter writing an article about the upcoming Ironman Triathlon Race. You are including the maps on pages 95 and 96 in your article. Write a paragraph that describes what the maps are for and how the reader can use them. Include a list of the estimated arrival times.

Earlier in this chapter, we wrote equations of lines in the form $y = mx + b$. The graph of the equation $y = -\frac{2}{3}x + 6$ is shown. The graph is a line with slope $-\frac{2}{3}$ and y-intercept 6.

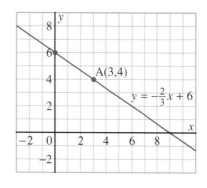

The coordinates of any point on the line satisfy the equation. For example, the graph shows that the point A(3, 4) lies on the line. We can check that these coordinates satisfy the equation. Substitute $x = 3$ and $y = 4$ into the equation $y = -\frac{2}{3}x + 6$.

Left side $= y$ Right side $= -\frac{2}{3}x + 6$
$\quad\quad = 4$ $= -\frac{2}{3}(3) + 6$
$\quad\quad\quad\quad\quad\quad\quad\quad\quad\quad\quad\quad = -2 + 6$
$\quad\quad\quad\quad\quad\quad\quad\quad\quad\quad\quad\quad = 4$

Since the left side equals the right side, the coordinates of A satisfy the equation $y = -\frac{2}{3}x + 6$.

Since $y = -\frac{2}{3}x + 6$ is an equation, we can use the rules for solving equations to write it in a different form.

$$y = -\frac{2}{3}x + 6 \qquad ①$$

Multiply each side by 3.
$$3y = 3\left(-\frac{2}{3}x + 6\right)$$
Use the distributive law.
$$3y = 3\left(-\frac{2}{3}x\right) + 3(6)$$
$$3y = -2x + 18$$
Add $2x$ to each side.
$$2x + 3y = 18$$
Subtract 18 from each side.
$$2x + 3y - 18 = 0 \qquad ②$$

The graph above is also the graph of this equation. The reason is that the coordinates of points that satisfy equation ① also satisfy equation ②. For example, we can check that the coordinates of A(3, 4) satisfy the equation. Substitute $x = 3$ and $y = 4$ into equation ②: $2x + 3y - 18 = 0$.

Left side $= 2x + 3y - 18$ Right side $= 0$
$\quad\quad = 2(3) + 3(4) - 18$
$\quad\quad = 6 + 12 - 18$
$\quad\quad = 0$

Hence, the coordinates of A satisfy equation ②.

When one equation can be changed into another equation using the rules for solving equations, the new equation will always produce the same points as the original equation.

When the equation $y = -\frac{2}{3}x + 6$ was changed to $2x + 3y - 18 = 0$, all the terms were collected on the left side of the equation. The equation has the form $Ax + By + C = 0$.

Any equation in the form $y = mx + b$ can be rearranged to the form $Ax + By + C = 0$.

Example 1

Write the equation $y = \frac{1}{2}x + 3$ in the form $Ax + By + C = 0$.

Solution

$$y = \frac{1}{2}x + 3$$

Multiply each side by 2.

$$2y = 2\left(\frac{1}{2}x + 3\right)$$

Use the distributive law.

$$2y = 2\left(\frac{1}{2}x\right) + 2(3)$$

$$2y = x + 6$$

Subtract x from each side.

$$2y - x = 6$$

Subtract 6 from each side.

$$2y - x - 6 = 0$$

Write the x-term first.

$$-x + 2y - 6 = 0$$

It is customary to write the equation with a positive x-term.
Multiply each side of the above equation by -1.

$$x - 2y + 6 = 0$$

Any equation in the form $Ax + By + C = 0$ can be written in the form $y = mx + b$.

Example 2

Write the equation $2x + 5y - 8 = 0$ in the form $y = mx + b$.

Solution

$2x + 5y - 8 = 0$
Solve the equation for y.
Subtract $2x$ from each side.

$\qquad 5y - 8 = -2x$

Add 8 to each side.

$\qquad 5y = -2x + 8$

Divide each side by 5.

$\qquad y = \frac{-2x + 8}{5}$

Write the right side as the sum of two fractions.

$\qquad y = \frac{-2}{5}x + \frac{8}{5}$

Example 3

Graph the equation $3x + 5y - 15 = 0$.

Solution

Method 1

> **Think ...**
>
> Write the equation in the form $y = mx + b$ to determine the slope and y-intercept.
> Use the slope and y-intercept to graph the line.

Solve $3x + 5y - 15 = 0$ for y.

$3x + 5y - 15 = 0$

Subtract $3x$ from each side.

$\qquad 5y - 15 = -3x$

Add 15 to each side.

$\qquad 5y = -3x + 15$

Divide each side by 5.

$\qquad y = \frac{-3x + 15}{5}$

Write as the sum of two fractions.

$\qquad y = -\frac{3}{5}x + \frac{15}{5}$

$\qquad y = -\frac{3}{5}x + 3$

The slope of the line is $-\frac{3}{5}$, and the y-intercept is 3.
Plot the point $(0, 3)$. Use run, 5, and rise, –3, to plot other points on the line. Draw a line through these points.

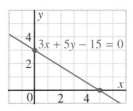

Method 2

> *Think ...*
>
> Determine pairs of values of x and y that satisfy the equation. Then plot the corresponding points.

Choose $x = 0$.
Substitute $x = 0$ in $3x + 5y - 15 = 0$, then solve for y.

$$3(0) + 5y - 15 = 0$$
$$5y - 15 = 0$$
$$5y = 15$$
$$y = 3$$

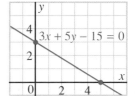

One point on the line has coordinates $(0, 3)$.
Choose $y = 0$.
Substitute $y = 0$ in $3x + 5y - 15 = 0$, then solve for x.

$$3x + 5(0) - 15 = 0$$
$$3x = 15$$
$$x = 5$$

Another point on the line has coordinates $(5, 0)$.
Plot these points on a grid, then draw a line through them.

Discussing the IDEAS

1. In *Example 2*, we obtained the equation $y = -\frac{2}{5}x + \frac{8}{5}$, which has the form $y = mx + b$. What other way is there to write the equation in this form?

2. In *Example 3*, Method 2, we graphed a line by determining the coordinates of the points where the line intersects the axes. Are there any disadvantages to this method of graphing a line? Explain.

3. Suppose the equation of a line is given. Explain how to find the coordinates of the points where the line intersects the axes.

Practise Your Skills

1. Expand.

   a) $2\left(-\frac{3}{2}x - 5\right)$ b) $4\left(-\frac{3}{4}x + 2\right)$ c) $5\left(-\frac{2}{5}x - 3\right)$

2. Divide each expression by the indicated number.

   a) $4x - 6$, by 2 b) $2x + 9$, by 3 c) $-3x + 2$, by 4

2.7 Exercises

A

1. Write each equation in the form $Ax + By + C = 0$.

   a) $y = -2x + 3$ b) $y = x + 1$ c) $y = 3x - 2$

2. Write each equation in the form $y = mx + b$.

   a) $x + y - 3 = 0$ b) $2x + y + 4 = 0$ c) $2x - y - 8 = 0$

B

3. Write each equation in the form $Ax + By + C = 0$.

   a) $y = \frac{1}{2}x - 4$ b) $y = -\frac{2}{3}x - 2$

   c) $y = \frac{3}{5}x - 8$ d) $y = 4x - \frac{5}{2}$

   e) $y = -\frac{3}{2}x + 4$ f) $y = -\frac{2}{3}x + \frac{5}{3}$

4. Write each equation in the form $y = mx + b$.

   a) $x + 2y + 8 = 0$ b) $x - 3y + 12 = 0$

   c) $2x + 3y - 18 = 0$ d) $3x - 2y + 6 = 0$

   e) $4x + 3y + 24 = 0$ f) $5x + 2y + 20 = 0$

5. The equation of each line is given in the form $Ax + By + C = 0$. Graph each line by determining its slope and y-intercept.

   a) $x + y - 4 = 0$ b) $2x - y + 8 = 0$

   c) $x - 2y + 6 = 0$ d) $3x + 2y - 6 = 0$

   e) $3x + 2y + 12 = 0$ f) $2x + 3y + 18 = 0$

6. The equation of each line is given in the form $Ax + By + C = 0$. Graph each line by determining the coordinates of two points on the line.

   a) $x + y - 8 = 0$ b) $2x + y - 10 = 0$

   c) $x + 2y - 4 = 0$ d) $5x + 2y + 10 = 0$

   e) $3x - 5y + 15 = 0$ f) $2x - 3y - 12 = 0$

7. The equations of six lines are given.
Suppose you graphed these lines on the graph at the right.

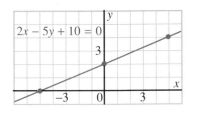

L_1 $2x - 5y + 15 = 0$ L_2 $2x + 5y + 20 = 0$
L_3 $5x - 2y + 10 = 0$ L_4 $4x - 10y - 10 = 0$
L_5 $2x + 5y + 10 = 0$ L_6 $2x + 5y - 10 = 0$

Which lines would have the same slope as the line on
the graph? Explain.

8. Which of these equations represent the same line? Explain.

L_1 $y = -\dfrac{1}{3}x + 4$ L_2 $y = -\dfrac{2}{5}x + 2$ L_3 $y = \dfrac{2}{7}x + 1$

L_4 $2x - 7y + 7 = 0$ L_5 $x + 3y - 12 = 0$ L_6 $2x + 5y - 10 = 0$

Ⓒ 9. The patterns below were produced on a graphing calculator.

a) The equation of one line in the first pattern is $x + y - 4 = 0$.
What are the equations of the other lines in this pattern?

b) The equation of one line in the second pattern is $2x + 3y - 12 = 0$.
What are the equations of the other lines in this pattern?

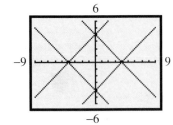

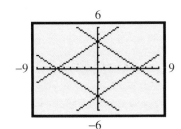

Communicating the IDEAS

Suppose you know the equation of
a line in the form $Ax + By + C = 0$.
Write to explain how to graph
the line. Illustrate
your explanation
with an example.

Algebra Tools

Direct Variation

- y varies directly as x; that is, y is a multiple of x.
- The equation has the form $y = mx$.
- The graph is a straight line through the origin (below left).

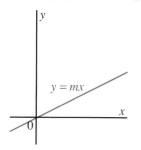
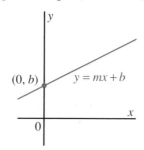

Partial Variation

- y varies partially as x; that is, y is a multiple of x, plus a constant.
- The equation has the form $y = mx + b$.
- The graph is a straight line that does not pass through the origin (above right).

Linear Function

- The graph of a linear function is a straight line.

Equation of a Linear Function

- $y = mx + b$, where m is the slope of the line and b is its y-intercept
- $y = m(x - p) + q$, where m is the slope of the line, and (p, q) are the coordinates of a point on the line
- $Ax + By + C = 0$, where all the terms are on the left side

Piecewise Linear Function

- A function composed of two or more linear functions.

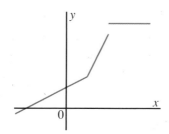

1. A grade 10 class is experimenting with a graphing calculator and a calculator-based ranger (CBR). A student holding a CBR walked toward and away from a wall. This screen was created, where the Xscl = 1.

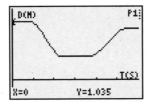

a) Describe the motion of the student holding the CBR.

b) Write the time period for each part of the walk.

2. The graph below shows how the international postage rate for air mail letters up to 20 g has changed since 1960.

a) How much did it cost to send an international air mail letter in 1980? In 1990?

b) Between which two years did the cost to send an international air mail letter increase the most?

c) Explain why this situation is represented by a piecewise linear function.

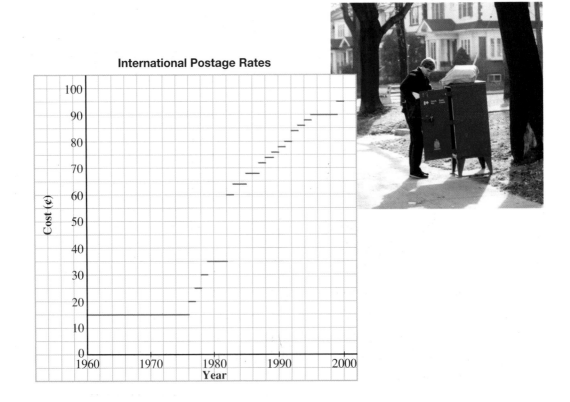

International Postage Rates

3. A photocopy centre advertises these rates:

From between 1 and 100 copies:	5¢ per copy
From between 101 and 250 copies:	4¢ per copy
From between 251 and 500 copies:	3¢ per copy
For 501 copies or more:	2¢ per copy

a) Sketch a graph to show the cost to make from zero to 600 copies.

b) To make 90 copies, it is cheaper to pay for 101 copies. Calculate the most cost effective schedule. Sketch a graph to show the cheapest way to make from zero to 600 copies.

FOCUS
CONSUMER

Buying a Pizza

4. A local pizza parlour advertises 64 different toppings. Their prices in dollars are shown in the table below.

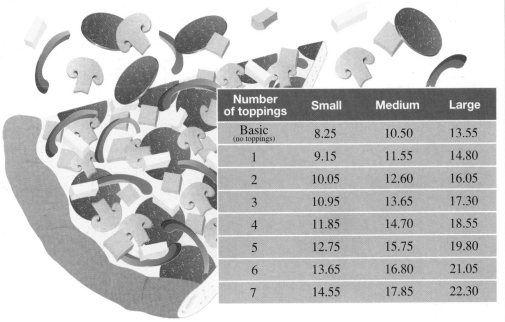

Number of toppings	Small	Medium	Large
Basic (no toppings)	8.25	10.50	13.55
1	9.15	11.55	14.80
2	10.05	12.60	16.05
3	10.95	13.65	17.30
4	11.85	14.70	18.55
5	12.75	15.75	19.80
6	13.65	16.80	21.05
7	14.55	17.85	22.30

a) How much does one topping cost for a small pizza?

b) How much does one topping cost for a medium pizza? for a large pizza?

c) Let n represent the number of toppings. Let P dollars represent the price of each size of pizza. For each pizza, write an equation for P as a function of n.

5. In January, 2000, the postage rates for air mail letters to international destinations were as follows.

0-20 g	$0.95
20-50 g	$1.45
50-100 g	$2.35
100-250 g	$5.35
250-500 g	$10.45

a) Create a graph to show the international postage rates for air mail letters up to 500 g.

b) Sketch the graph. Label the axes and the scales.

c) Explain why this situation involves a piecewise linear function.

6. Mathematical Modelling On pages 93–100, you estimated the arrival times of the Ironman Triathlon competitors at different locations along the bike and run routes. Your models were based on the 1999 winners' times from the Ironman Canada web site. The information below was also obtained from this site.

Course Records

Event	Men's time (h:min)	Women's time (h:min)
Swim	0:43	0:49
Bike	4:28	4:59
Run	2:39	2:50

Suppose you had used these data instead of the data on page 93.

a) How would you modify the assumptions on page 93?

b) Use grid paper.
   i) Draw distance-time graphs similar to that on page 94, but based on the assumptions in part a.
   ii) Use the graphs to predict the arrival times at the locations in the tables on pages 94 and 97.

c) i) Create graphs similar to that on page 97, but based on the assumptions in part a.
   ii) Trace to predict the arrival times at the locations in the tables on page 98.

d) Which times do you think would be closer to the actual times: those predicted using the winners' times from the 1999 race, or those predicted using the course records? Explain.

7. An object is thrown downward with a speed of 19 m/s. Its speed, s metres per second, after t seconds is given by the equation $s = 19 + 9.8t$.

a) Calculate the speed of the object after 3 s.

b) How long does it take the speed to reach 72.9 m/s?

c) Solve the equation for t. Determine how long it takes for the speed to reach 130.72 m/s.

8. Determine the equation of the line with the given slope through the given point.

   a) 3; (3, 4)   b) −1; (2, −3)   c) $\frac{1}{2}$; (−3, 2)

9. Determine the equation of the line through the points A(3, 5) and B(−1, 2).

10. Write each equation in the form $y = mx + b$.

    a) $2x + y − 3 = 0$ b) $x − 2y − 4 = 0$ c) $3x + 4y + 1 = 0$

11. Write each equation in the form $Ax + By + C = 0$.

    a) $y = 3x − 2$ b) $y = -\frac{2}{3}x − 5$ c) $y = \frac{3}{4}x + 2$

12. The equation of each line is given in the form $Ax + By + C = 0$. Graph each line by determining its slope and y-intercept.

    a) $6x + 2y − 3 = 0$ b) $3x + y − 2 = 0$ c) $x − 2y − 1 = 0$

13. The equation of each line is given in the form $Ax + By + C = 0$. Graph each line by determining the coordinates of two points on the line.

    a) $2x + y + 4 = 0$ b) $3x − 2y − 6 = 0$ c) $5x + 2y + 10 = 0$

MATHEMATICAL MODELLING

When Will the Skydivers Meet?

Skydivers sometimes arrange themselves to make formations. Some large formations can include dozens of people. To make a large formation, the skydivers will need to jump from different planes. They must plan how to meet at the same point during their dives.

1. Suppose two skydivers jump from two planes at the same time. One plane is higher than the other. How is it possible for the skydivers to meet at the same point during their dives?

2. What factors affect how fast a skydiver falls?

You will return to this problem in Section 3.5.

FYI Visit www.awl.com/canada/school/connections

For information related to the above problem, click on <u>MATHLINKS</u>, followed by Foundations of Mathematics 10. Then select a topic under When Will the Skydivers Meet?

Preparing
for the Chapter

This section reviews these concepts:

- Solving equations in one variable
- Solving an equation for x or for y
- Substituting

1. Solve each equation for x.

   a) $3x = 9$ **b)** $-4x = 20$ **c)** $\frac{1}{2}x = 5$

   d) $-\frac{1}{2}x = 2$ **e)** $\frac{1}{4}x = -6$ **f)** $-\frac{1}{4}x = 3$

2. Solve each equation for x.

   a) $x + 3 = 5$ **b)** $3(x + 2) = 4$ **c)** $5x + 2(x - 1) = 5$

   d) $2x + 2(3x - 1) = 6$ **e)** $3x - 4(x + 2) = 7$ **f)** $-2x + 4(8 - 3x) = 4$

   g) $6x + 2(8 + 5x) = 0$ **h)** $-4x + 3(-4 + 2x) = 4$ **i)** $2x - 3(2 - x) = 9$

3. Solve each equation for x.

   a) $x + y = 3$ **b)** $x - y = 7$ **c)** $x + 2y = -8$

   d) $-x + y = 10$ **e)** $-x - 3y = 5$ **f)** $x + 4y = 13$

4. Solve each equation for y.

   a) $x + y = 4$ **b)** $x - y = -6$ **c)** $3x + y = 5$

   d) $-x - y = -7$ **e)** $-2x + y = 1$ **f)** $4x + y = -3$

5. Substitute each value of x to determine y.

   a) Substitute $x = 2$ in $y = 2x + 3$.   **b)** Substitute $x = -2$ in $y = 3x - 4$.

   c) Substitute $x = 0$ in $y = -3x + 5$.   **d)** Substitute $x = 3$ in $y = -x - 7$.

   e) Substitute $x = -1$ in $y = -2x - 1$.   **f)** Substitute $x = -5$ in $y = x + 11$.

6. Substitute each value of x to determine y.

   a) Substitute $x = 2$ in $x + y = 5$.   **b)** Substitute $x = 0$ in $x - y = -2$.

   c) Substitute $x = 6$ in $2x + y = 7$.   **d)** Substitute $x = -3$ in $x - 3y = 9$.

   e) Substitute $x = 1.5$ in $4x - 2y = 8$.   **f)** Substitute $x = -0.5$ in $3x + y = 7$.

7. Substitute each value of y to determine x.

   a) Substitute $y = 0$ in $x + y = -3$.   **b)** Substitute $y = -2$ in $x - y = 4$.

   c) Substitute $y = 5$ in $x + 2y = 7$.   **d)** Substitute $y = -1.5$ in $2x - y = 6.5$.

   e) Substitute $y = 2.5$ in $2x + 3y = -2.5$.   **f)** Substitute $y = 0.5$ in $-3x + 2y = -7$.

Two banquet halls are considered for a wedding reception. Here are their charges:

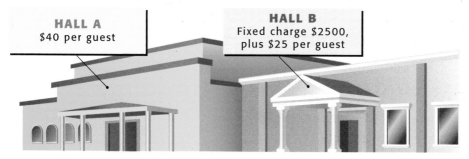

HALL A
$40 per guest

HALL B
Fixed charge $2500,
plus $25 per guest

Which hall is cheaper?

We shall consider the cost of each hall for different numbers of guests.

Number of guests	Cost for Hall A	Cost for Hall B
50	50 × $40 = $2000	50 × $25 + $2500 = $3750
100	100 × $40 = $4000	100 × $25 + $2500 = $5000
200	200 × $40 = $8000	200 × $25 + $2500 = $7500

From the table:

Hall A is cheaper for 100 guests or fewer.
Hall B is cheaper for 200 guests.

When the number of guests is between 100 and 200, it is not clear which hall is cheaper.

For how many guests would the cost for the halls be the same? To find out,
we can plot the data on a grid.

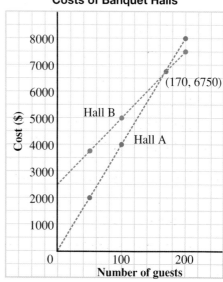

Costs of Banquet Halls

(170, 6750)

Hall B

Hall A

We plotted *Number of guests* horizontally. We chose a scale of 1 square to 20 people.

We plotted *Cost in dollars* vertically. We chose a scale of 1 square to $500, to be able to plot costs of $3750 and $7500.

From the graph, the lines intersect at a point that represents approximately 170 guests and a cost of $6750.

Hall A is cheaper when there are fewer than 170 guests.
Hall B is cheaper when there are more than 170 guests.

The cost for each hall can be represented by an equation.

For Hall A, the cost, C dollars, is $40 \times$ the number of guests.
Let the number of guests be represented by n.
Then the equation is $C = 40n$.

For Hall B, the cost, C dollars, is $25 \times$ the number of guests + $2500.
The equation is $C = 25n + 2500$.

Each equation above is a linear equation. A pair of linear equations considered together is a *linear system*.

To solve a linear system means to determine the coordinates of the point of intersection of the two lines. The coordinates of this point satisfy both equations.

From the graph, the solution of the linear system
$C = 40n$
$C = 25n + 2500$
is approximately (170, 6750).

We can use a graphing calculator to determine a more accurate solution of this linear system.

Example 1

Solve this linear system.
$C = 40n$
$C = 25n + 2500$

Solution

Rewrite each equation in terms of x and y.
$y = 40x$ ①
$y = 25x + 2500$ ②

Press ⌊ Y= ⌋.

To enter equation ①, press 40 ⌊X,T,θ,n⌋.

Use ⌊▼⌋ to move the cursor next to Y2.

To enter equation ②, press 25 ⌊X,T,θ,n⌋ ⌊ + ⌋ 2500.

Use ⌊◄⌋ to move the cursor to the left of Y2. Press ⌊ENTER⌋ to select a heavy line. This helps distinguish the graphs on the screen.

Press ⌊WINDOW⌋. Use the values in the table on page 115 to determine what to input.

Input Xmin = 0, Xmax = 200, Xscl = 50, Ymin = 0, Ymax = 8000, Yscl = 1000.

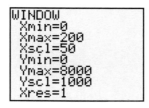
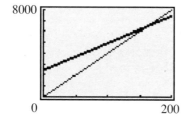

Press ⌊GRAPH⌋ to see the screen above right.

To determine the coordinates of the point of intersection, press ⌊ 2nd ⌋ ⌊TRACE⌋ for ⌊ CALC ⌋. Then press **5** to select **intersect**.

Use ⌊►⌋ or ⌊◄⌋ to move the cursor close to the point of intersection, and to the left of the point. Press ⌊ENTER⌋ to see a screen similar to that below left.

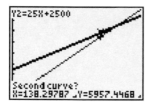
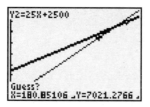

Use ⌊►⌋ to move the cursor to the right of the point of intersection and close to the point. Press ⌊ENTER⌋ to see a screen similar to that above right.

Press ⌊ENTER⌋ again. The calculator displays the screen below, with the coordinates of the point of intersection.

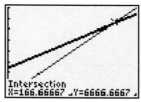

The solution of the linear system is (166.666 67, 6666.6667).

In *Example 1*, the solution of the linear system contained decimals. However, the situation described by the linear system involves numbers of guests and costs in dollars.

When we use a graphing calculator to solve a linear system, we must consider the situation. In this case, we round the number of guests to 167.

Substitute $n = 167$ in each equation to determine the cost.

For $C = 40n$:

When $n = 167$, $C = 40 \times 167$
$$C = 6680$$

For $C = 25n + 2500$:

When $n = 167$, $C = 25 \times 167 + 2500$
$$C = 6675$$

For Hall A, 167 guests cost $6680.
For Hall B, 167 guests cost $6675.

Example 2

Solve this linear system. Determine the solution to 2 decimal places.
$$y = 2x + 2$$
$$y = -1.5x - 3$$

Solution

$y = 2x + 2$ ①
$y = -1.5x - 3$ ②

Press [Y=]. Clear any equations on the screen.
Use [▲] to move the cursor next to Y1.

To enter equation ①, press 2 [x,T,θ,n] [+] 2.
Use [▼] to move the cursor next to Y2.

To enter equation ②, press [(-)] 1.5 [x,T,θ,n] [-] 3.
Use [◄] to move the cursor to the left of Y2. Press [ENTER] to select a heavy line.

Press [WINDOW]. Input Xmin = -9.4, Xmax = 9.4, Xscl = 1, Ymin = -6, Ymax = 6, Yscl = 1.

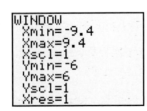
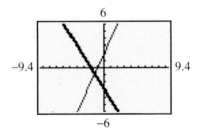

Press [GRAPH] to see the screen above right.

To determine the point of intersection, press [2nd] [TRACE] for [CALC].
Then press **5** to select **intersect**.

Use [▶] or [◀] to move the cursor close to the point of intersection and to the
left of the point. Press [ENTER] to see a screen similar to that below left.

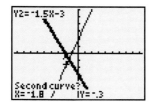
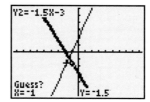

Use [▶] to move the cursor close to the point of intersection and to the right
of the point. Press [ENTER] to see a screen similar to that above right.

Press [ENTER] again. The calculator displays the screen below, with the
coordinates of the point of intersection.

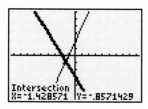

From the screen, to 2 decimal places, the solution of the linear system is
(−1.43, −0.86).

Discussing *the* IDEAS

1. For the graph on page 115, why were the points joined with a broken line?

2. In *Example 1*

   a) Which equation represents a direct variation? Explain.

   b) Which equation represents a partial variation? Explain.

3. In the solution of *Example 1*, why were the equations rewritten in terms of x and y?

4. In the solution of *Example 1*, explain the choice of numbers input for the WINDOW.

5. Following *Example 1*, we rounded the first coordinate to get 167. Why can we not round
   the second coordinate to 6667 to determine the corresponding cost?

6. For *Example 2*, explain how the coordinates of the point of intersection were determined
   by rounding.

A 1. Each graph compares the rental costs of two halls. For each graph:

    i) Describe what the point of intersection represents.

    ii) For how many guests is each hall cheaper? Explain.

a)   **Hall Rental Costs**   b)   **Hall Rental Costs**

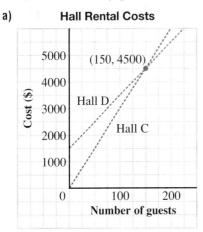
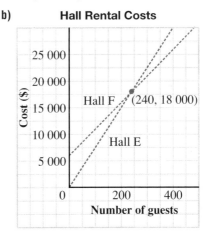

B 2. This graph represents the costs to produce and sell pizza. Graph A represents the daily cost to produce pizza. Graph B represents the daily income from the sale of pizzas.

    a) Describe what the point of intersection represents.

    b) How many pizzas must be sold before there is a profit (that is, the income is greater than the cost)? Explain.

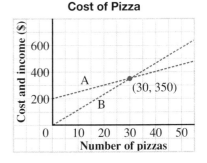

3. Belleville and St. Catharines are two towns in Ontario. They are 300 km apart.

    Graph A represents a car travelling from Belleville to St. Catharines at an average speed of 80 km/h.

    Graph B represents a car travelling from St. Catharines to Belleville at an average speed of 90 km/h.

    The approximate coordinates of the point of intersection are shown.

    Describe what the point of intersection represents.

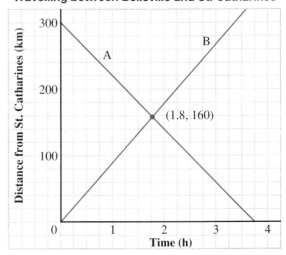

4. A company produces compact discs. Each disc sells for $8. The income, C dollars, from the sale of x discs is given by $C = 8x$.
The cost to produce x discs is given by $C = 4x + 48\ 000$.

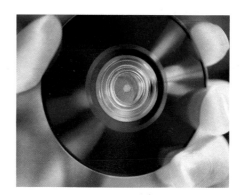

a) Solve this linear system:

$C = 8x$

$C = 4x + 48\ 000$

b) What does the point of intersection represent?

c) How many discs must be sold before there is a profit (that is, the income is greater than the cost)?

5. To clear snow from a driveway, a student charges $5 per hour. A snow-removal company charges an annual fee of $160.
The cost, C dollars, to employ the student for x hours is given by $C = 5x$.
The cost, C dollars, to employ the company is given by $C = 160$.

a) Solve the linear system formed by the two equations.

b) What does the point of intersection represent?

c) When is it cheaper to employ the student? Explain.

6. Gander and Corner Brook are two towns in Newfoundland. They are 350 km apart.
Car A travels from Corner Brook to Gander at an average speed of 70 km/h. Its journey is described by the equation $d = 350 - 70t$.
Car B travels from Gander to Corner Brook at an average speed of 80 km/h. Its journey is described by the equation $d = 80t$.

For each car, d kilometres represents its distance from Gander after driving for t hours.

a) Solve the linear system formed by the two equations.

b) What does the point of intersection represent?

7. The equations and graphs of a linear system are shown. Estimate the solution of each system.

a) $y = -3x + 5$
$y = 0.5x + 2$

b) $y = -2x + 6$
$y = -0.75x + 1$

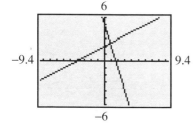

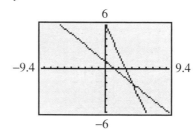

c) $y = 2x + 1$
$y = -1.5x + 8$

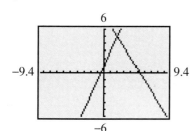

d) $y = 0.75x - 2.5$
$y = -\frac{2}{3}x + \frac{1}{3}$

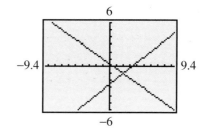

8. Solve each linear system.

a) $y = x + 1$
$y = -2x + 4$

b) $y = -x + 4$
$y = x - 2$

c) $y = -x + 1$
$y = -2x + 5$

9. Solve each linear system. Write each coordinate to 2 decimal places where necessary.

a) $y = 0.5x - 3$
$y = 2.5x - 1$

b) $y = 0.3x + 1$
$y = -1.2x - 3$

c) $y = 2.4x - 5$
$y = 0.45x + 1$

FOCUS
TECHNOLOGY

Using a Table to Solve a Linear System

10. To solve the system: $y = 6x + 1$
$y = -10x + 97$

- Press [Y=]. With the cursor next to Y1=, enter 6 [X,T,θ,n] [+] 1.
  Move the cursor next to Y2=, then enter [(-)] 10 [X,T,θ,n] [+] 97.

- Press [2nd] [WINDOW] for [TBLSET].
  Use [▼] and [ENTER] to produce the window below.

```
TABLE SETUP
 TblStart=0
 ΔTbl=1
Indpnt: Auto Ask
Depend: Auto Ask
```

The x-values in the table will start at 0 and increase by 1.

- Press [2nd] [GRAPH] for [TABLE]. The table below is displayed.

X	Y1	Y2
0	1	97
1	7	87
2	13	77
3	19	67
4	25	57
5	31	47
6	37	37

X=0

- Use ▼ to scroll down the table to find where Y1 = Y2.
- What is the solution of this system?

11. Use the method of exercise 10 to solve each system.

a) $y = x - 4$
$y = -4x + 46$

b) $y = -8x + 10$
$y = -2x + 40$

c) $y = x + 5$
$y = 2x + 15$

C **12.** A cellular phone company offers two plans.

Plan A: a monthly fee of $28, 30 minutes free, then $0.55 for each additional minute

Plan B: a monthly fee of $40 and $0.25 per minute

The monthly cost, C dollars, for t minutes of calls is:

Plan A: $C = 0.55(t - 30) + 28$
Plan B: $C = 0.25t + 40$

a) Solve the linear system formed by the two equations.

b) What does the point of intersection represent?

c) Suppose a person uses the phone for more than 100 minutes each month. Which is the better plan? Explain.

d) Suppose a person uses the phone for 50 minutes each month. Which is the better plan? Explain.

e) Explain when plan A is better than plan B.

Communicating
the IDEAS

Write to describe a situation that could be represented by a linear system. Explain what the solution of the system represents. Include a sketch of a graph in your explanation.

In Section 3.1, we solved linear systems graphically. We can also solve a linear system algebraically.

Recall the linear system from *Example 1*, page 116.

$C = 40n$

$C = 25n + 2500$

In each equation, C dollars is the cost to rent a hall for n guests.

This system can be solved algebraically.

$C = 40n$ ①

$C = 25n + 2500$ ②

Since the left sides of equations ① and ② are equal, the right sides must also be equal.

$40n = 25n + 2500$

Solve for n.

Subtract $25n$ from each side.

$40n - 25n = 25n + 2500 - 25n$

$\qquad 15n = 2500$

Divide each side by 15. Use a calculator.

$$n = \frac{2500}{15}$$

$$n = 166.\overline{6}$$

Do not clear the calculator screen.

Substitute $n = 166.\overline{6}$ in equation ① to determine C.

$C = 40 \times 166.\overline{6}$

Use a calculator. Press $\boxed{\times}$ 40 $\boxed{=}$.

$C = 6666.\overline{6}$

The solution is $(166.\overline{6},\ 6666.\overline{6})$.

The solution represents the number of guests for which the halls cost the same.

However, the number of people is a whole number, so round $166.\overline{6}$ to 167 people.
The halls cost approximately the same for 167 guests.

Example 1

Solve this linear system.

$y = 2x + 1$

$y = -3x + 11$

Solution

$y = 2x + 1$ ①
$y = -3x + 11$ ②

Since the left sides of equations ① and ② are equal, the right sides must also be equal.

$2x + 1 = -3x + 11$

Solve for x.

Add $3x$ to each side.

$2x + 1 + 3x = -3x + 11 + 3x$
$\qquad 5x + 1 = 11$

Subtract 1 from each side.

$5x + 1 - 1 = 11 - 1$
$\qquad\quad 5x = 10$

Divide each side by 5.

$\qquad\quad x = 2$

Substitute $x = 2$ in equation ① to determine y.

$y = 2(2) + 1$
$y = 5$

The solution of the system is (2, 5).

We can use a graphing calculator to check the solution in *Example 1*.

Input Y1= 2 [X,T,θ,n] [+] 1 and Y2= [(-)] 3 [X,T,θ,n] [+] 11.

Press [WINDOW] and set it as shown below left.

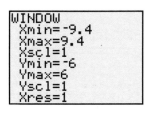
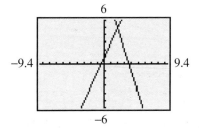

Press [GRAPH] to see the screen above right.

From the screen, the solution is (2, 5).

Alternatively, press [2nd] [TRACE] for [CALC], then use **intersect** to see the screen below.

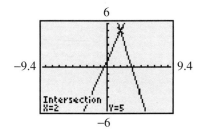

We can check a solution algebraically, as shown in *Example 2*.

Example 2

Solve, then check this linear system.
$x + y = 6$
$x - y = -3$

Solution

$x + y = 6$ ①
$x - y = -3$ ②

> **Think ...**
>
> Rearrange each equation to solve for y.

$x + y = 6$ ①
Subtract x from each side.
$x + y - x = 6 - x$
$y = 6 - x$ ③

$x - y = -3$ ②
Subtract x from each side.
$x - y - x = -3 - x$
$-y = -3 - x$
Multiply each side by -1.
$y = 3 + x$ ④

Since the left sides of equations ③ and ④ are equal, the right sides must also be equal.
$6 - x = 3 + x$
Solve for x.
Subtract x from each side.
$6 - x - x = 3 + x - x$
$6 - 2x = 3$
Subtract 6 from each side.
$6 - 2x - 6 = 3 - 6$
$-2x = -3$
Divide each side by -2.
$x = 1.5$

Substitute $x = 1.5$ in equation ① to determine y.

$y = 6 - 1.5$
$y = 4.5$
The solution is (1.5, 4.5).

Check.
Substitute $x = 1.5$ and $y = 4.5$ in equation ①: $x + y = 6$
Left side $= x + y$ Right side $= 6$
        $= 1.5 + 4.5$
        $= 6$
Since the left side equals the right side, (1.5, 4.5) satisfies equation ①.

Substitute $x = 1.5$ and $y = 4.5$ in equation ②: $x - y = -3$
Left side $= 1.5 - 4.5$ Right side $= -3$
        $= -3$
Since the left side equals the right side, (1.5, 4.5) satisfies equation ②.

Since the solution satisfies both equations, the solution is correct.

Example 3

Solve, then check this linear system.
$3x - 2y = 8$ ①
$4x + y = 7$ ②

Solution

> **Think ...**
>
> Solve equation ② for y. Then substitute that value of y in equation ①.

Solve equation ② for y.
     $4x + y = 7$
Subtract $4x$ from each side.
     $4x + y - 4x = 7 - 4x$
           $y = 7 - 4x$
Substitute this expression for y in equation ①: $3x - 2y = 8$
    $3x - 2(7 - 4x) = 8$
    $3x - 14 + 8x = 8$
      $11x - 14 = 8$
Add 14 to each side.
          $11x = 22$

Divide each side by 2.
$$x = 2$$
Substitute $x = 2$ in equation ② to determine y.
$$4(2) + y = 7$$
$$8 + y = 7$$
$$y = -1$$
The solution is $(2, -1)$.

Check.
Substitute $x = 2$ and $y = -1$ in equation ①: $3x - 2y = 8$

Left side = $3x - 2y$ Right side = 8
$$= 3(2) - 2(-1)$$
$$= 6 + 2$$
$$= 8$$
Since the left side equals the right side, $(2, -1)$ satisfies equation ①.

Substitute $x = 2$ and $y = -1$ in equation ②: $4x + y = 7$

Left side = $4x + y$ Right side = 7
$$= 4(2) - 1$$
$$= 8 - 1$$
$$= 7$$
Since the left side equals the right side, $(2, -1)$ satisfies equation ②.

Since the solution satisfies both equations, the solution is correct.

Discussing the IDEAS

1. Look at the solution of the linear system on page 124. Explain why the method is called "solving by substitution."

2. In the solution of the linear system on page 124, may we round $6666.\overline{6}$ to 6667 and say that 167 guests cost $6667? Explain.

3. In the solution of *Example 1*, we substituted $x = 2$ in equation ① to determine y. Could we have substituted $x = 2$ in equation ② instead? Explain.

4. In the *Check* of the solution in *Example 2*, why did we substitute for x and y in equations ① and ② instead of equations ③ and ④?

5. In the solution of *Example 2*, could the first step have been to rearrange each equation to solve for x? Try it. Explain the result.

1. Solve each linear system.

a) $y = x + 1$
$y = -x + 3$

b) $y = -x + 3$
$y = x - 1$

c) $y = x + 1$
$y = -x + 5$

2. Solve each linear system.

a) $y = 2x - 4$
$y = -2x + 8$

b) $y = -2x + 8$
$y = 2x - 8$

c) $y = -2x + 5$
$y = 2x + 1$

3. Solve each linear system.

a) $y = -3x + 1$
$y = 3x + 7$

b) $y = 3x + 1$
$y = -3x - 11$

c) $y = -3x + 5$
$y = 3x - 13$

B **4.** Solve, then check each linear system.

a) $y = x + 5$
$y = 2x + 5$

b) $y = 3x - 13$
$y = x - 5$

c) $y = 2x + 3$
$y = -x - 6$

5. Solve, then check each linear system.

a) $y = 3x - 10$
$y = -3x + 8$

b) $y = 2x$
$y = -2x + 8$

c) $y = 3x + 17$
$y = -2x - 8$

6. a) Solve, then check each linear system.

i) $x + y = 4$
$x - y = 0$

ii) $x - y = -5$
$x + y = -1$

iii) $x + y = -2$
$x - y = 6$

iv) $x - y = -3$
$x + y = 5$

v) $x + y = -7$
$x - y = 1$

vi) $x - y = 8$
$x + y = 2$

b) Choose one linear system from part a. Write to explain how you solved it.

7. Pyramid Stables charges $20/h, including insurance, for a trail ride.
Sara's Stables charges $16/h, and $12 for insurance.
Let C dollars represent the cost of a trail ride for h hours.
A linear system that represents the costs of the rides is:

$C = 20h$
$C = 16h + 12$

a) Write to explain each equation.

b) Solve the linear system.

c) Explain what the solution represents.

d) Suppose a person rides for 2 h. Which stable should he choose? Explain.

8. A fishing boat sails to and from the fishing grounds at an average speed of 10 knots. One knot is 1 nautical mile per hour. While the people are fishing, the average speed of the boat is 3 knots. The total distance travelled during an 11-h trip is 61 nautical miles.

Let x hours represent the time travelling to and from the fishing grounds. Let y hours represent the time spent fishing.

A linear system that represents the trip is:

$x + y = 11$
$10x + 3y = 61$

a) Write to explain each equation. **b)** Solve the linear system.

c) Explain what the solution represents. **d)** How much time was spent fishing?

9. Solve, then check each linear system.

a) $x = y + 4$
  $x = 2y + 8$

b) $x = y - 5$
  $x = 3y + 13$

c) $x = -y + 6$
  $x = y - 3$

d) $x = 2y - 3$
  $x = 3y - 3$

e) $x = -2y + 11$
  $x = 4y - 13$

f) $x = -3y - 14$
  $x = 4y + 42$

10. Solve each linear system.

a) $3x + y = 2$
  $2x + 3y = 13$

b) $2x + 3y = 6$
  $2x - y = -2$

c) $x - 2y = -8$
  $3x + y = -3$

11. Solve, then check each linear system.

a) $3x - y = -2$
  $2x + y = -8$

b) $x - 2y = 9$
  $3x - y = 17$

c) $2x - y = 2$
  $x + 4y = 28$

d) $x - y = 5$
  $2x + 4y = -5$

e) $2x - 2y = -9$
  $x + 4y = 18$

f) $3x - y = 10$
  $4x - 2y = 11$

C **12.** Solve, then check each linear system.

a) $3x - 2y = -4$
  $3x + 2y = 16$

b) $2x - 3y = -7$
  $2x + 3y = 11$

c) $3x - 2y = 4$
  $2x + 3y = 7$.

d) $3x - 5y = 22$
  $3x + 5y = 2$

e) $2x + 2y = 3$
  $2x + 3y = 3$

f) $8x + 2y = -11$
  $4x + 2y = -5$

Communicating the IDEAS

You know two methods to solve a linear system: using a graphing calculator and using substitution. Write to explain when you would use each method. Include examples of linear systems in your explanation.

3.3 Solving Linear Systems by Elimination

In the solution of *Example 1*, page 125, we solved a linear equation using this property:

- Equal terms can be added to both sides of an equation without changing the equality.
  That is: to solve $2x + 1 = -3x + 11$

  Add $3x$ to each side.
  $$2x + 1 + 3x = -3x + 11 + 3x$$
  $$5x + 1 = 11$$
  Add -1 to each side.
  $$5x + 1 - 1 = 11 - 1$$
  $$5x = 10$$
  $$x = 2$$

We can extend this property to combine two equations by adding the left sides, then the right sides. This creates another equation. When we combine equations in this way, we *add* the equations.

We can solve certain linear systems by adding the equations. When we do this, we may *eliminate* one variable.

Example 1

Solve this linear system.
$$x + y = 6$$
$$x - y = 4$$

Solution

$x + y = 6$ ①
$x - y = 4$ ②

> **Think ...**
>
> We can add two equations without affecting the solution. Since $+y$ and $-y$ are opposites, when we add the equations, we eliminate y.

$$
\begin{array}{ll}
x + y = 6 & \quad ① \\
\underline{x - y = 4} & \quad ② \\
\text{Add.} \quad 2x = 10 & \\
x = 5 &
\end{array}
$$

Substitute $x = 5$ in equation ① to determine y.
$$5 + y = 6$$
$$y = 1$$
The solution of the system is $(5, 1)$.

On page 114, exercise 1d, you solved a linear equation using this property:

- Both sides of an equation can be multiplied by a constant without changing the equality.

  That is: to solve $-\frac{1}{2}x = 2$

  Multiply each side by –2.

$$(-2)\left(-\frac{1}{2}x\right) = (-2)(2)$$
$$x = -4$$

We can extend this property to solve certain linear systems, as illustrated by *Examples 2* and *3*.

___

Example 2

Solve, then check this linear system.
$2x + 5y = 16$
$x - y = 1$

Solution

$2x + 5y = 16$ ①
$x - y = 1$ ②

> **Think ...**
>
> If we multiply equation ② by 5, the y-coefficients will be opposites. We can then add the equations to eliminate y.

Copy ①. $2x + 5y = 16$ ①
Multiply ② by 5. $5x - 5y = 5$ ③
Add. $\overline{7x = 21}$
                      $x = 3$

Substitute $x = 3$ in equation ② to determine y.
       $3 - y = 1$
         $-y = -2$
          $y = 2$
The solution is (3, 2).

Check.
Substitute $x = 3$ and $y = 2$ in equation ①: $2x + 5y = 16$
Left side = $2x + 5y$ Right side = 16
       = $2(3) + 5(2)$
       = $6 + 10$
       = 16
Since the left side equals the right side, (3, 2) satisfies equation ①.

Substitute $x = 3$ and $y = 2$ in equation ②: $x - y = 1$

Left side $= x - y$ Right side $= 1$
$$= 3 - 2$$
$$= 1$$

Since the left side equals the right side, (3, 2) satisfies equation ②.

Since the solution (3, 2) satisfies both equations, the solution is correct.

Example 3

Solve, then check this linear system.

$x - 4y = -9$

$2x - 3y = -8$

Solution

$x - 4y = -9$ ①

$2x - 3y = -8$ ②

> **Think ...**
>
> If we multiply equation ① by -2, the x-coefficients will be opposites. We can then add the equations to eliminate x.

Multiply equation ① by -2. $-2x + 8y = 18$ ③

Copy ②. $\underline{2x - 3y = -8}$ ②

Add. $5y = 10$

                                      $y = 2$

Substitute $y = 2$ in equation ① to determine x.

$$x - 4(2) = -9$$
$$x - 8 = -9$$
$$x = -1$$

The solution is $(-1, 2)$.

Check.

Substitute $x = -1$ and $y = 2$ in equation ①: $x - 4y = -9$

Left side $= x - 4y$ Right side $= -9$
$$= -1 - 4(2)$$
$$= -1 - 8$$
$$= -9$$

Since the left side equals the right side, $(-1, 2)$ satisfies equation ①.

Substitute $x = -1$ and $y = 2$ in equation ②: $2x - 3y = -8$

Left side $= 2x - 3y$ Right side $= -8$

$$= 2(-1) - 3(2)$$
$$= -2 - 6$$
$$= -8$$

Since the left side equals the right side, $(-1, 2)$ satisfies equation ②.

Since the solution $(-1, 2)$ satisfies both equations, the solution is correct.

For some linear systems, we multiply both equations by constants to eliminate one variable.

Example 4

Solve this linear system.
$$5x - 3y = -5$$
$$3x - 2y = -4$$

Solution

$5x - 3y = -5$ ①
$3x - 2y = -4$ ②

Think ...

To eliminate x, multiply equation ① by 3 and equation ② by -5. Then the x-coefficients will be opposites.

Multiply equation ① by 3. $15x - 9y = -15$ ③
Multiply equation ② by -5. $\underline{-15x + 10y = 20}$ ④
Add. $y = 5$

Substitute $y = 5$ in equation ① to determine x.
$$5x - 3(5) = -5$$
$$5x - 15 = -5$$
$$5x = 10$$
$$x = 2$$

The solution is $(2, 5)$.

1. In the solution of *Example 1*, we substituted $x = 5$ in equation ① to determine y. Could we have substituted $x = 5$ in equation ② instead? Explain.

2. To solve the system in *Example 2*, we eliminated y. Could we have solved the system by eliminating x? Explain.

3. In the solution of *Example 2*, why did we substitute $x = 3$ in equation ② instead of in equation ①?

4. In the solution of *Example 4*, how did we know to multiply equation ① by 3 and equation ② by –5?

5. In the solution of *Example 4*, what would we multiply each equation by to eliminate y? Solve the system this way to check.

Practise Your Skills

1. Add.

a) $\quad 6$
$\quad\underline{3}$

b) $\quad 6$
$\quad\underline{-\ 3}$

c) $\quad -6$
$\quad\underline{3}$

d) $\quad -6$
$\quad\underline{-\ 3}$

e) $\quad 4$
$\quad\underline{10}$

f) $\quad 4$
$\quad\underline{-\ 10}$

g) $\quad -4$
$\quad\underline{10}$

h) $\quad -4$
$\quad\underline{-\ 10}$

2. Add.

a) $\quad 2x$
$\quad\underline{5x}$

b) $\quad -2x$
$\quad\underline{+\ 5x}$

c) $\quad 2y$
$\quad\underline{-\ 5y}$

d) $\quad -2y$
$\quad\underline{-\ 5y}$

e) $\quad 8x$
$\quad\underline{3x}$

f) $\quad -8x$
$\quad\underline{+\ 3x}$

g) $\quad 8y$
$\quad\underline{-\ 3y}$

h) $\quad -8y$
$\quad\underline{-\ 3y}$

3.3 Exercises

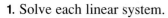

A 1. Solve each linear system.

a) $x + y = 3$
$x - y = 5$

b) $x + y = -7$
$x - y = 9$

c) $x - y = 6$
$x + y = -6$

d) $x - y = -4$
$x + y = 10$

e) $x + y = -1$
$x - y = 1$

f) $x - y = 0$
$x + y = 0$

2. Solve each linear system.

a) $x + y = 1$
$3x - y = 15$

b) $2x + y = 3$
$3x - y = 7$

c) $x + 3y = -6$
$-x - 2y = 3$

d) $x - y = -5$
$2x - y = -7$

e) $x - y = 2$
$x + 2y = 8$

f) $x - y = -5$
$x - 3y = 13$

3. Solve each linear system.

a) $2x + 3y = 12$
$2x - 3y = -12$

b) $5x - 7y = 24$
$3x + 7y = -8$

c) $-3x + 4y = 7$
$3x + y = -2$

d) $x + 3y = 12$
$5x - 3y = 6$

e) $-2x - y = 5$
$2x - 2y = 7$

f) $4x + 3y = -9$
$-4x + y = -19$

4. Solve each linear system.

a) $5x - 4y = -7$
$5x + 3y = 14$

b) $x - 3y = -2$
$2x - 3y = -4$

c) $2x - 3y = -23$
$2x + 5y = 17$

d) $x - 4y = -30$
$3x - 4y = -26$

e) $3x + 2y = -18$
$3x - 4y = 18$

f) $3x - 5y = 26$
$2x - 5y = 19$

5. Choose one linear system from exercise 3 or 4. Write to explain how you solved it.

6. Solve, then check each linear system.

a) $x - y = 2$
$2x + 3y = 14$

b) $3x - 2y = -19$
$x + y = -3$

c) $x + y = -1$
$3x - 4y = 25$

d) $5x + 3y = -36$
$x - y = -4$

e) $x + y = 5$
$2x - 5y = 17$

f) $4x - 3y = -35$
$x + y = 0$

7. In hockey, the points a player scores is the sum of the goals scored and the assists.

In one season, Mats Sundin scored 83 points.

Sundin scored 21 fewer goals than assists.
Let x represent the number of goals scored.
Let y represent the number of assists.

A linear system that represents this situation is:

$y + x = 83$
$y - x = 21$

a) Write to explain each equation.

b) Solve the linear system.

c) Explain what the solution represents.

8. Solve, then check each linear system.

a) $x + 2y = 2$
$2x - 3y = 4$

b) $3x + 4y = -21$
$x - 3y = 6$

c) $2x + y = -3$
$3x + 4y = -7$

d) $3x - 11y = 20$
$x - 3y = 5$

e) $3x - 10y = 16$
$2x + y = 3$

f) $x - 3y = 10$
$2x + 5y = -13$

9. Solve each linear system.

a) $2x - 8y = 17$
$4x + 6y = -21$

b) $3x - 4y = 7$
$4x + 5y = -32$

c) $3x + 4y = -1$
$5x - 3y = 8$

d) $4x + 3y = 9$
$2x - 7y = 13$

e) $2x - 5y = 26$
$5x - 4y = 31$

f) $5x + 8y = -23$
$8x - 5y = -19$

10. Point P lies on the lines with equations $2x + y = 5$ and $3x + 2y = 2$. On which of the following lines does P also lie?

a) $5x + 3y = 7$

b) $x + y = -3$

c) $4x + 3y = -1$

d) $y = -11$

Ⓒ **11.** In the equation $2x + 5y = 8$, the coefficients form a pattern. That is, each coefficient is 3 more than the previous one.

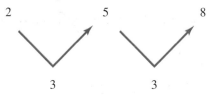

a) Write two different equations with coefficients that form the pattern above.

b) Solve the linear system in part a.

c) Write a linear equation with a different pattern in its coefficients. Repeat parts a and b for your equation. Write to explain what you notice.

Communicating *the* IDEAS

You have learned 3 methods to solve a linear system.

a) Write to describe each method.

b) Is one method always easier than another? Write to explain your answer. Refer to examples from any preceding exercises.

3.4 Applications of Linear Systems

You have learned how to solve a linear system with a graphing calculator and algebraically. Certain problems can be described using linear systems, as illustrated in 3.1, 3.2, and 3.3 Exercises. The solution of the problem can be determined from the solution of the linear system.

Example 1

Eight thousand people attended a rock concert. The ticket prices were $50 and $30. The total revenue from the ticket sales was $250 000. How many tickets of each price were sold? Check the solution.

Solution

Let x represent the number of $50-tickets sold.
Let y represent the number of $30-tickets sold.
The total number of tickets sold is 8000.
Use the preceding 3 sentences to write an equation: $x + y = 8000$ ①

The revenue from x $50-tickets is $50x$ dollars.
The revenue from y $30-tickets is $30y$ dollars.
The total revenue is $250 000.
Use the preceding 3 sentences to write an equation: $50x + 30y = 250\ 000$ ②

Equations ① and ② form a linear system.

Solve the system by substitution.
Solve equation ① for y.
$y = 8000 - x$

Substitute this expression for y in equation ②:

$$50x + 30y = 250\ 000$$
$$50x + 30(8000 - x) = 250\ 000$$
$$50x + 240\ 000 - 30x = 250\ 000$$
$$20x + 240\ 000 = 250\ 000$$
$$20x = 250\ 000 - 240\ 000$$
$$20x = 10\ 000$$
$$x = 500$$

Substitute $x = 500$ in equation ① to determine y.

$$500 + y = 8000$$
$$y = 8000 - 500$$
$$y = 7500$$

500 tickets were sold at $50 each and 7500 tickets were sold at $30 each.

Check.
500 tickets at $50 each is $500 \times \$50 = \$25\ 000$
7500 tickets at $30 each is $7500 \times \$30 = \$225\ 000$
Total revenue is $\$25\ 000 + \$225\ 000 = \$250\ 000$
Since this is the given revenue, the solution is correct.

The solution of *Example 1* illustrates how to use a linear system to solve a problem. For similar problems, follow these steps:

- Identify two unknown quantities. Let them be represented by two variables, such as x and y.
- Use the given information to write two equations in x and y.
- Solve the linear system.
- Use the solution of the linear system to solve the problem.
- Check the solution(s) to the problem.

Example 2

A golf club charges its members an annual fee, and a greens fee for each golf game played.

In one year, member A played 12 games and paid $814.

In the same year, member B played 29 games and paid $1188.

a) What is the annual fee?

b) What is the greens fee?

Solution

> **Think ...**
>
> There are two quantities to be determined. Let them be represented by x and y. Write a linear system.

Let x dollars represent the annual fee.
Let y dollars represent the greens fee.

Member A paid x dollars annual fee and $12y$ dollars greens fee.
Member A paid $814.
Use this information to write an equation:

$x + 12y = 814$ ①

Member B paid x dollars annual fee and $29y$ dollars greens fee.
Member B paid $1188.
Use this information to write an equation:

$x + 29y = 1188$ ②

Equations ① and ② form a linear system.
Solve the system by elimination.

Copy ②. $x + 29y = 1188$ ②
Multiply ① by −1. $\underline{-x - 12y = -814}$ ③
Add. $17y = 374$
                            $y = 22$

Substitute $y = 22$ in equation ① to determine x.

$x + 12(22) = 814$
$x + 264 = 814$
$x = 814 - 264$
$x = 550$

a) The annual fee is $550.

b) The greens fee is $22.

Check.
Member A paid $12 \times \$22 + \$550 = \$814$.
Member B paid $29 \times \$22 + \$550 = \$1188$.
Since this matches the given information, the solution is correct.

Example 3

A salesperson works at an electronics store.
She has a choice of two payment plans.

Plan A: a commission of 11% of her monthly sales

Plan B: a monthly salary of $1500 plus a commission
of 2% of her monthly sales

Which is the better payment plan for the salesperson?

Solution

Consider the person's pay for different monthly sales.

Monthly sales ($)	Plan A payment ($)	Plan B payment ($)
5 000	$5\,000 \times 0.11 = 550$	$1500 + 0.02 \times 5\,000 = 1600$
10 000	$10\,000 \times 0.11 = 1100$	$1500 + 0.02 \times 10\,000 = 1700$
15 000	$15\,000 \times 0.11 = 1650$	$1500 + 0.02 \times 15\,000 = 1800$
20 000	$20\,000 \times 0.11 = 2200$	$1500 + 0.02 \times 20\,000 = 1900$

From the table:

The salesperson earns more with plan B, for sales up to approximately $15 000.

The salesperson earns more with plan A, for sales above approximately $15 000.

Each payment plan can be represented by an equation.

Let the monthly sales be represented by m dollars.

Let the monthly payment be represented by P dollars.

For plan A, the payment is $P = 0.11m$.

For plan B, the payment is $P = 1500 + 0.02m$.

These two equations form a linear system.

Use a graphing calculator to solve the linear system.

The variable P is graphed as Y1, then Y2.

The variable m is graphed as X.

Press ⬚ Y= ⬚ 0.11 ⬚X,T,θ,n⬚. Use ⬚▼⬚ to move the cursor next to Y2.

Press 1500 ⬚ + ⬚ 0.02 ⬚X,T,θ,n⬚. Use ⬚◄⬚ to move the cursor to the left of Y2. Press ⬚ENTER⬚.

Press ⬚WINDOW⬚. Input Xmin = 0, Xmax = 20000, Xscl = 5000, Ymin = 0, Ymax = 2500, Yscl = 500, to see the screen below left.

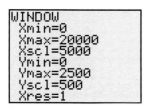

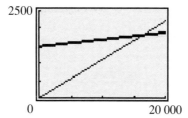

Press ⬚GRAPH⬚ to see the screen above right.

To determine the point of intersection, press ⬚ 2nd ⬚ ⬚TRACE⬚ for ⬚ CALC ⬚.
Press **5** to select **intersect**.

Use ⬚►⬚ or ⬚◄⬚ to move the cursor close to the point of intersection and to its left.
Press ⬚ENTER⬚.

Use ⬚►⬚ to move the cursor close to the point of intersection and to its right.
Press ⬚ENTER⬚.

Press [ENTER] again to see the screen below.

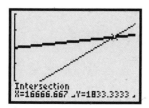

From the screen, the solution of the linear system is (16 666.667, 1833.3333).
This means that for monthly sales of approximately $16 667, the monthly
payment is approximately the same for both payment plans.

From the screen, the heavier line represents plan B.
If the sales person expects her monthly sales to be less than $16 667,
she should choose plan B.

If the sales person expects her monthly sales to be greater than $16 667,
she should choose plan A.

Discussing *the* IDEAS

1. In the solution of *Example 1*, we solved equation ① for y. Could we have completed the example by solving equation ① for x instead? Explain.

2. In the solution of *Example 1*, we substituted $x = 500$ in equation ①. Could we have substituted in equation ② instead? Explain.

3. In the solution of *Example 2*, we used elimination to solve the linear system. Could we have used substitution instead? Explain.

4. In the solution of *Example 3*, explain the choice of the numbers input for the WINDOW.

Practise Your Skills

1. Calculate each percent.

   a) 10% of $360

   b) 1% of $360

   c) 15% of $4500

   d) 1.5% of $4500

   e) 20% of $1600

   f) 2% of $1600

   g) 35% of $230

   h) 3.5% of $230

A 1. Solve each linear system.

a) $x + y = 7$
$x - y = 11$

b) $2x - y = 10$
$x + y = 8$

c) $-x + 2y = 1$
$x - 3y = 2$

d) $3x + y = 4$
$5x + y = 6$

e) $6x - 3y = 6$
$2x + 3y = -6$

f) $3x + 5y = 12$
$7x + 5y = 8$

B 2. Solve, then check each linear system.

a) $3x + y = 18$
$x + 2y = 11$

b) $5x + 2y = 5$
$3x - 4y = -23$

c) $3x + y = 4$
$5x + 2y = 5$

d) $6x - 5y = -2$
$2x + 3y = 18$

e) $5x + 3y = 5$
$3x + 2y = 4$

f) $2x - 5y = -4$
$3x + 4y = -29$

3. An airplane travels at an average speed of 740 km/h with a tail wind, and 560 km/h with a head wind. Let s kilometres per hour represent the speed of the plane with no wind. Let w kilometres per hour represent the wind speed. This situation can be described by a linear system:

$s + w = 740$
$s - w = 560$

a) Solve the linear system.

b) Explain what the solution represents.

c) What is the wind speed?

4. At a sale, each CD is one price and each tape is another price. Three CDs and two tapes cost $67. One CD and three tapes cost $48. Let c dollars represent the price of one CD. Let t dollars represent the price of one tape. This situation can be described by a linear system:

$3c + 2t = 67$
$c + 3t = 48$

a) Write to explain each equation.   b) Solve the linear system.

c) What does one CD cost?   d) What does one tape cost?

5. A salesperson works at a computer store. He has a choice of two payment plans.
Plan A: a commission of 10% of his monthly sales
Plan B: a monthly salary of $1750 plus a commission of 1.5% of his monthly sales
Let m dollars represent the monthly sales.
Let P dollars represent the monthly payment.

This situation can be described by a linear system:

$P = 0.10m$

$P = 1750 + 0.015m$

a) Write to explain each equation.

b) Solve the linear system.

c) Which is the better payment plan for the salesperson? Explain.

6. A regular light bulb costs $0.55 and uses $6 of electricity every 1000 h.
This is a rate of $\frac{\$6}{1000 \text{ h}}$, or $0.006/h.

An energy-saver light bulb costs $1.55 and uses $4 of electricity every 1000 h.
This is a rate of $\frac{\$4}{1000 \text{ h}}$, or $0.004/h.

Let h hours represent the time for which the light bulbs burn.
Let c dollars represent the cost of electricity.
This situation can be described by a linear system:

$c = 0.55 + 0.006h$

$c = 1.55 + 0.004h$

a) Write to explain each equation.

b) Solve the linear system.

c) Explain what the solution represents.

d) Which bulb is cheaper to use? Explain.

7. Three footballs and one soccer ball cost $155.
Two footballs and three soccer balls cost $220.
What are the costs of one football and one soccer ball?

8. A tennis club charges an annual fee and
an hourly fee for court time. In one year,
member A played for 39 h and paid $384.
In the same year, member B played for 51 h
and paid $456.

a) What is the annual fee?

b) What is the hourly fee?

9. A salesperson has a choice of two payment plans.
Plan A is a commission of 7% of her weekly sales.
Plan B is a weekly salary of $200 plus a commission of 2.5% of her weekly sales.
Which is the better payment plan for the salesperson? Explain.

10. Greenthumb Garden Products Ltd. manufactures wheelbarrows and carts.
A wheelbarrow has one wheel. A cart has two wheels. The company has
500 wheels with which to make 300 vehicles.

a) Write an equation to describe the total number of vehicles.

b) Write an equation to describe the total number of wheels.

c) Solve the linear system formed by the equations in parts a and b.

d) How many wheelbarrows can be made?

e) How many carts can be made?

11. A sports club is planning an awards banquet. It will provide two main courses: vegetarian lasagna for $8 and turkey dinner for $10. The budget is $980 for 100 people.

   a) Write an equation to describe the total number of meals.

   b) Write an equation to describe the total cost.

   c) Solve the linear system formed by the equations in parts a and b.

   d) How many vegetable lasagnas can be provided?

   e) How many turkey dinners can be provided?

12. A taxi ride costs $2.50 plus $0.35 per kilometre. A limousine charges a flat rate of $35.

   a) Write an equation to describe the cost of a taxi ride.

   b) Write an equation to describe the cost of a limousine ride.

   c) Solve the linear system formed by the equations in parts a and b.

   d) Which is the cheaper way to travel: taxi or limousine? Explain.

13. Two companies quote prices to print a school's yearbook.
   Company A: a fee of $3500 plus $3.25 per book
   Company B: a fee of $1700 plus $4.75 per book

   a) For how many yearbooks do the two companies charge the same amount?

   b) The yearbook committee thinks it can sell 1200 copies. Which company should the committee use? Explain.

C 14. The fuel consumption for a particular car is 11 L/100 km for city driving and 8 L/100 km for highway driving. These fuel consumptions are 0.11 L/km and 0.08 L/km, respectively. In one week, a car used 62 L of fuel to travel 600 km.

   a) How far did the car drive in the city?

   b) How far did the car drive on the highway?

Communicating *the* IDEAS

Write to explain how the solution of a linear system can be used to solve a problem. Include three examples of linear systems in your explanation.

3.5 When Will the Skydivers Meet?

On page 112, you considered the problem of how two skydivers can meet when they jump from planes at different altitudes. When a skydiver jumps from a plane, she soon reaches a downward speed that is approximately constant. Air resistance prevents her from falling faster. This constant speed depends on the body position.

Spread stable position

Diving or standing position

In this position, a skydiver falls at speeds from 50 m/s to 60 m/s.

In this position, a skydiver can reach speeds of 80 m/s to 90 m/s.

Allison and Beth jump at the same time from planes at different altitudes. This graph shows how their heights above the ground change while they fall.

Note: To construct this graph, a major assumption was made — that the skydivers fall at constant speeds from the moment they jump from their planes. In practice, the skydivers' initial downward speed is 0 m/s, and it takes a few seconds to reach the constant speeds on the graph. See exercise 4.

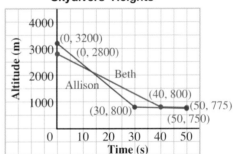

Skydivers' Heights

For exercises 1 to 4, use the graph on page 146.

1. **a)** What were the heights of Allison's and Beth's planes when they jumped?

   b) How far did Allison fall in 30 s? What was her speed in metres per second?

   c) How far did Beth fall in 40 s? What was her speed?

d) Who used the spread stable position? Who used the dive position?

e) How many seconds after jumping were they at the same height?

f) What did they do when they reached a height of 800 m? Explain.

2. Let h metres represent each person's height above the ground t seconds after she jumped.

a) Write an equation to represent Allison's height.

b) Write an equation to represent Beth's height.

c) The equations in parts a and b form a linear system. Solve the system algebraically. Compare the result with your answer to exercise 1e.

3. Suppose Allison and Beth plan to stay together when they meet.

a) What should Allison do when she meets Beth?

b) How many seconds after jumping will they be at the same height?

c) For how many seconds will they be able to stay together?

d) Suppose a graph were drawn to represent this situation. Describe how it would be different from the graph on page 146.

4. Suppose the graph on page 146 is redrawn to show that the downward speeds are initially 0 m/s.

a) Draw a sketch to show how you think the graph would change.

b) Do you think there would be a significant change in your estimate of the time for the skydivers to meet? Explain.

5. Two skydivers plan to meet and stay together during their dives. One person jumps from a plane at 2950 m and falls at 55 m/s in spread stable position. At the same time, the other person jumps from a plane at 3350 m and falls at 85 m/s in diving position.

a) Write equations representing their heights after t seconds.

b) How many seconds after jumping will they meet?

c) They will open their parachutes at 800 m. For how many seconds will they be able to stay together?

Communicating *the* IDEAS

Suppose two skydivers plan to jump from different planes and stay together during their dives. Write to explain how a linear system can be used to estimate the number of seconds after jumping until they meet. Include a graph with your explanation.

Mathematics | **Toolkit**

Algebra Tools

- A linear system is a pair of linear equations.
  For example, $C = 40n$
  $$C = 25n + 2500$$

- A linear system can be solved in different ways.

  – by graphing

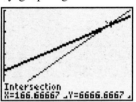

Intersection
X=166.66667 Y=6666.6667

  The solution is (166.666 67, 6666.6667).

  – by substitution
  Solve one equation for y. Then substitute that value for y in the other equation.
  Solve the resulting equation for x. Substitute that value for x in one of the original
  equations to determine y.
  See *Example 3*, pages 127, 128.

  – by elimination
  Eliminate x by multiplying the two equations by numbers that produce opposite
  x-coefficients. Add the equations. Solve the resulting equation for y. Substitute
  that value for y in one of the original equations to determine x.
  See *Example 4*, page 134.

1. Two banquet halls are considered for a wedding reception.
Hall X charges a fixed cost of $5000, plus $125 per guest.
Hall Y charges a fixed cost of $7500, plus $100 per guest.
This situation can be described by a linear system:
$$C = 5000 + 125g$$
$$C = 7500 + 100g$$

a) In each equation:
   i) What does C represent? **ii)** What does g represent?

b) Solve the linear system.

c) What does the point of intersection represent?

d) Which hall is cheaper? Explain.

2. Kirkland Lake and Thunder Bay are two towns in Ontario. They are 880 km apart.

   Car A travels from Kirkland Lake to Thunder Bay at an average speed of 75 km/h. Its journey is described by $d = 880 - 75t$.

   Car B travels from Thunder Bay to Kirkland Lake at an average speed of 85 km/h. Its journey is described by $d = 85t$.

   a) Write to explain each equation.

   b) Solve the linear system formed by the two equations.

   c) What does the point of intersection represent?

3. Refer to the *Technology Focus*, pages 122, 123. Use a table to solve each linear system.

   a) $y = 10x - 30$
   $y = 7x + 15$

   b) $y = 2x - 283$
   $y = -3x + 282$

   c) $y = -2x - 89$
   $y = 3x + 91$

4. Solve each linear system by substitution. Check each solution.

   a) $x + y = 1$
   $3x - 2y = 13$

   b) $3x + y = 7$
   $5x + 2y = 13$

   c) $4x + y = -5$
   $2x + 3y = 5$

   d) $3x + 2y = 16$
   $2x - y = -1$

   e) $3x - 2y = 4$
   $2x + y = 5$

   f) $3x + 2y = 48$
   $x + y = 48$

5. A radio program has 12 commercial breaks in a 1-h show. Each break is 30 s or 60 s long. The total time for commercials in the show is 10 min. This situation can be described by a linear system:

   $x + y = 12$
   $0.5x + y = 10$

   a) Write to explain each equation.

   b) Solve the linear system.

   c) What does the solution of the system represent?

   d) How many 30-s commercial breaks were there?

6. Solve each linear system by elimination. Check each solution.

   a) $x - 2y = 1$
   $x + 2y = 3$

   b) $2x + 3y = 5$
   $2x + 9y = 11$

   c) $6x + 7y = 10$
   $2x - 3y = 14$

   d) $2x + 3y = 6$
   $x + 2y = 4$

   e) $3x + 2y = 5$
   $2x + 3y = 0$

   f) $2x + y = -5$
   $3x + 5y = 3$

Testing Your Knowledge

7. A company manufactures two types of T-shirts.
The cost of 1 deluxe shirt is $8 for materials and $30 for labour.
The cost of 1 standard shirt is $5 for materials and $12 for labour.
In one month, the company spends $1700 for materials and $5700 for labour.
This situation can be described by a linear system:

$8d + 5s = 1700$

$30d + 12s = 5700$

a) Write to explain each equation.

b) Solve the linear system.

c) What does the solution of the system represent?

d) How many standard shirts were produced in 1 month?

8. Solve, then check each linear system.

a) $x + 5y = -17$
$3x - 4y = 6$

b) $x + 5y = -11$
$4x - 3y = 25$

c) $3x + 2y = 6$
$3x - 2y = -12$

d) $9x + 2y = 2$
$4x + y = 1$

e) $2x + y = 3$
$7x + 2y = 0$

f) $x = 2y - 5$
$x = -5y + 2$

9. A company manufactures two types of snow boards.

The cost for a racing board is $120 for materials and $180 for labour.

The cost for a freestyle board is $100 for materials and $120 for labour.

The company's budget is $49 000 for materials and $64 200 for labour.

a) Write an equation to describe the total cost for materials.

b) Write an equation to describe the total cost for labour.

c) Solve the linear system formed by the equations in parts a and b.

d) How many of each board can be produced?

10. A local fitness club has two payment plans.

Plan A: a monthly membership fee of $30, plus a fee of $1 per visit

Plan B: a fee of $5 per visit

Which is the cheaper plan for one month? Explain.

11. Part-time sales staff at a store are offered a choice of two payment plans.

Plan A: $1500 per month, plus a commission of 4% of monthly sales

Plan B: $1700 per month, plus a commission of 2% of monthly sales

Which is the better plan? Explain.

12. Suppose a hockey team offers players two salary packages. Package A has a base salary of x dollars and a $1000 bonus for each goal scored. Package B has a base salary $10 000 less than that in Package A, but pays a bonus of $1500 for each goal scored.

a) Let y represent the number of goals a player scores in a season. Write a linear system to represent the two salary packages.

b) Solve the linear system to determine the number of goals for which the salaries are equal.

13. Mathematical Modelling

Two skydivers jump from two planes at the same time. The jump of skydiver A is modelled by the equation $d = 2700 - 53t$.

The jump of skydiver B is modelled by the equation $d = 3000 - 70t$.

a) Explain what d and t represent in each equation.

b) At what altitude was each skydiver when she jumped? Explain.

c) Which skydiver is in the spread stable position? Explain.

d) Which skydiver is in the dive position? Explain.

e) At what time will one skydiver pass the other? Explain.

f) Each skydiver will open her parachute at an altitude of 1000 m. At this time, will one skydiver have passed the other? Explain.

Cumulative Review

1. A company produces software for the preparation of income tax returns. The monthly sales for one year are shown.

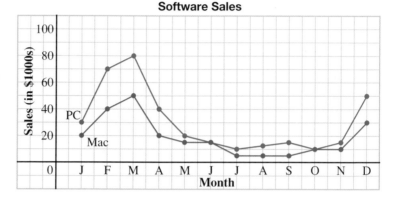

Software Sales

Sales (in $1000s) vs **Month** (J F M A M J J A S O N D)

a) Explain what each broken-line graph represents.

b) Which month had the greatest sales?

c) Determine the sales for the PC software for January and February.

d) What is the difference between the sales of PC software for January and February?

e) What is the difference between the sales of PC software and Mac software in March?

2. Graph each function. If you use a graphing calculator, sketch the graph. If you use a grid, make a table of values first.

a) $y = 2x + 2$ b) $y = 2x - 5$ c) $y = 0.2x + 2$

d) $y = 0.2x^2 + 1$ e) $y = 0.2x^2$ f) $y = 0.2x^2 - 3$

3. Explain why all the graphs in exercise 2 represent functions.

4. The time, t seconds, to put x litres of fuel in a car is given by $t = 35 + 0.5x$.

a) Graph the function. Use values of x from 0 to 60. Choose suitable scales.

b) Determine the time needed to put 40 L of fuel in the car.

c) Suppose it took 49 s to fill a car with fuel. How much fuel was used?

5. A candle is 8.5 cm long. When lit, its length decreases by 0.5 cm each minute.

a) Make a table of values to show the relationship between the time, t minutes, the candle is alight and its length, L centimetres.

b) When is the candle 5.5 cm long?

c) How long is the candle after 11 minutes?

d) For how many minutes will the candle stay alight? Explain.

e) Write the equation that relates L and t.

6. a) Predict which functions will have straight-line graphs.

   i) $y = 0.2x + 3$　　　　**ii)** $y = 2 - x$　　　　**iii)** $y = \frac{3}{x}$

   iv) $y = x^2 + 2x$　　　　**v)** $y = \frac{1}{3}x + \frac{2}{5}$　　　**vi)** $y = -\frac{2}{3}x$

   b) Use a graphing calculator to check your predictions in part a.

7. State the slope and y-intercept of the graph of each function.

   a) $y = \frac{1}{4}x - 5$　　　　**b)** $y = 3(x - 2) + 1$　　　　**c)** $y = -2(x + 5) - 6$

8. Write the equation of the line passing through each point with the given slope.

   a) P(-2, 3), slope 3　　　　**b)** Q(1, 4), slope $\frac{2}{5}$

9. Determine the equation of the line through each pair of points.

   a) A(2, 5) and B(5, 20)　　　　**b)** C(-6, 5) and D(-3, -7)

10. A class published a cookbook using recipes obtained from family members. It costs $2400 to print 500 copies. The class plans to sell each book for $8.50. The profit, P dollars, depends on the printing costs and on n, the number of books sold. An equation that relates these quantities is $P = 8.5n - 2400$.

   a) What is the profit when 300 books are sold?

   b) What is the greatest possible profit?

   c) Solve the equation for n. How many books must be sold to make a profit of $1000?

11. Solve each equation.

   a) $4y - 5 = 5(1 - y) + 3$　　　**b)** $8 - 3(a + 1) = 2a - 7$　　　**c)** $1 - \frac{x}{3} = 2$

   d) $\frac{x}{2} + 3 = 8 - 2x$　　　　**e)** $3 - 4x = \frac{1}{3}(x - 4)$　　　**f)** $\frac{2}{3}(2x - 1) = 1 + 4x$

12. Xpress Smart delivers parcels locally. The graph shows the amounts paid for delivering parcels to various distances.

   a) What is the cost to deliver a parcel 15 km?

   b) Explain what the horizontal segment of the graph represents.

   c) Suggest why the segment on the right is less steep than the middle segment.

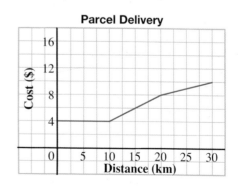

Parcel Delivery

13. Xpress Smart also delivers parcels to the airport for a fee that depends on the mass of the parcel. The table lists the functions used to calculate the costs for various sized parcels. Choose several values of m in each range. Calculate each corresponding cost, C. Draw a graph to represent the data.

Mass, m (kg)	0 to 10	between 10 and 20	20 and greater
Cost, C ($)	$C = 0.4m + 10$	$C = 1.6m - 2$	$C = 0.5m + 20$

14. A developer advertises these prices for building lots. Basic cost is $125/m^2$.
For lots between 200 m^2 and 400 m^2, there is a surcharge of $50/m^2$ on the area over 200 m^2.
For lots larger than 400 m^2, there is a discount of $20/m^2$ on the area over 400 m^2.

a) Calculate the cost of buying a 300-m^2 lot.

b) Calculate the cost of buying a 1000-m^2 lot.

c) Draw a graph to show the costs of buying lots up to 1000 m^2 in area.

15. Write each equation in the form $Ax + By + C = 0$.

   a) $y = 2x - 8$

   b) $y = \frac{2}{3}x + 5$

   c) $y = \frac{1}{2}x - \frac{2}{3}$

16. Write each equation in the form $y = mx + b$.

   a) $5x - y + 4 = 0$

   b) $3x + y - 10 = 0$

   c) $7x - 2y + 12 = 0$

17. Graph each equation.

   a) $x - 2y - 6 = 0$

   b) $3x + y - 5 = 0$

   c) $2x + 5y - 20 = 0$

18. The graph represents the cost to operate a car rental agency. Line A represents the cost to own the cars. Line B represents the income from renting the cars to customers.

   a) What does the point of intersection represent?

   b) How many days elapse before a profit is earned?

   c) What is the profit after 30 days?

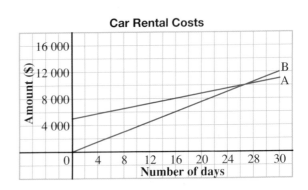

19. Solve each linear system. Write each coordinate to 2 decimal places.

a) $y = 0.4x + 1$
$y = 2.2x - 1$

b) $y = 1.2x + 2$
$y = -0.5x - 3$

20. Solve each linear system.

a) $y = x - 1$
$y = 2x - 3$

b) $y = -x + 3$
$y = 2x + 6$

c) $y = x - 6$
$y = -x + 10$

d) $y = -2x + 1$
$y = x - 5$

e) $y = 3x + 5$
$y = -x + 1$

f) $y = 2x + 4$
$y = x + 7$

21. Solve each linear system, then check.

a) $y = 5x + 5$
$y = -3x + 21$

b) $y = 10x$
$y = 5x + 50$

c) $y = -3x + 2$
$y = x + 26$

22. Solve each linear system.

a) $x + y = 6$
$x - y = 2$

b) $2x + y = 9$
$2x - y = -5$

c) $x + 2y = -4$
$x + y = -7$

d) $2x + y = -4$
$x - y = 1$

e) $x + y = 11$
$-x + y = 1$

f) $7x + y = 9$
$2x + y = -1$

23. A movie theatre charges $5 for a youth ticket and $7 for an adult ticket.
The theatre seats 300 people. On an evening when the theatre was full, the receipts showed $1724 from ticket sales.
The situation can be described by this linear system:
$x + y = 300$
$5x + 7y = 1724$

a) Write to explain each equation.

b) Solve the linear system. What does the solution represent?

c) How many youth tickets were sold?

d) How many adult tickets were sold?

24. A parking garage charges a flat rate plus an hourly fee. One day, a person parked for 7 h and paid $11.25. The next day, she parked for 3 h and paid $6.25.

a) What was the flat rate?

b) What was the hourly fee?

Project

Indoor Skiing

Suggested Group Size: 2

Materials: grid paper, graphing software (optional)

This project highlights a real situation. You will apply some of the mathematics you learned in the preceding chapters while working as part of a team. In the workplace, companies rely on their employees to work well together. Plan together how to tackle this project. Respect each other and allow your partner to have her or his say.

THE TASK

You will work with a partner to design an indoor ski hill. You will use piecewise linear functions, slope, and lines of best fit.

ASSESSMENT

There are many valid responses to this project. Your teacher may choose to use the rubric on page 159 to assess your work.

BACKGROUND

Many giant malls have parks. For example, the West Edmonton Mall in Alberta has a water park. Visitors can enjoy the wave pool, water slides, and bungee jumping, or they can simply lie by the pool.

- Why do you think indoor theme parks are popular?
- What kind of indoor theme park would be popular in southern Ontario?

The 1998 winter Olympics helped to make the sport of mogul skiing popular. The developers of a mall in southern Ontario want to include an indoor ski hill with a mogul run on one part of the hill. You have been asked to submit a sample design.

MEMO

To: Ski Hill Designer
From: Mall Developers Incorporated

Thank you for agreeing to submit a design for
our proposed ski hill. Please consider the
following criteria in your design.

- An indoor ski hill must be smaller than an
  actual ski hill, yet still be fun to use.
  Suggested measurements are 35 m high,
  100 m long, and 40 m wide.

- The moguls should be 1 m high and
  3 m apart. They should be shaped like
  rounded pyramids.

A ski hill usually has sections with steep slopes and
sections with less steep slopes, particularly near the
bottom of the hill, so the skier can slow down and stop
safely. You can draw a side view, or elevation profile,
of a ski hill and give the coordinates of the points
where the slope changes.

Horizontal distance (m)	Vertical height (m)
0	60
31	40
50	35
70	20
95	10
115	5
125	0
150	0

1. Graph the data in the table. Join adjacent points
   with a line segment. This creates a simplified
   elevation profile of a ski hill.

2. Calculate the slope of each segment of the ski hill.

3. Which part of the ski hill has the steepest slope? Which part has the least
   steep slope?

4. Suppose the hill had constant slope. Draw the line of this hill on your graph.
   What is the equation of this line?

5. Estimate the length of the ski run. Do you think your estimate is higher or
   lower than the actual length? Explain.

Project

Apply what you learned in Getting Started. Work with your partner to develop a proposal for an indoor ski hill for Mall Developers Incorporated. Your final project will be a poster display of your design. It should include:

- An accurate plan of the entire hill showing the positions of the moguls
- An elevation profile of the ski hill

Work together to decide the locations of the ski run and the moguls. An escalator or lift will be needed to transport skiers to the top.

Show by calculation that the ski run is close to the requested length. Calculate the slope of each section of the ski hill. Prepare an elevation profile of the ski run for the poster display. Make a table to show the slope of each section. Design the moguls area of the run with several moguls. Draw an elevation profile of the moguls area for display.

Career Opportunities

Designers' skills are used in many fields. If you like the outdoors, there is work in landscape design and leisure area design. Many designs are now produced using computer aided design (CAD) software. These programs require you to be able to work in the coordinate plane. Once the design is complete, many skilled workers are needed for the project. For example, a golf course requires bulldozer operators, grader operators, and underground pipe layers to prepare the course before sod is laid.

Rubric for the Indoor Skiing project

Level 1	Level 2	Level 3	Level 4
Knowledge/Understanding			
The student:			
• for Getting Started, constructs a graph, compares slopes, and draws a constant slope line with a few parts correct.	• for Getting Started, constructs a graph, compares slopes, draws a constant slope line, and writes an equation with some parts correct.	• for Getting Started, constructs a graph, compares slopes, draws a constant slope line, and writes an equation with most parts correct.	• for Getting Started, constructs a graph, compares slopes, draws a constant slope line, and writes an equation correctly.
• calculates a few slopes and distances accurately.	• calculates some slopes and distances accurately.	• calculates most slopes and distances accurately.	• calculates the slopes and distances accurately.
Thinking/Inquiry/Problem Solving			
The student:			
• designs a hill that has little similarity with the suggested measurements.	• designs a hill with measurements close to some of those suggested.	• designs a hill with measurements close to those suggested.	• designs a hill to fit the suggested measurements.
• designs a hill with moguls.	• designs a hill with moguls matching some requirements.	• designs a hill with moguls matching most requirements.	• designs a hill to fit mogul requirements of size and shape.
• states a connection between the design and the requirements.	• explains connections between the design and the requirements.	• explains that the design follows the requirements.	• explains how the design follows the requirements exactly.
Communication			
The student:			
• draws a view of the hill.	• draws a view of the hill that could be an elevation profile.	• draws a clear elevation profile.	• draws a clear, detailed elevation profile.
• presents a poster that can be interpreted by someone familiar with the project.	• presents a fairly organized poster that can be interpreted.	• presents an effective, organized poster that is easy to interpret and promotes the proposal.	• presents an effective, organized poster that shows why the proposal is perfect.
• communicates with proper form for some slopes and/or units.	• communicates fairly clearly with proper form for most slopes and units.	• communicates clearly with proper form for slopes and units.	• communicates clearly in a convincing manner with proper form for slopes and units.
Application			
The student:			
• creates a design and a display that correctly follow some of the instructions.	• creates a design and a display that correctly follow many of the instructions.	• creates a design and a display that correctly follow almost all instructions.	• creates a design and a display that correctly follow the instructions.
• demonstrates little understanding of an indoor ski hill with moguls.	• demonstrates some understanding of an indoor ski hill with moguls.	• demonstrates an understanding of an indoor ski hill with moguls.	• demonstrates excellent understanding of an indoor ski hill with moguls.

MATHEMATICAL MODELLING

Food for a Healthy Heart

You probably know about the importance of reducing fat in our diet. The Heart and Stroke Foundation recommends we eat food that contains less than 30% of its total energy as fat.

Many packaged foods claim to contain reduced fat. But it is not always clear whether the food meets the Heart and Stroke Foundation's guideline.

How can we tell if food contains less than 30% of its total energy as fat?

Here are some things to consider.

1. Why do we need food energy?

2. List some foods you think would satisfy the Heart and Stroke Foundation's guideline.

3. List some foods you think might not satisfy the Heart and Stroke Foundation's guideline.

You will return to this problem in Section 4.3.

 FYI Visit www.awl.com/canada/school/connections

For information related to the above problem, click on <u>MATHLINKS</u>, followed by Foundations of Mathematics 10. Then select a topic under Food for a Healthy Heart.

Nutrition Information

per serving 30 g = 125 mL

Energy	79 Cal
	330 kJ
Protein	4.1 g
Fat	1.2 g
Polyunsaturates	0.7 g
Monounsaturates	0.2 g
Saturates	0.2 g
Cholesterol	0 mg
Carbohydrate	22 g
Sugars	2.5 g
Starch	1 g
Dietary	g
Sodium	

Recommendation

Eat food containing less than 30% of its total energy as fat.

Heart and Stroke Foundation

Preparing for the Chapter

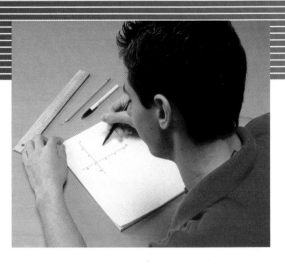

This section reviews these concepts:

- Graphing linear functions
- Relating multiplication and division
- Percent
- Ratio
- Solving equations

1. A store gives its customers a 20% discount coupon. The table shows the regular price and the discount price.

Regular price ($)	Discount price ($)
0	0
20	16
40	32
60	48
80	64
100	80

   a) Graph the data in the table. Plot *Regular price* horizontally and *Discount price* vertically.

   b) Use the graph to calculate the discount price for each regular price.

   i) $10   **ii)** $30

   iii) $50   **iv)** $85

   c) Use the graph to calculate the regular price for each discount price.

   i) $12   **ii)** $20

   iii) $40   **iv)** $60

2. Students in a grade 10 class participated in the Terry Fox Run. The pledges collected by five students are shown.

   a) Each student jogged 10 km. How much money did each student raise?

   b) Copy this table. Insert your answers to part a.

Amount pledged per kilometre (¢)	Amount of money raised ($)
40	
75	
50	
60	
25	

   c) Use the data in the table to draw a graph. Describe the graph.

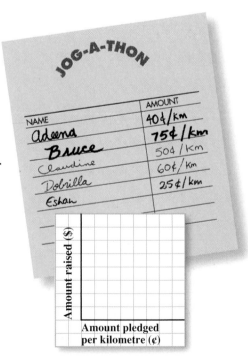

3. **Estimation** Terry Fox ran 5373 km. Since 1980, the money raised in Terry Fox's name has exceeded $250 million. Estimate the mean amount of money raised per kilometre.

4. a) You know many multiplication facts such as $7 \times 6 = 42$. Use these three numbers to:

   i) write another multiplication fact   **ii)** write two division facts

   b) The four facts in part a form a "family of facts." Choose a different multiplication fact. Write the four facts that form its "family."

5. a) Since $6 \times 16 = 96$, what does $96 \div 6$ equal?

   b) Use a calculator to determine 32×1.45. Without using the calculator, determine $46.4 \div 1.45$.

6. Reduce to lowest terms.

   a) $\frac{3}{9}$ b) $\frac{100}{60}$ c) $\frac{21}{49}$ d) $\frac{36}{24}$ e) $\frac{36}{64}$

7. Write each percent as a fraction.

   a) 10% b) 20% c) 40% d) 70% e) 90%

   f) 110% g) 120% h) 140% i) 170% j) 190%

8. Write each percent in exercise 7 as a decimal.

9. Write each fraction as a percent.

   a) $\frac{1}{2}$ b) $\frac{1}{4}$ c) $\frac{3}{4}$ d) $\frac{1}{3}$ e) $\frac{2}{3}$

   f) $\frac{7}{10}$ g) $\frac{4}{5}$ h) $\frac{4}{6}$ i) $\frac{3}{8}$ j) $\frac{5}{12}$

10. Write each fraction in exercise 9 as a decimal.

11. Write each ratio as a fraction in lowest terms.

    a) $2 : 10$ b) $10 : 2$ c) $12 : 4$ d) $4 : 12$

    e) $50 : 70$ f) $70 : 50$ g) $25 : 60$ h) $60 : 25$

12. Solve each equation.

    a) $2x = 10$ b) $3x = 12$ c) $2x = 7$ d) $5x = 12$

    e) $4x = 10$ f) $3x = 10$ g) $5x = 70$ h) $10x = 14$

13. Solve each equation.

    a) $\frac{x}{3} = 4$ b) $\frac{x}{2} = 10$ c) $\frac{x}{10} = 2$ d) $\frac{x}{4} = 3$

    e) $\frac{x}{3} = \frac{4}{6}$ f) $\frac{x}{2} = \frac{10}{5}$ g) $\frac{x}{10} = \frac{2}{3}$ h) $\frac{x}{4} = \frac{3}{5}$

Comparing Linear Relationships

Part A Production Costs

To produce a booklet, there is a fixed cost of $3000 to set up the press. There is also a variable cost of $2 to print and bind each booklet.

1. Copy and complete the table.

Number of booklets	Cost ($)
0	
200	
400	
600	
1200	
1600	

2. Find patterns in the table. Describe each pattern.

3. a) Graph the data in the table. Plot *Number of booklets* horizontally and *Cost* vertically.

    b) Describe the graph.

    c) Use the graph to determine the cost to produce 800 booklets.

4. Use *C* dollars to represent the cost and *n* to represent the number of booklets. Write an equation to describe the relationship between the cost and the number of booklets.

Part B Currency Conversion

Canadians who travel to the United States convert Canadian dollars to US dollars. In the fall of 1999, it cost $1.50 Cdn to buy $1.00 US.

5. Copy and complete the table.

Amount ($US)	Amount ($Cdn)
0	
200	
400	
600	
1200	
1600	

6. Find patterns in the table. Describe each pattern.

7. a) Graph the data in the table. Plot *Amount ($US)* horizontally and *Amount ($Cdn)* vertically.

   b) Describe the graph.

   c) Use the graph to determine the cost in Canadian dollars to buy $800 US.

8. Use *C* to represent the amount in Canadian dollars and *U* to represent the amount in US dollars. Write an equation to describe the relationship between the Canadian dollar and the US dollar.

Part C

9. Compare your work in Part A and Part B. You should have found more patterns in the table in Part B than in the table in Part A. Explain why.

In Part A of the *Investigation*, you should have obtained this table and graph.

Number of booklets	Cost ($)
0	3000
200	3400
400	3800
600	4200
1200	5400
1600	6200

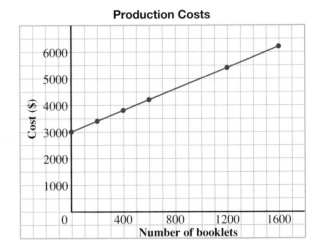

The patterns in the table are:

- In the first column, constant differences of 200 in the first 4 rows
- In the second column, constant differences of $400 in the first 4 rows

For the equation, use C dollars to represent the cost for n booklets.
The equation is $C = 2 \times$ number of booklets $+ 3000$, or $C = 2n + 3000$.

In Part B of the *Investigation*, you should have obtained this table and graph.

Amount ($US)	Amount ($Cdn)
0	0
200	300
400	600
600	900
1200	1800
1600	2400

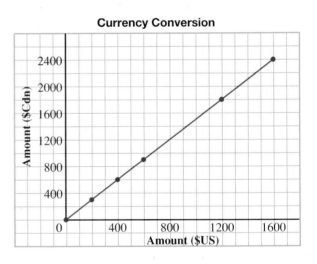

Since the line on the graph passes through (0, 0), there are more patterns in this table than in the previous table.

Pattern 1

Amount ($US)	Amount ($Cdn)
0	0
200	300
400	600
600	900
1200	1800
1600	2400

$0 \times 1.5 = 0$

$200 \times 1.5 = 300$

$400 \times 1.5 = 600$

$600 \times 1.5 = 900$

$1200 \times 1.5 = 1800$

$1600 \times 1.5 = 2400$

We see: In any row of the table, when the amount in US dollars is multiplied by 1.5, the result is the amount in Canadian dollars.

We say: The conversion rate is $1.50 Canadian for every $1 US.

Using C to represent an amount in Canadian dollars and U to represent the amount in US dollars, the equation is $C = 1.5U$.

Pattern 2

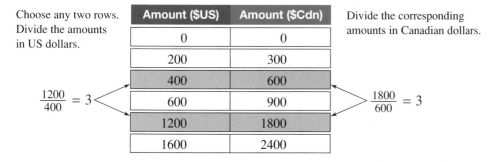

Choose any two rows. Divide the amounts in US dollars.

$\frac{1200}{400} = 3$

Amount ($US)	Amount ($Cdn)
0	0
200	300
400	600
600	900
1200	1800
1600	2400

Divide the corresponding amounts in Canadian dollars.

$\frac{1800}{600} = 3$

We see: When the amount in US dollars is tripled, the amount in Canadian dollars is tripled. In general, when the amount in US dollars is multiplied by any number, the amount in Canadian dollars is multiplied by the same number.

We say: The growth factor is the same for both US dollars and Canadian dollars.

The relationship between the amount in US dollars and the amount in Canadian dollars is an example of a proportional relationship. Its equation is $C = 1.5U$.

On page 166, the relationship between the number of booklets and the cost is an example of a *non-proportional* relationship. Its equation is $C = 2n + 3000$.

Proportional Relationships

- A relationship that has an equation of the form $y = mx$ is called a *proportional relationship*.
- The graph of a proportional relationship is a straight line through the origin.
- Corresponding quantities are related by multiplication or division.

The squares represent the numbers in two rows of a table of values.

These patterns can only be used with proportional relationships. You can use either the horizontal pattern or the vertical pattern, whichever seems easier.

Example 1

Four apples cost $1.20.

a) At this price, how much would 24 apples cost?

b) How many apples could you buy for $6?

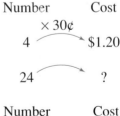

Solution

a) 4 apples cost $1.20.

1 apple costs $\frac{\$1.20}{4} = 30¢$.

24 apples cost $24 \times 30¢ = 720¢$, or $7.20.

b) One apple costs 30¢.

For $6, the number of apples you can buy is $\frac{\$6.00}{\$0.30} = 20$.

Number   Cost

$\times 30¢$

4     $1.20

24   ?

Number   Cost

$\times 30¢$

? $6.00

The reasoning in the solution of *Example 1* is an example of *proportional reasoning*.

Example 2

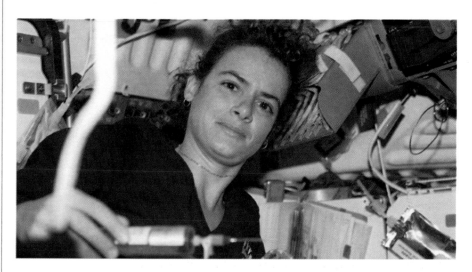

On May 27, 1999, the space shuttle *Discovery* was launched. Julie Payette, a Canadian scientist, was one astronaut on the 10-day mission. After 2 days, she had travelled 1.4 million km in space.

a) How far had Julie travelled after 10 days?

b) How much time would it take Julie to travel 5 million km?

Solution

a) In 2 days, Julie travelled 1.4 million km.
Divide to find how 10 days are related to
2 days: $\frac{10}{2} = 5$
In 10 days, Julie travels 5 times as far as in 2 days.
Multiply 1.4 million km by 5:
1.4 million km × 5 = 7 million km
In 10 days, Julie travelled 7 million km.

Days   Distance

$\times 5 \begin{pmatrix} 2 \\ \\ 10 \end{pmatrix}$ $\times 5 \begin{pmatrix} 1.4 \\ \\ ? \end{pmatrix}$

b) In 2 days, Julie travelled 1.4 million km.
Divide to find how 5 million km are related
to 1.4 million km: $\frac{5}{1.4} \doteq 3.57$
To travel 5 million km takes about 3.57
times as long as to travel 1.4 million km.
Multiply 2 days by 3.57: $2 \times 3.57 = 7.14$
It would take a little more than 7 days to travel 5 million km.

Days   Distance

$\begin{pmatrix} 2 \\ \\ ? \end{pmatrix} \times 3.57$ $\begin{pmatrix} 1.4 \\ \\ 5 \end{pmatrix} \times 3.57$

1. What is proportional reasoning?

2. Why do you use multiplication or division, and not addition or subtraction, to solve problems in proportional situations?

3. Explain why the relationship in *Example 1* is a proportional relationship.

4. What other way is there to solve *Example 1*? Explain.

5. Here is a variation of the problem in *Example 1a*. How would the solution to *Example 1* change for this problem?
   Three apples cost 79¢. How much would 11 apples cost?

6. Explain why the relationship in *Example 2* is a proportional relationship.

7. What other way is there to solve *Example 2*? Explain.

8. Here is a variation of the problem in *Example 2a*. How would the solution to *Example 2* change for this problem?
   After 3 days, Julie had travelled 2.2 million km. How far had she travelled after 8 days?

Practise Your Skills

1. Divide.

   a) $\dfrac{1.60}{2}$ b) $\dfrac{4.25}{5}$ c) $\dfrac{120}{4}$

   d) $\dfrac{30}{5}$ e) $\dfrac{40}{3}$ f) $\dfrac{1.89}{4}$

2. Multiply.

   a) 25×2 b) 0.80×3 c) 0.85×4

   d) 30×12 e) 6×20 f) 0.47×5.25

3. Calculate each percent.

   a) 2% of $500   b) 4% of $500

   c) 6% of $500   d) 8% of $2000

   e) 10% of $2000   f) 11% of $2000

4. Calculate each percent.

   a) 4% of $200   b) 6% of $400

   c) 4% of $600   d) 6% of $800

   e) 4% of $1000   f) 6% of $1200

A 1. One apple costs 25¢. Determine each cost.

    a) 2 apples   **b)** 3 apples   **c)** 4 apples

    d) 6 apples   **e)** 9 apples   **f)** 12 apples

2. Two candy bars cost $1.60. Determine each cost.

    a) 1 candy bar   **b)** 3 candy bars   **c)** 4 candy bars

    d) 8 candy bars   **e)** 16 candy bars   **f)** 20 candy bars

3. **a)** Harvest Mix costs $4.25 for 500 g. Determine the cost of each mass of
    Harvest Mix.

      i) 100 g   **ii)** 200 g   **iii)** 300 g   **iv)** 400 g

      v) 600 g   **vi)** 700 g   **vii)** 800 g   **viii)** 900 g

    b) Describe the pattern in the answers for part a.

4. One Norwegian kroner, 1 k, is worth
$0.19 Cdn. What does it cost to buy
each number of kroner?

    a) 100 k   **b)** 2000 k

    c) 3500 k   **d)** 9000 k

5. One English pound, £1, is worth
$2.42 Cdn. What does it cost to buy
each number of pounds?

    a) £5   **b)** £40

    c) £300   **d)** £5000

B 6. In 4 min, a student typed 120 words. Suppose he
continues to type at this rate.

    a) How many words could he type in 12 min?

    b) How long would it take him to type 600 words?

7. In 5 h of part-time work, a student earned $30.

    a) At this rate, how much would she earn in 20 h?

    b) How long would it take her to earn $150?

8. Scientists estimate that about one person in 9 is
left-handed. The population of Canada is about
31 million. About how many left-handed people
are there in Canada?

9. In 2 min, a laser printer printed 8 pages of text.

   a) How many pages would it print in 6 min?

   b) How long would it take to print 20 pages?

   c) Which arithmetic operations did you use to complete parts a and b?

   d) Is this a proportional situation? Explain.

10. A taxi charges a basic fare of $2 plus 80¢/km.

   a) How much would it cost to travel 5 km?

   b) How far could you travel for $10?

   c) Which arithmetic operations did you use to complete parts a and b?

   d) Is this a proportional situation? Explain.

11. Rosanna cycled 40 km in 3 h.

   a) At this speed, how far could she cycle in 5 h?

   b) How long would it take her to cycle 100 km?

12. On her mission aboard the space shuttle *Discovery*, Julie Payette completed 153 Earth orbits in 235.2 h.

   a) How long did it take to complete each orbit?

   b) Julie travelled 4 million km. How far did she travel on each orbit?

13. Two problems and their solutions are shown.

   a) Are the solutions correct? Explain.

   b) For each incorrect solution, provide a correct solution.

Problem 1

Lesley and Sarah can run equally fast around a track. Lesley started first. After Sarah had run 2 laps, Lesley had run 6 laps. When Sarah had run 10 laps, how many laps had Lesley run?

Solution

Sarah	Lesley
2	6
×5	×5
10	?

Since $6 \times 5 = 30$, Lesley ran 30 laps.

Problem 2

Kurt has a part-time job at a supermarket. He earned $56 by working for 8 h. At this rate, how much would he earn by working for 20 h?

Solution

Hours	Earnings
8	20
×7	×7
56	?

Since $20 \times 7 = 140$, he would earn $140.

---

14. a) Solve each problem below.

b) How are the problems similar? How are they different?

c) Did you use the same method to solve each problem? If your answer was no, complete part d.

d) Why did you not use the same method to solve each problem? Could you have used the same method to solve each problem? Explain.

Problem 1

In 4 games, Janet had 12 hits. At this rate, how many hits would she have in 20 games?

Problem 2

In 5 games, Mark had 11 hits. At this rate, how many hits would he have in 30 games?

Problem 3

In 6 games, Rhonda had 14 hits. At this rate, how many hits would she have in 44 games?

15. Sue invested $800 in Canada Savings Bonds. After one year, she received $40 interest. Suppose she had invested other amounts. How much interest would she have received for each amount?

a) $400   **b)** $1200   **c)** $100   **d)** $1500

16. Choose one principal to invest: $500 or $2000

a) Copy and complete each table for your investment.

i)

Interest rate (%)	Interest earned after one year ($)
0	
2	
4	
6	
8	

ii)

Interest rate (%)	Total amount after one year ($)
0	
2	
4	
6	
8	

b) Use the data in the tables to draw two graphs.

c) Which graph represents a proportional situation? Why does it represent a proportional situation? Why does the other graph not represent a proportional situation?

d) Compare your graphs with those of someone who chose the other amount to invest. How are the graphs similar? How are they different?

17. Choose one interest rate: 4% or 6%

a) Copy and complete each table for your interest rate.

i)

Principal invested ($)	Interest earned after one year ($)
200	
400	
600	
800	
1000	

ii)

Principal invested ($)	Total amount after one year ($)
200	
400	
600	
800	
1000	

b) Use the data in the tables to draw two graphs.

c) Why are both these situations proportional, while only one situation in exercise 16 is proportional?

d) Compare your graphs with those of someone who chose the other amount to invest. How are the graphs similar? How are they different?

18. A 400-g box of corn flakes costs $1.89. A 525-g box of the same cereal costs $2.29. Which box is the better buy? Explain.

19. A brand of liquid detergent is sold in 2 sizes — $1.99 for 500 mL and $2.39 for 950 mL. Which is the better buy? Explain.

20. Choose either exercise 18 or exercise 19. Solve the problem in a different way.

21. The following item was obtained from the CN Tower's website in January, 2000 (www.cntower.ca/ll_calc.html).

- If the Tower were lying on its side, it would take Donovan Bailey, the world's fastest man, 54.45 s to run the distance. (Remember? He broke the 100-m world's record in Atlanta in 1996.)

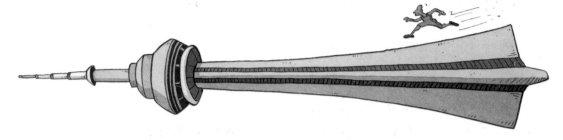

a) The person who wrote this item knew that Donovan Bailey ran 100 m in 9.84 s, and the height of the CN Tower is 553 m. Explain how the person determined that it would take Donovan 54.45 s to run the distance.

b) Is the reasoning in part a a valid application of proportional reasoning? Explain.

C **22.** In some provinces, police stop speeders by measuring the speed of a car from a helicopter. Marks are painted 0.5 km apart at the edge of a highway. The police in the helicopter measure the time it takes for a car to travel from one mark to the next.

a) Suppose a car is travelling at 80 km/h. How long would it take to go from one mark to the next?

b) Suppose a car takes 20 s to go from one mark to the next. At this rate, how far would it go in one hour? What is its speed in kilometres per hour?

Communicating the IDEAS

Write to explain what is meant by a proportional situation. Your explanation should be understood by someone who has not heard this term. Use an example of a proportional situation and an example of a non-proportional situation to illustrate your explanation.

Comparing Heights on a Photograph with Actual Heights

Use the photograph of the CN Tower on page 177.

1. Copy and complete the table. Measure and record each height on the photograph, to the nearest tenth of a centimetre.

	Height on photograph (cm)	Actual height (m)
	0.0	0
Glass floor		
Sky Pod		
Top of CN Tower		

2. a) Graph the data. Plot *Height on photograph* horizontally and *Actual height* vertically.

b) Does it make sense to join the points? Explain.

c) Describe the graph.

Comparing actual heights with heights on the photograph

3. Use a calculator to determine each ratio, then compare the results. Explain.

a) $\dfrac{\text{Actual height of Sky Pod}}{\text{Height of Sky Pod on photo}}$ **b)** $\dfrac{\text{Actual height of Glass floor}}{\text{Height of Glass floor on photo}}$

c) $\dfrac{\text{Actual height of CN Tower}}{\text{Height of CN Tower on photo}}$

4. What can you conclude from the results of exercises 2 and 3?

Comparing height ratios on the photograph and on the tower

5. Consider the Glass floor and the Sky Pod. Use a calculator to determine each ratio, then compare the results. Explain.

a) $\dfrac{\text{Height of Sky Pod on photo}}{\text{Height of Glass floor on photo}}$ **b)** $\dfrac{\text{Actual height of Sky Pod}}{\text{Actual height of Glass floor}}$

6. a) Use the top of the tower and the Glass floor. Repeat exercise 5.

b) Use the top of the tower and the Sky Pod. Repeat exercise 5.

7. What can you conclude from the results of exercises 5 and 6?

Note: Save your data and your graph for use in 4.2 Exercises.

Top of Tower, 553 m

Sky Pod, 447 m

Glass floor, 342 m

Ground Level

In the *Investigation*, you probably obtained these results.

Comparing actual heights with heights on the photograph

	Height on photograph (cm)	Actual height (m)	$\dfrac{\text{Actual height}}{\text{Height on photo}}$
Glass floor	11.6	342	$\frac{342}{11.6} \doteq 29.5$
Sky Pod	15.2	447	$\frac{447}{15.2} \doteq 29.4$
Top of Tower	18.8	553	$\frac{553}{18.8} \doteq 29.4$

We see: In any row of the table, when the actual height in metres is divided by the height on the photograph in centimetres, the result is the same, approximately 29.4.

We say: The actual heights in metres are 29.4 times the heights on the photograph in centimetres.

Comparing height ratios on the photograph and on the tower

	Height on photograph (cm)	Actual height (m)
Glass floor	11.6	342
Sky Pod	15.2	447
Top of Tower	18.8	553

$$\frac{15.2}{11.6} \doteq 1.31 \qquad \frac{447}{342} \doteq 1.31$$

We see: Choose any two rows. Divide the heights on the photograph, and divide the actual heights. The results will be the same. The results shown above are for the Glass floor and the Sky Pod.

We say: The Sky Pod is 1.31 times as high as the Glass floor.

There is a proportional relationship between the heights on the photograph and the corresponding heights on the tower. The heights on the photograph are proportional to the actual heights.

The above example shows that a proportional relationship can be expressed in terms of ratios.

Proportional Relationship

• Ratios relating corresponding quantities are equal.

In a problem involving a proportional relationship, three facts are usually given or easily determined. We use equal ratios to determine a fourth fact.

Example 1

The basketball player, Wilt Chamberlain, dwarfs the famous jockey, Eddie Arcaro, in this photograph. Chamberlain was 2.16 m tall. Calculate Arcaro's height.

Solution

Measure the heights on the photograph.
Arcaro's height is 7.0 cm.
Chamberlain's height is 9.4 cm.

Write the heights in a table.
Let Eddie's height be represented by x metres.

	Height on photograph (cm)	Actual height (m)
Arcaro	7.0	x
Chamberlain	9.4	2.16

Method 1

Write equal vertical ratios, beginning with x.

$$\frac{x}{2.16} = \frac{7.0}{9.4}$$

$$\frac{x}{2.16} \doteq 0.7447$$

Multiply each side by 2.16.

$$x \doteq 2.16 \times 0.7447$$
$$x \doteq 1.61$$

Arcaro's height is about 1.61 m.

Photo   Actual

7.0 x

9.4   2.16

Method 2

Write equal horizontal ratios, beginning with x.

$$\frac{x}{7.0} = \frac{2.16}{9.4}$$

$$\frac{x}{7.0} \doteq 0.2298$$

Multiply each side by 7.0.

$$x \doteq 7.0 \times 0.2298$$

$$x \doteq 1.61$$

Arcaro's height is about 1.61 m.

Photo	Actual
7.0	x
9.4	2.16

In *Example 1*:

Method 1 shows that Arcaro is about 0.75 times as tall as Chamberlain. Hence, Arcaro's height is about 0.75 times Chamberlain's height.

Method 2 shows that Chamberlain's height in metres is about 0.23 times his height on the photograph in centimetres. The same relationship applies to Arcaro's height.

Example 2

The fuel consumption for a 1999 Toyota Paseo is quoted as 6.7 L/100 km for highway driving.

a) At this rate, how many litres of fuel would be used to drive 1715 km from Windsor to Thunder Bay?

b) How far could you drive on the highway on 55 L of fuel?

Solution

a) Let x litres represent the volume of fuel needed.

	Volume of fuel (L)	Distance (km)
Consumption	6.7	100
Trip	x	1715

Method 1

Write equal vertical ratios, beginning with x.

$$\frac{x}{6.7} = \frac{1715}{100}$$

$$\frac{x}{6.7} = 17.15$$

Fuel	Distance
6.7	100
x	1715

Multiply each side by 6.7.

$x = 6.7 \times 17.15$

$x = 114.905$

About 115 L of fuel would be used.

Method 2

Write equal horizontal ratios, beginning with x.

Fuel	Distance

6.7 → 100

$\frac{x}{1715} = \frac{6.7}{100}$

$\frac{x}{1715} = 0.067$

Multiply each side by 1715.

x → 1715

$x = 1715 \times 0.067$

$x = 114.905$

About 115 L of fuel would be used.

b) Let x kilometres represent the distance you could drive on 55 L of fuel.

	Volume of fuel (L)	Distance (km)
Consumption	6.7	100
Tank	55	x

Method 1

Write equal vertical ratios, beginning with x.

Fuel	Distance

6.7 → 100
55 → x

$\frac{x}{100} = \frac{55}{6.7}$

$\frac{x}{100} \doteq 8.2090$

Multiply each side by 100.

$x \doteq 100 \times 8.2090$

$x \doteq 820.90$

You could drive about 821 km on 55 L of fuel.

Method 2

Write equal horizontal ratios, beginning with x.

Fuel	Distance

6.7 → 100
55 → x

$\frac{x}{55} = \frac{100}{6.7}$

$\frac{x}{55} \doteq 14.9254$

Multiply each side by 55.

$x \doteq 55 \times 14.9254$

$x \doteq 820.90$

You could drive about 821 km on 55 L of fuel.

1. On page 178, the ratios of actual height (in metres) to height on the photograph (in centimetres) were approximately 29.5, 29.4, and 29.4.

   a) Explain why these are not equal.

   b) Are heights on the tower 29.4 times the heights on the photograph? Explain.

2. On page 178, the height ratios for the Glass floor and the Sky Pod on the photograph and on the tower were approximately 1.31. Are these ratios equal? Explain.

3. Refer to the solutions of *Example 1*. Why did we begin with *x* when we wrote the equal ratios?

4. Suppose the photograph on page 179 is enlarged. It becomes twice as wide and twice as high. How would the solution of *Example 1* change?

5. Choose part a or part b of *Example 2*. Choose one of the two methods of solving this part. Explain the solution without using an equation.

Practise Your Skills

1. Solve each equation.

   a) $\frac{x}{10} = 3$

   b) $\frac{x}{5} = 2.5$

   c) $\frac{x}{7} = 0.3$

   d) $\frac{x}{100} = \frac{4}{7}$

   e) $\frac{x}{60} = \frac{10}{3}$

   f) $\frac{x}{3.4} = \frac{2.7}{3.2}$

2. Write each fraction as a percent.

   a) $\frac{1}{4}$

   b) $\frac{1}{2}$

   c) $\frac{3}{10}$

   d) $\frac{4}{5}$

   e) $\frac{3}{4}$

   f) $\frac{1}{5}$

   g) $\frac{4}{9}$

   h) $\frac{3}{8}$

   i) $\frac{4}{3}$

   j) $\frac{5}{8}$

   k) $\frac{5}{4}$

   l) $\frac{9}{10}$

4.2 Exercises

B 1. **Estimation** Look at the measurements from the photograph on page 179.

   a) Determine $\frac{\text{Arcaro's height}}{\text{Chamberlain's height}}$.

   b) Estimate Arcaro's height as a percent of Chamberlain's height.

   c) Check your answer to part b using the statement on page 180, following *Example 1*.

2. **Estimation** Look at the measurements on page 178 from the photograph on page 177.

   a) Determine $\dfrac{\text{height of Sky Pod}}{\text{height of the tower}}$.

   b) Use part a to estimate the height of the Sky Pod as a percent of the height of the tower.

   c) Determine $\dfrac{\text{height of the Glass floor}}{\text{height of the tower}}$.

   d) Use part c to estimate the height of the Glass floor as a percent of the height of the tower.

3. Use the photograph on page 177. The large white object on the left is the SkyDome.

   a) Measure the height of the top of SkyDome above ground level.

   b) Calculate the actual height of the top of SkyDome above ground level.

4. According to a newspaper report, 4 out of every 5 Canadians buy lottery tickets. The population of Canada is about 31 million. How many Canadians buy lottery tickets?

5. At the bulk food store, 120 g of candies cost $1.65.

   a) At this price, how much would 70 g cost?

   b) What mass could be bought for $2.50?

6. At the bulk food store, 188 g of mixed nuts cost $2.61.

   a) At this price, how much would 450 g cost?

   b) What mass could be bought for $5.00?

7. In the summer, a student worked part-time for a landscaping company. In 3.5 h, she earned $28.70.

   a) At this rate, how much would she earn in 9 h?

   b) How many hours would it take her to earn $350?

8. A person had 12 hits in 30 at-bats. At this rate, how many at-bats would she need to have 100 hits?

9. A person jogged 3 km in 25 min.

   a) At this rate, how far could she jog in 40 min?

   b) How long would it take her to jog 20 km?

10. This photograph shows a parachutist about to leap from the arm of a statue above Rio de Janeiro, Brazil. The entire statue is 30 m high. The photograph shows the top half of the statue. Estimate the height of the parachutist.

11. a) In the 1996 baseball season, Sammy Sosa had 40 home runs in 124 games. Suppose Sosa continued hitting home runs at this rate. There were 162 games in the 1999 season. How many home runs would Sosa have had at the end of the season?

b) Use the Internet or a sports almanac. Find Sosa's home runs for the 1999 season. Did Sosa's hitting improve or decline from 1996 to 1999? Explain.

12. Two problems are given. A possible solution is shown for each problem.

    a) Are the solutions correct? Explain.

    b) For each incorrect solution, provide a correct solution.

    Problem 1

    A car travels 178 km in 2.25 h. At this rate, what time would it take to travel 420 km?

    Solution

    $$\frac{x}{420} = \frac{178}{2.25}$$
    $$\doteq 79.11$$
    $$x \doteq 420 \times 79.11$$
    $$\doteq 33\ 227$$

    It would take 33 227 h.

    Problem 2

    In January 2000, one Canadian dollar was worth $0.70 US. How much would it cost in Canadian dollars to obtain $200 US?

    Solution

    $0.70 US is equal to $1 Cdn, which is 30¢ more. Therefore, $1 US is equal to $1.30 Cdn, which is also 30¢ more. To obtain $200 US, it would cost 200 × $1.30 Cdn, or $260 Cdn.

13. The following facts were obtained from SkyDome's web site in January, 2000. (www.skydome.com/welcome/facts.html)

    Fun Facts

    - If one were to line up all the hot dogs served by McDonald's at SkyDome in one year, the hot dogs would cover the distance of 3241 stolen bases.

    - Since the opening of SkyDome, the number of hot dogs served at SkyDome would stretch close to 800 km — or the distance from Toronto to Ottawa and back again.

    a) Measure the length of one hot dog. One stolen base is a distance of 27.4 m. According to the first fact, how many hot dogs are served in one year?

    b) According to the second fact, how many hot dogs had been served at SkyDome since its opening?

14. The fuel consumption for a 1999 Pontiac Firefly is quoted as 4.5 L/100 km for highway driving.

    a) At this rate, how many litres of fuel would be used for 225 km of highway driving?

    b) How far could you drive on the highway on 30 L of fuel?

15. The photograph below appeared in a newspaper in 1998 with the caption "Hard to park." Suppose the person standing in the middle of the photograph is 1.8 m tall.

a) Copy and complete this table.

		Length on photograph (cm)	Actual length (m)
i)	Height of one person		1.8
ii)	Height of one tire		
iii)	Height of the truck		
iv)	Width of the truck		
v)	Length of a staircase		
vi)	Width of a staircase		

b) Graph the data. Plot *Length on photograph* horizontally and *Actual length* vertically.

c) Measure the width of a tire on the photograph. Use the graph to determine the width of the tire on the truck.

d) There are two lines forming an X on the grill at the front of the truck. Measure the length of one of these lines on the photograph. Use the graph to determine the length of the line on the truck.

C

The Hubble Space Telescope

The Hubble Space Telescope was launched in 1990. Since it is high above Earth's atmosphere, it can gather more data than ground-based observatories.

In 1999, this diagram appeared in a newspaper article about the Hubble Space Telescope. The diagram shows the floppy disks needed to hold all the data gathered by the telescope since it was launched. The building on the left is the Empire State Building in New York, which was once the tallest building in the world. The height of this building is 381 m.

16. a) Estimate the height of the pile of floppy disks.

b) Three floppy disks form a pile 1 cm high. About how many floppy disks are in the pile?

17. Suppose the data were put on CDs instead of floppy disks. One CD can hold the same amount of data as 500 floppy disks. In its case, one CD is 1 cm high. Calculate the height of the pile of CDs.

18. The CN Tower is 553 m high. Draw a scale diagram to compare the height of the pile of CDs with the height of the CN Tower.

19. Do you think the diagram at the right is misleading? Explain.

20. Suppose the diagram at the right were a scale diagram of a pile of floppy disks. A floppy disk is about 9 cm wide.

a) Measure the width and the height of the pile on the diagram.

b) Calculate the height of the pile of floppy disks.

c) Compare the answer in part b with the answer to exercise 16a.

d) Suppose a scale diagram was drawn to show the floppy disks needed to hold all the data gathered by the Hubble telescope. How many times as high as the diagram at right should it be?

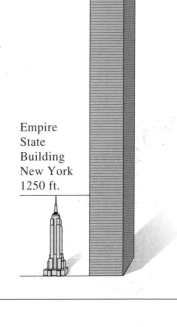

Empire
State
Building
New York
1250 ft.

21. The amount of gold in jewellery is measured in karats (K), with 24 K representing pure gold. The mark 10 K on a ring means that $\frac{10}{24}$ of the material in the ring is gold. Since the second term in this ratio is always 24, it is omitted when describing the gold content. A 30-g gold pin is marked 18 K. What is the mass of the gold in the pin?

22. Most television sets are rectangular with a width-to-height ratio of 4:3. The screens in many movie theatres have width : height ratio = 16 : 9. These different ratios create a problem when a movie is shown on TV. When the movie image is reduced to fit the TV screen, it will not be high enough to fill the screen.

40 cm

30 cm

h

a) A TV screen is 40 cm wide and 30 cm high. Calculate the height of the movie image shown.

b) How wide are the black bars at the top and bottom of the screen?

23. This statement was heard on a television program about dinosaurs.

If the time since Earth was formed is compressed into a single year, the dinosaurs would have appeared on December 13. They would have become extinct on December 25.

Carry out calculations to check this statement. Use this information:

Earth is 4 billion years old.
The dinosaurs appeared about 200 million years ago and became extinct 65 million years ago.

24. Refer to exercise 23. The first humans appeared about 3 million years ago. Suppose the time since Earth was formed is compressed into a single year. When would the first humans appear?

Communicating *the* IDEAS

Write to explain what is meant by proportional reasoning. To illustrate your explanation, use one of the exercises as an example.

4.3 Food for a Healthy Heart

On page 160, you considered the problem of how to determine if food meets the Heart and Stroke Foundation's guideline:

Food should contain less than 30% of its total energy as fat.

Nutrition information is printed on the packages of many foods. This example is from a granola cereal. We need to know the amount of energy and the amount of fat. The data show that one serving of this cereal contains:

Energy: 139 Cal
Fat: 5.5 g

NUTRITION INFORMATION	
per 30 g serving (1/3 cup)	
Energy	139 Calories
	580 kilojoules
Protein	2.8 g
Fat	5.5 g
Carbohydrate	20.0 g
Sugars	9.3 g
Starch	8.3 g
Dietary Fibre	2.0 g
Sodium	23 mg
Potassium	170 mg

We shall use these numbers to determine if less than 30% of the energy in one serving is fat. However, energy is measured in Calories and fat is measured in grams. To relate these quantities, we need to know that 1 g of fat contains 9 Cal of energy.

1. In 1 g of fat, there are 9 Cal of energy.

   a) One serving of cereal contains 5.5 g of fat. How many Calories is this?

   b) One serving of cereal contains 139 Cal. The answer in part a is the number of these Calories that are fat.
   i) What fraction of the 139 Cal are fat?
   ii) What percent of the 139 Cal are fat?

2. Does this brand of granola cereal satisfy the guideline? Explain.

3. Use the method of exercise 1. Determine whether each food satisfies the guideline.

a)

Frozen waffles	
78 g serving (1 waffle)	
Energy	230 Cal
Fat	5 g

b)

Crackers	
20 g = 4 crackers	
Energy	90 Cal
Fat	3.0 g

c)

Salmon fillets	
1 fillet serving	
Energy	228 Cal
Fat	12 g

4. Describe a rule to determine whether packaged food meets the recommended guideline of the Heart and Stroke Foundation.

The calculations in exercises 1 and 3 may be difficult to remember when shopping. We need a more efficient model that involves fewer steps.

Suppose a food is labelled as follows:

Energy: e Calories
Fat: f grams

We repeat the steps in exercise 1.

One serving of the food contains f grams of fat.
1 g of fat contains 9 Cal.
So, f grams of fat contain $f \times 9$ Cal, or $9f$ Cal.

The fraction of the total energy that is fat is $\dfrac{9f}{e}$.

Suppose a food just satisfies the guideline. Then 30% of the total energy is fat.

30% is $\dfrac{30}{100}$, or $\dfrac{3}{10}$.

$$\frac{9f}{e} = \frac{3}{10}$$

Multiply each side by e.

$$9f = \frac{3}{10}e$$

Multiply each side by 10.

$$90f = 3e$$

Divide each side by 3.

$$30f = e$$

For a food that just satisfies the guideline, this equation tells us that the energy is the number of grams of fat multiplied by 30.

For example, consider the crackers in exercise 3b:

Energy $= 3.0 \times 30 = 90$

grams of fat

Since the result, 90, is equal to the energy on the label, the crackers just satisfy the guideline.

For the frozen waffles in exercise 3a:

$$5 \times 30 = 150$$

grams of fat

Since 150 is less than the energy on the label, the frozen waffles satisfy the guideline.

For the salmon fillets in exercise 3c:

$$12 \times 30 = 360$$

grams of fat

Since 360 is greater than the energy on the label, the salmon fillets do not satisfy the guideline

This is the rule for determining if a food satisfies the Heart and Stroke Foundation's guideline.

- Multiply the number of grams of fat by 30.
- If the result is less than or equal to the energy in Calories on the label, the food satisfies the guideline.

5. Use this rule to determine whether each food satisfies the guideline.

a)
Homogenized milk	
250-mL serving (1 cup)	
Energy	157 Cal
Fat	8.6 g

b)
1% milk	
250-mL serving (1 cup)	
Energy	108 Cal
Fat	2.6 g

c)
Skim milk	
250-mL serving (1 cup)	
Energy	91 Cal
Fat	0.5 g

d)
Half and Half cream	
15-mL serving	
Energy	19 Cal
Fat	1.6 g

e)
Bran flakes	
30-g serving	
Energy	105 Cal
Fat	0.7 g

f)
Apple sauce	
125-mL serving	
Energy	105 Cal
Fat	0.1 g

g)
Shortening	
15-mL serving	
Energy	72 Cal
Fat	8.0 g

h)
Ice cream	
125-mL serving	
Energy	161 Cal
Fat	9.2 g

i)
Frozen pie shell	
220-g serving ($\frac{1}{8}$ shell)	
Energy	110 Cal
Fat	6.8 g

j)
Tortillas	
57 g (1 tortilla)	
Energy	160 Cal
Fat	2.7 g

k)
Sole fillets	
1 fillet serving	
Energy	275 Cal
Fat	16 g

l)
Chicken & vegetable pie	
225-g serving ($\frac{1}{8}$ pie)	
Energy	467 Cal
Fat	27 g

6. Record the nutrition information from two foods in your home. Determine whether each food meets the guideline for fat energy content.

7. a) Can you find any foods in your home that indicate the percent of energy that is fat?

 b) Why do some food manufacturers not disclose this information?

Communicating *the* IDEAS

Write a report of your findings so that someone who did not complete these exercises could understand. Describe the problem you investigated, the conclusions you reached, and how you arrived at your conclusions. Use appropriate written explanations, tables, equations, or calculations.

Interpreting scale diagrams and maps involves proportional reasoning. Scales can be shown in three different ways. Each *Investigation* involves one of these ways.

Working with a Ratio Scale with Units

Basketball was invented in 1891 by James Naismith, a Canadian. It is one of the most popular professional sports in North America. Here is a scale diagram of the basketball court in the Air Canada Centre, Toronto.

1. Explain what the scale "1 cm to 2.9 m" means.

2. Measure the length and the width of the drawing. Use the scale to determine the length and the width of a basketball court, in metres.

3. Use the scale.

   a) Determine the distance from the three-point line to the centre of the basket.

   b) Determine the distance from the free throw line to the centre of the basket.

   c) Determine the diameters of the centre circle and the restraining circle.

4. The basket is 3.05 m above the floor. How high would the basket be on a scale drawing that has the same scale as this one?

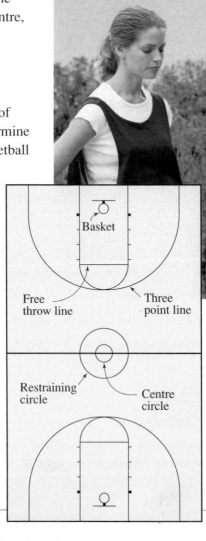

Basket

Free throw line

Three point line

Restraining circle

Centre circle

Scale: 1 cm to 2.9 m

Working with a Ratio Scale without Units

Orienteering is an international sport. Each competitor carries a map that indicates a number of circled locations that must be visited in order. The challenge is to find and follow the fastest route between locations.

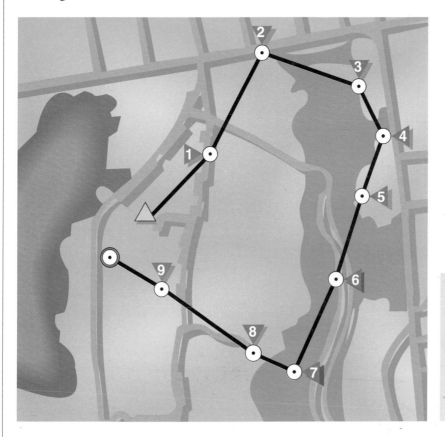

Legend

Scale 1 : 20 000

△ Start

◉ Finish

⊙ Control Location

1. What do you think the scale "1:20 000" means?

2. **a)** Decide how you could use the ratio in exercise 1 to determine the distance from location 2 to location 3.

   b) Use your method to determine the distance.

3. Determine the total distance from start to finish.

Working with a Bar Scale

In 1999, two balloonists became the first people to fly a balloon, named *Breitling Orbiter 3*, non-stop around the world. This scale diagram about the flight appeared in a magazine article.

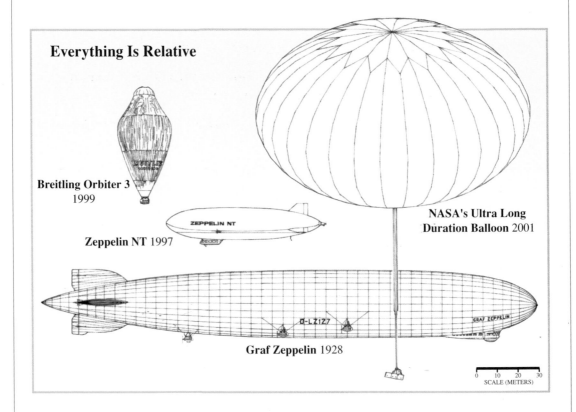

Everything Is Relative

Breitling Orbiter 3 1999

Zeppelin NT 1997

NASA's Ultra Long Duration Balloon 2001

Graf Zeppelin 1928

0 10 20 30
SCALE (METERS)

1. The scale at the bottom is called a *bar scale*. What do you think this means?

2. a) Decide how you could use the bar scale to determine the width and the height of *Breitling Orbiter 3*.

   b) Determine the width and the height using your method.

3. Determine the width and the height of NASA's Ultra Long Duration balloon.

4. Determine the length and the height of each Zeppelin balloon.

1. Three different scales were involved in the *Investigations*. Which scale do you think is easiest to use? Explain.

2. When you use the ratio scale without units, does it matter which units you use in the calculations? Explain.

3. What are some different ways to use the bar scale?

4. When you use the bar scale, does it matter which units you use? Explain.

5. A ratio scale contains two numbers or quantities. How can you tell which number represents the diagram and which number represents the object?

6. Choose one *Investigation*. Explain how proportional reasoning is involved when you use the scale.

4.4 Exercises

A 1. Calculate each scale or length to complete this table.

	Length in drawing (cm)	Actual length (cm)	Scale
a)	5	50	
b)	5	10	
c)	10	5	
d)	10	50	
e)	4.5	225	
f)	3.7	92.5	
g)	38	3.8	
h)	420	8.4	
i)	5.2		1:10
j)	12.7		1:25
k)		370	1:10
l)		7500	1:250

2. Calculate each scale or length to complete this table.

	Actual distance (km)	Distance on map (cm)	Scale
a)	120	60	
b)	120	40	
c)	120	10	
d)	120	6	
e)	90	3.6	
f)		4.3	1:5000
g)		13.2	1:2500
h)	75		1:500 000
i)	150		1:50 000
j)	99	3.3	
k)	1.4	3.5	
l)	3.6		1:20 000
m)	56		1:200 000
n)		10.8	1:35 000
o)		15.4	1:4000

B 3. These are the scales for the drawings below, but they are not in the correct order.

| 1 cm to 85 cm | 1 cm to 3 cm | 1 cm to 24 m |

Match each drawing with the correct scale.

a)

b)

c)

4. Determine the length of each item.

a) Scale: 1 cm to 12 cm   **b)** Scale: 1 cm to 40 cm   **c)** Scale: 4 cm to 1 cm

5. A student drew this floor plan of a house.

    a) What are the dimensions of each room?

    b) What are the dimensions of the house?

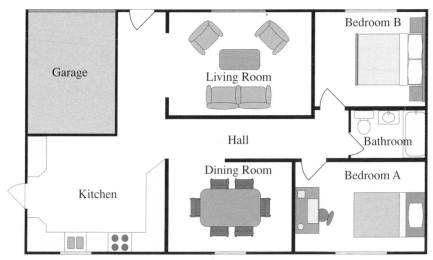

Scale 1 cm : 2 m

6. Determine the lengths of the zebra and the ladybug.

    a) 1 cm to 50 cm   **b)** 1 cm to 1.5 mm

7. Determine the approximate length of each part on the car.

    a) height of door   **b)** diameter of wheel   **c)** length of car

Scale: 1 cm to 36.7 cm

8. Determine the approximate length of each part of the mosquito.

a) abdomen   **b)** antennae   **c)** wings

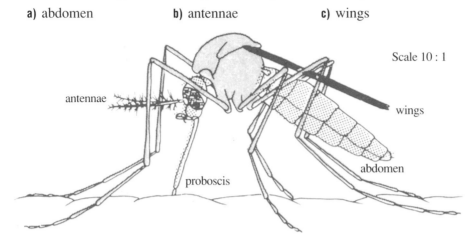

Scale 10 : 1

9. This map shows the swim course for the Ironman Canada Triathlon in Penticton, British Columbia. Use the scale to determine the total length of the swim course.

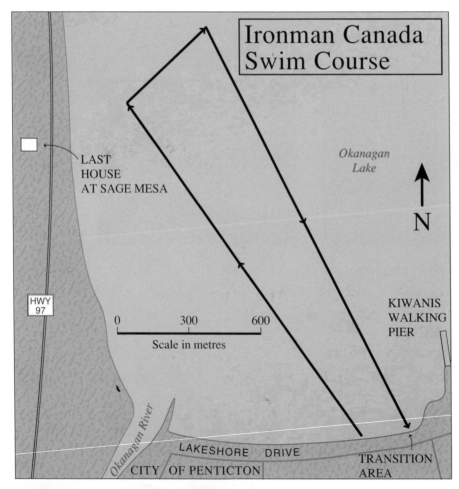

10. The scale drawing below shows the floor plan of the observation level in the Sky Pod of the CN Tower. The small rectangles around the circular rim represent the windows.

a) The floor of the observation level is circular. Determine the diameter of this circle.

b) Suppose you walk once around the observation level. Calculate how far you would walk.

c) Make up your own exercise about this scale diagram. Exchange exercises with a classmate. Complete your classmate's exercise.

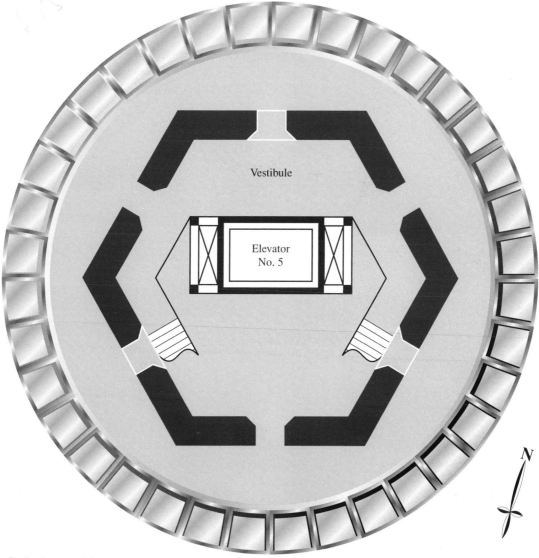

Scale: 1 cm to 1.1 m

11. This scale diagram shows the top view, front view, and side view of the Space Shuttle. Use the scale to determine each distance.

   a) the length   b) the height   c) the wing span

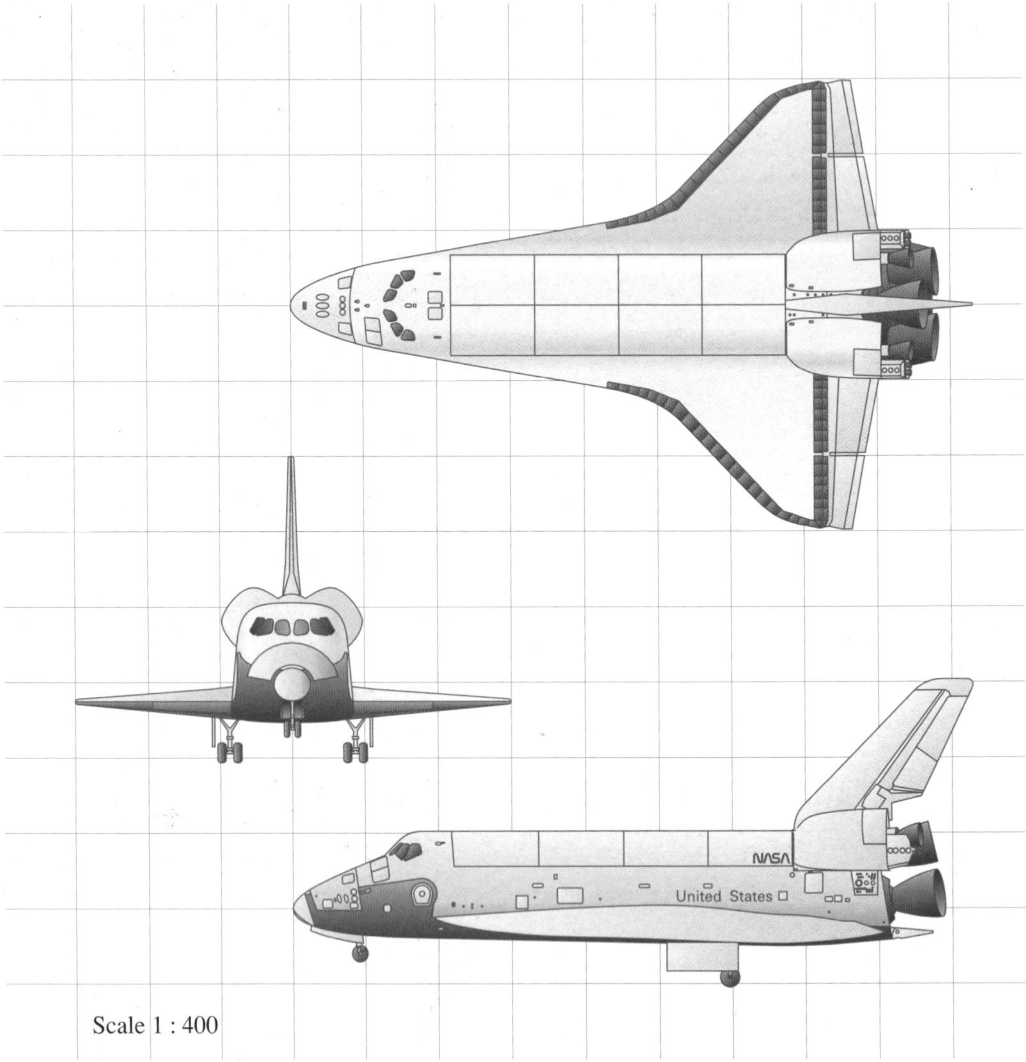

Scale 1 : 400

12. The satellite photograph on page 201 was obtained from the Internet. The area shown measures 560 km by 760 km. Point Pelee, near the bottom of the image, is the southernmost point in Canada.

   a) Choose three towns or cities on the photo. Calculate the distance from each town or city to Point Pelee.

   b) Graph the data. Plot *Image distance* horizontally and *Actual distance* vertically.

   c) Choose three other towns or cities on the photo. Use the graph to determine the distance from each town or city to Point Pelee.

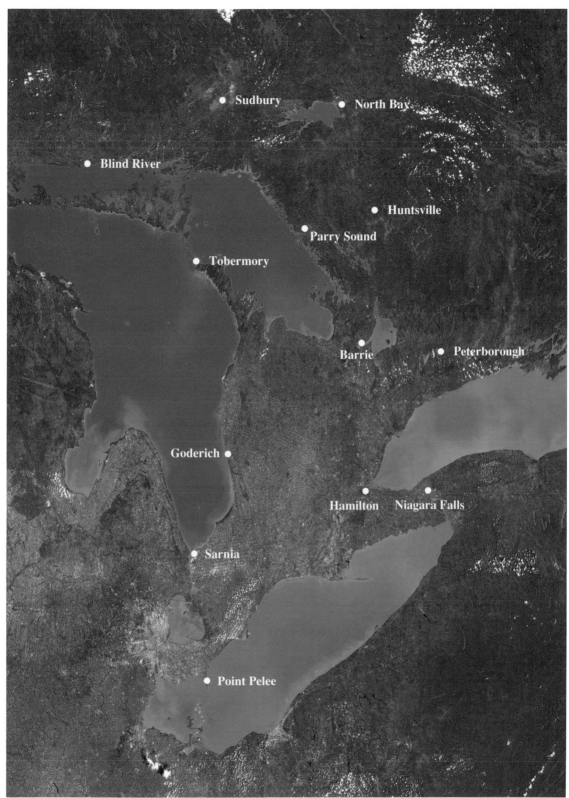

Sudbury

North Bay

Blind River

Huntsville

Parry Sound

Tobermory

Barrie

Peterborough

Goderich

Hamilton   Niagara Falls

Sarnia

Point Pelee

13. Estimate each distance.

    a) the diameter of the moon

    c) the width of the eye in the needle

    e) the width of the ball in a ball point pen

    b) the length of Prince Edward Island

    d) the thickness of a hair

    f) the width of a pollen grain

a)

1000 km

b)
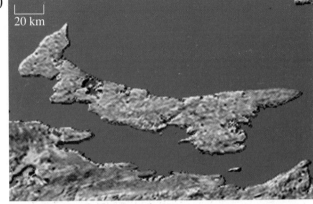
20 km

c)
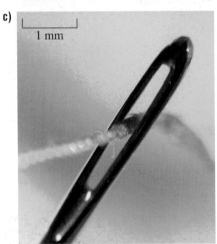
1 mm

d)

0.1 mm

e)
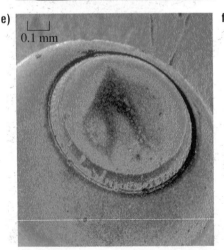
0.1 mm

f)

0.01 mm

14. In 1999, Teri Murden became the first woman to row alone across the Atlantic Ocean. She left the Canary Islands on September 13 and arrived in Guadeloupe after rowing for 82 days.

a) Use the map to determine the length of Murden's journey from the Canary Islands to Guadeloupe.

b) On average, how many kilometres did Murden row each day?

c) One day, Murden rowed 150 km. Suppose she could row at this rate for the entire trip. How many days would it take?

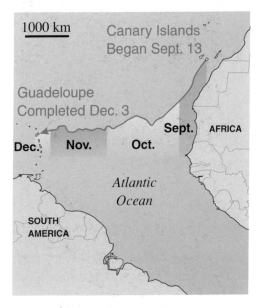

1000 km

Canary Islands
Began Sept. 13

Guadeloupe
Completed Dec. 3

Sept.   AFRICA

Dec.   Nov.   Oct.

Atlantic Ocean

SOUTH
AMERICA

The Importance of the St. Lawrence Seaway

The St. Lawrence Seaway permits ocean-going ships to enter the Great Lakes and sail to Thunder Bay. These ships pass through the Welland Canal, between Lake Ontario and Lake Erie. The signpost in the map on the next page is at the Welland Canal.

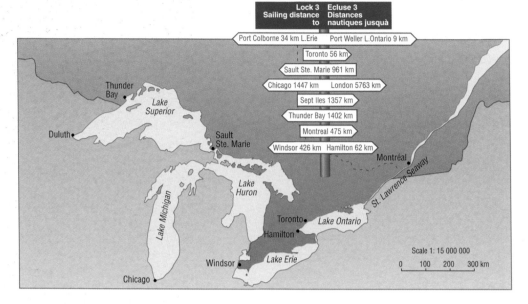

15. Use the map to estimate each distance. Check your result by using the signpost to determine the distance.

a) the sailing distance from Sault Ste. Marie to Thunder Bay

b) the sailing distance from Windsor to Sault Ste. Marie

16. The diagram below is called a profile. It is a side view that shows how the water level changes as the ships sail from one lake to another.

a) What is the change in water level from Montreal to Lake Ontario? What is the mean change in water level through the locks in the St. Lawrence Seaway?

b) What is the change in water level from Lake Ontario to Lake Erie? What is the mean change in water level through the locks in the Welland Canal?

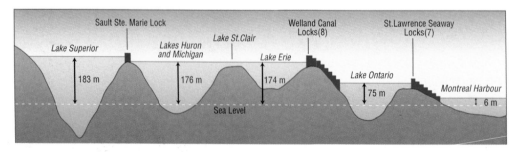

17. a) Explain how you can tell from the profile that the horizontal scale is not the same as the vertical scale.

b) Do you think this diagram is misleading? Explain.

18. This photograph appeared in a newspaper in 1997. Louis Weiss is shown with his scale model of the Parliament Buildings in Ottawa. The central tower is the Peace Tower.

a) Weiss worked 6000 hours in eight-hour days to make the model. How many days did he work?

b) Suppose Weiss worked 5 days a week. About how many months did it take him to make the model?

c) An adult male's head is about 18 cm wide. Measure the width of Weiss' head and the height of the Peace Tower in the photograph. Use the results to calculate the height of the model.

d) The Peace Tower is about 92 m high. How many times as tall as the tower in the model is the Peace Tower?

19. This graph appeared in a newspaper article in 1999. It compares the number of riders using public transit in Toronto (the TTC), with the number of riders using the GO train. The caption under the graph stated "While the number of TTC riders has fallen by about 20% over the last 10 years, the number of GO train passengers has risen by about 25%."

a) Measure the vertical distance that represents a decrease of 20% in the number of TTC riders.

b) Estimate the vertical distance that represents an increase of 25% in the number of GO train riders.

c) Is the graph misleading? Explain.

d) If your answer to part c is yes, describe how you would redraw the graph so that it is not misleading.

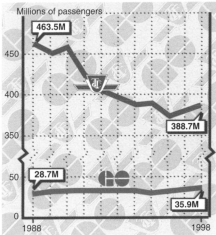

Communicating the IDEAS

Obtain a used envelope from another city or country. Use a map or an atlas to determine how far it travelled. Write to explain how you determined the distance.

To draw a scale diagram, we must consider the space available for the diagram.

Example

The diameter of a basketball is approximately 24.5 cm. The diameter of the hoop that holds that basket is 45.0 cm. Draw a scale diagram to represent the ball and the hoop.

Solution

> **Think ...**
>
> The hoop is the larger of the two objects. We decide how large we want this to be on the scale diagram. This establishes the scale. For the space available here, a diameter of 4.5 cm was chosen.

Choose 4.5 cm to represent 45 cm, the diameter of the hoop.

Divide 45 by 4.5 to determine what 1 cm represents.

1 cm represents $\frac{45 \text{ cm}}{4.5} = 10$ cm.

Hence, the scale is 1 cm to 10 cm, or 1 : 10.

To represent the hoop:
The diameter of the hoop is 45.0 cm.
The diameter of the circle representing
the hoop is $\frac{45 \text{ cm}}{10} = 4.5$ cm.

The radius of the circle is $\frac{4.5 \text{ cm}}{2} = 2.25$ cm.
Construct a circle with radius 2.25 cm.

To represent the ball:
The diameter of the ball is 24.5 cm.
The diameter of the circle representing
the ball is $\frac{24.5 \text{ cm}}{10} = 2.45$ cm.

The radius of this circle is $\frac{2.45 \text{ cm}}{2} = 1.225$ cm.
Construct a circle with radius 1.2 cm.

Write the scale on the diagram.

Scale: 1 cm to 10 cm

1. In the solution of the *Example*, why was 4.5 cm chosen as the diameter, and not 5.0 cm?

2. In the solution of the *Example*, what other way is there to determine the radius of the circle to represent the hoop? Explain.

4.5 Exercises

A

1. The scale on a diagram is 1 cm to 5 cm. What is the actual distance for each measured distance?

   a) 5 cm   b) 10 cm   c) 15 cm

   d) 20 cm   e) 25 cm   f) 30 cm

2. The scale on a diagram is 1 cm to 1000 m. What is the actual distance for each measured distance?

   a) 3 cm   b) 7 cm   c) 10 cm

   d) 14 cm   e) 25 cm   f) 2.5 cm

3. A diagram is to be drawn to a scale of 1 cm to 20 km. What is the distance on the diagram for each actual distance?

   a) 20 km   b) 10 km   c) 40 km

   d) 50 km   e) 60 km   f) 100 km

4. A diagram is to be drawn to a scale of 1 cm to 10 m. What is the distance on the diagram for each actual distance?

   a) 10 m   b) 19 m   c) 29 m

   d) 290 m   e) 350 m   f) 35 m

B

5. Draw a scale diagram of each ball. Use the same scale for each ball.

Sport	Diameter of ball (cm)
Baseball	7.4
Golf	4.3
Table tennis	3.7
Volleyball	20.9

6. A hockey puck is 7.6 cm in diameter and 2.5 cm thick. Make a scale diagram to show a top view and a side view of a hockey puck. Use the same scale as in exercise 5.

7. A table tennis table is 2.7 m long and 1.5 m wide. The playing surface is divided into two halves by the net, and also a centre line perpendicular to the net. Draw a scale diagram of a table tennis table to show the net and the centre line.

8. A volleyball court is 18 m long and 9 m wide. The centre line under the net divides the court into two equal squares. On each side of the net, the attack line is 3 m from the centre line. Draw a scale diagram of a volleyball court to show the centre line and the two attack lines.

9. In women's gymnastics, the uneven bars are 160 cm apart. The lower bar is 160 cm above the floor and the upper bar is 240 cm above the floor.

a) Draw a scale diagram to show a side view of the uneven bars.

b) The bars are 200 cm long. Draw a scale diagram to show a front view of the uneven bars.

10. A Canadian football field is 65 yards wide and measures 110 yards between the goalposts. Each end zone adds 25 more yards to the field's length. Draw a scale diagram of a Canadian football field.

11. In Olympic diving there are two boards, the springboard and the platform. The springboard is 5 m long and 3 m above the water. The platform is 6 m long and 10 m above the water. Each board projects 1.5 m beyond the water's edge. The water is 5 m deep. Draw a scale diagram to show a side view of the boards and the water. Make any reasonable assumptions about the thickness of the boards and the dimensions of the supports.

Distorting a Drawing

The drawing of a truck (below left) is clip art obtained from a computer. A drawing program was used to make the other drawings. The original drawing was stretched in one direction or in two directions.

Original drawing

Stretched horizontally

Stretched vertically

Stretched horizontally and vertically

12. Measure the drawings. Use the results to describe how each stretched drawing was obtained from the original drawing.

13. Explain why the drawings at the right on page 209 and above left are distorted, but the drawing above right is not distorted.

14. a) Make a scale drawing of your classroom.

b) Compare your scale drawing with the scale drawing of another student. Did both of you use the same scale? Explain.

C **15.** A home plate in baseball has 5 sides. The longest side side is 43.2 cm long. The two sides next to this side are one-half as long. The other two sides are 30.5 cm long. The angles formed by the sides measure either 90° or 135°. Make a scale diagram of a home plate.

16. Explain why the photograph of home plate is misleading. How would you take the photograph so the picture is not misleading?

Communicating *the* IDEAS

Choose one exercise you completed. Write to explain how you determined the scale for the diagram.

Proportional Reasoning Tools

Proportional Relationship

- The equation has the form $y = mx$.
- The graph is a straight line through the origin.
- Corresponding quantities are related by multiplication and division.
- Ratios relating corresponding quantities are equal.

Scale Diagrams

- A ratio scale may have units; for example, 1 cm to 2.5 m.
- A ratio scale may not have units; for example, 1 : 20 000.
- A bar scale looks like part of a ruler, with equal distances marked.

1. About 3 in 10 Canadians are nearsighted.
   About 6 in 10 are farsighted. The
   population of Canada is about 31 million.

   a) How many Canadians are nearsighted?

   b) How many are farsighted?

2. A laser printer can print 15 pages of text per minute. Ignore the time it takes
   to refill the paper tray.

   a) How many pages can the printer print in each time?
   i) 1 h   **ii)** 1 day

   b) How long would it take to print each number of pages?
   i) 1000   **ii)** 10 000

3. An amount of $42 500 was raised in the first 20 min of the Red Shield
   Telethon.

   a) At this rate, how much would be raised in 3 h?

   b) How long would it take to raise $1 000 000?

4. The following fact was obtained from SkyDome's web site in January, 2000
   (www.skydome.com/welcome/facts.html). Suppose the fact is true. How
   many seats does each person wash per hour?

   - SkyDome is the only facility in the world to wash 37 000 seats of the
     stadium after each event. It takes 14 people 8 hours to complete.

5. The following fact was obtained from the CN Tower's web site in January, 2000 (www.cntower.ca/ll_calc.html). Shaquille O'Neal is a basketball player. Use the information to determine his mass. One ton is about 907 kg.

   • The Tower has a mass of 130 000 tons. That's the same as 852 459 Shaquille O'Neals.

6. **Estimation** On Canada Day, about 2400 people created a human flag. Each person held a large red or white card above her or his head.

   a) Estimate the number of people holding a red card and the number holding a white card.

   b) Suppose all the students in your school were to create a human flag like this. Estimate the number of red cards and the number of white cards needed.

7. Determine the length and the width of this can. The scale is 1 cm to 15 mm.

The Trans-Canada Trail

When completed, the Trans-Canada Trail will be the world's longest nature trail.

—— Land Trail
—— Water Trail

Trail facts
Total length	15 000 km
Total cost	$42 million
Amount raised by January, 2000:	$6.7 million

8. Estimation Use the information above. Estimate the length of the part of the trail that lies in Ontario.

9. What percent of the total cost was raised by January, 2000?

10. By January, 2000:

    65% of the amount raised represented donations from individuals.

    30% represented donations from corporations.

    5% represented government grants.

    How much was raised by each group?

    a) individuals   **b)** corporations   **c)** government grants

11. A person can buy 1 m of the trail for $40. The person's name is inscribed on a panel in a pavilion along the route. A pavilion can hold up to 12 000 names.

    a) How much money does a pavilion represent?

    b) What length of trail does this represent?

12. Use the information in exercises 10 and 11. How many name panels would be needed to inscribe the names of the individuals who had contributed by January, 2000?

13. The following item appeared in a newspaper in January, 2000.

Each smoke shortens life 11 minutes, study says

Every cigarette smoked shaves 11 minutes off your life, according to a new study.

- The 11 minutes gained from not smoking a single cigarette could be used for a brisk walk or reading a newspaper.
- The three hours and 40 minutes of living gained from not smoking a pack of cigarettes is enough time to run a marathon, or watch the movie *Titanic*.
- The 1.5 days gained from not smoking a carton of 200 cigarettes could be used to enjoy a flight around the world.

Dr. Mary Shaw found that a moderate male smoker, who smokes 16 cigarettes a day from 17 to 71, will consume 5,772 cigarettes a year, or 311,688 in a lifetime.

That average smoker will also live 6 1/2 years less than an average non-smoker. (The 11 minutes a cigarette calculation comes from dividing this shorter life span — 3.4 million minutes — by the number of cigarettes smoked over 54 years, 311,688).

a) Assume a 17-year-old smokes 16 cigarettes per day.

   i) How many cigarettes does the 17-year-old smoke in one year?

   ii) How many cigarettes does the 17-year-old smoke by age 71?

b) How many minutes are there in each time period?

   i) 1 h   **ii)** 1 day

   iii) 1 year   **iv)** $6\frac{1}{2}$ years?

c) Do the results of parts a and b support the statement that every cigarette smoked shaves 11 min off your life? Explain.

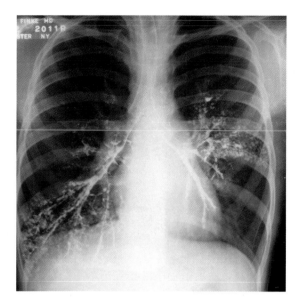

14. The map (below left) has no scale. It is 58 km from Nanaimo to Vancouver.

   a) Take measurements to determine the distance from Vancouver to Victoria.

   b) Express the scale of the map as a ratio.

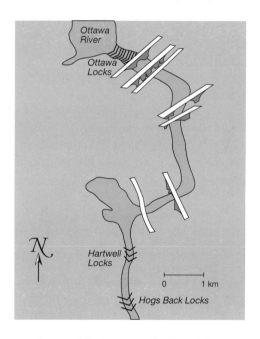

15. In winter, the Rideau Canal (above right) becomes the world's longest skating rink. Estimate the length of the canal from the Ottawa Locks to the Hartwell Locks.

16. The scale drawing below shows the Sky Pod of the CN Tower. Calculate each measurement.

    a) the height of the Sky Pod

    b) the diameter of the Sky Pod

    c) the diameter of the part of the CN Tower below the Sky Pod

    d) the diameter of the part of the CN Tower above the Sky Pod

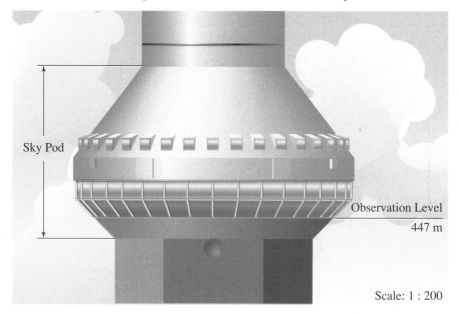

Sky Pod

Observation Level
447 m

Scale: 1 : 200

17. Suppose Canada's coastline were straightened out and represented by the width of this page. On this scale, the distance from coast to coast across Canada would be represented by this dash —. It is about 5000 km across Canada. How long is Canada's coastline?

18. **Mathematical Modelling** Determine if these foods satisfy the recommendation of the Heart and Stroke Foundation, that food should contain less than 30% of its total energy as fat.

a)

Rice Dream drink box amount per serving	
Energy	120 Cal
Fat	2 g

b)

Peanut butter 14 g serving	
Energy	86 Cal
Fat	7.5 g

c)

Chocolate cake mix per 1/12 cake	
Energy	175 Cal
Fat	4.5 g

19. A scanning electron microscope was used to take the photograph of an ant (below left).

   a) Take measurements from the photograph. Determine approximately:

      i) the width of the ant's head

      ii) the diameter of the ant's eye

   b) Which scale could be used for this photograph? Explain.

      i) 1 cm to 70 cm   **ii)** 70 cm to 1 cm

      iii) 1 mm to 70 mm   **iv)** 70 mm to 1 mm

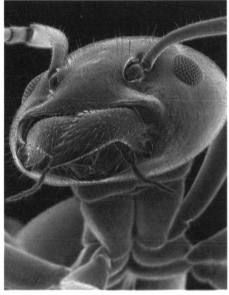

Scale 70 : 1

20. The photograph (above right) shows workers lowering a dome over a dish antenna near Winnipeg. The antenna is used to warn of approaching storms.

   a) Estimate the radius of the dish antenna.

   b) Estimate the diameter of the dome.

21. a) Explain why the second term in the scale on a map of a large area is a large number with several zeros.

   b) Describe a way to convert a scale in this form to one in the form "1 cm to _ km".

5 Similar Triangles and Trigonometry

Measuring Inaccessible Heights

The Great Pyramid, at Giza, near Cairo, Egypt is one of the Seven Wonders of the World. It was built around 2700 BC as the tomb of the Pharaoh Khufu.

When built, the Great Pyramid measured about 146 m high.

Mount Everest is about 8848 m high.

Have you ever wondered how people know these measurements? The word *trigonometry* means "triangle measurement." You can use trigonometry to calculate the lengths of sides and the measures of angles in any triangle, if you have enough information to draw the triangle.

In this chapter, you will learn about trigonometry. You will use it to determine the height of an inaccessible object, such as a tree. Here are some things to consider.

1. What instruments will you need to measure distances?

2. How can you measure angles?

3. How do you think you could measure the height of a tree?

You will return to this problem in Section 5.11.

FYI Visit www.awl.com/canada/school/connections

For information related to the above problem, click on <u>MATHLINKS</u>, followed by Foundations of Mathematics 10. Then select a topic under Measuring Inaccessible Heights.

Preparing
for the Chapter

This section reviews these concepts:

- The Pythagorean Theorem
- Angle measures in triangles
- Angle measures with parallel lines
- Geometry terms

1. In each diagram, the lengths of two sides of a right triangle are given. Use the Pythagorean Theorem to calculate the length of the third side, to the nearest tenth of a centimetre.

a)

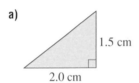

b)

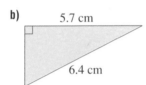

2. An extension ladder extends to 10.0 m. The ladder leans against a house. The foot of the ladder is 2.0 m away from the house. How high will the ladder reach?

3. Determine the angle measure indicated by each letter.

a)

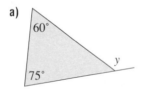

b)

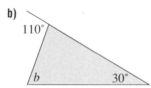

c)

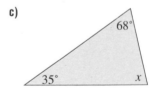

4. Determine the angle measure indicated by each letter.

a)

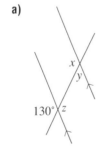

b)

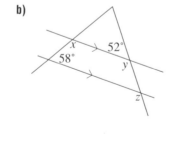

c)

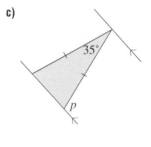

5. Determine the measure of each angle represented by *a*. Explain your answers.

a)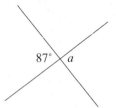

b)

c)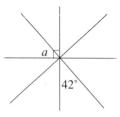

6. Explain each term. Draw a diagram to illustrate your explanation.

a) parallel lines

b) a vertex of a triangle

c) an angle

d) perpendicular lines

e) an obtuse angle

f) an acute angle

g) a leg of a right triangle

h) the hypotenuse of a right triangle

7. Draw a triangle of each type.

a) scalene

b) isosceles

c) equilateral

8. Use a protractor to measure each angle.

a) ∠ABM

b) ∠AMB

c) ∠BAM

d) ∠BCE

e) ∠ECB

f) ∠ACH

g) ∠AHC

h) ∠JHM

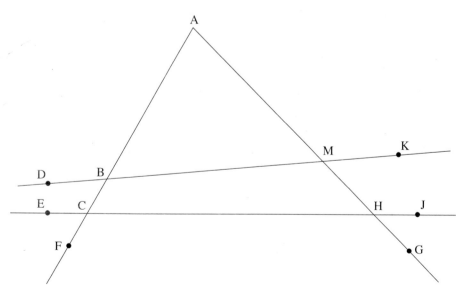

5.1 Using *The Geometer's Sketchpad*: Additional Tools

Chapter 1 introduced *The Geometer's Sketchpad*. In this chapter, you will need additional skills to explore trigonometry.

As you extend your skills with *The Geometer's Sketchpad,* you will review work from grade 7 on reflections and rotations.

Use *The Geometer's Sketchpad*.

INVESTIGATION 1

Constructing Scale Diagrams

The Dilate command is used to construct a scale diagram of a figure or segment.

1. *Construct a scale diagram of a segment.*

   SET-UP

   • From the File menu, choose New Sketch.
   • From the Display menu, choose Preferences. Make sure that Autoshow Labels for Points is checked. This automatically labels points. Recall that you can change any label by using the Text tool. Double-click on the label and type a new letter.

Click on this tool …	*… and do this:*
	a) Draw line segment AB.
	b) Select point A. Double-click on A. The point flashes to show that it is a centre of movement.
	c) Shift-click segment AB and point A. From the Transform menu, choose Dilate. When the dialog box appears, type 2.00 in the New box and 1.00 in the Old box. Click OK. You have dilated segment AB about A.
	d) What do you notice?
	e) Draw another line segment CD. Mark C as the centre of movement. In the dialog box, type 0.5 in the New box. Click OK. What do you notice?
	f) Find out what happens if you select D to be the centre of movement.

2. *Construct a scale diagram of a pentagon.*

SET-UP

- From the File menu, choose New Sketch.
- From the Display menu, choose Preferences. Make sure that Autoshow Labels for Points is checked.

Click on this tool … … and do this:

a) Draw points A, B, C, D, and E.

b) Shift-click points A, B, C, D, and E, in order.

c) From the Construct menu, choose Segment.

d) Double-click on A to make it a centre of movement.

e) Shift-click each vertex and side to select the pentagon.

f) From the Transform menu, choose Dilate. When the dialog box appears, type 2.00 in the New box and 1.00 in the Old box. Click OK.

g) What do you notice?

h) Construct another figure. Use the Dilate command, but this time type 0.5 in the New box. Write to explain what happens.

Recall that reflections and rotations are two types of transformations.

Reflecting Figures

The Reflect command is used to reflect a figure or segment. Before using this command, you must draw a mirror line.

1. *Reflect a line segment.*

SET-UP

- From the File menu, choose New Sketch.
- Make sure that Autoshow Labels for Points is checked.

Click on this tool … … and do this:

a) Draw line segment AB.

b) Draw a second segment CD that does not intersect AB.

c) Select segment CD. Double-click on CD. The little black squares flash to show that CD is now a mirror line.

d) Select segment AB. From the Transform menu, choose Reflect.

e) Drag point A. What do you notice?

f) What happens when segment AB is dragged to intersect segment CD? Explain.

2. *Reflect a pentagon.*

SET-UP

- From the File menu, choose New Sketch.
- Make sure that Autoshow Labels for Points is checked.

Click on this tool … … and do this:

a) Draw points A, B, C, D, and E.

b) Shift-click points A, B, C, D, and E, in order.

c) From the Construct menu, choose Polygon Interior. The pentagon will be grey. The stripes show that it is selected. Change the colour: from the Display menu, choose Color, then select the red rectangle.

d) Draw a line segment beside the pentagon.

e) Double-click the segment to mark it as a mirror line.

f) Select the pentagon by clicking anywhere on the red area.

g) From the Transform menu, choose Reflect. A striped image pentagon is produced. Change its colour to blue.

h) Drag a vertex of the red pentagon and watch the effect on the image pentagon.

Rotating Figures

1. *Rotate a segment.*

SET-UP

- From the File menu, choose New Sketch.
- Make sure that Autoshow Labels for Points is checked.
- Make sure the Line tool is in the Toolbox.

Click on this tool … … and do this:

a) Draw line AB.

b) Double-click on A to mark it as a centre of rotation.

c) Select line AB. From the Transform menu, choose Rotate.

d) When the dialog box appears, type 60 and click OK. What happened?

e) Drag points A and B. What do you notice?

f) Select line AB. From the Transform menu, choose Rotate. When the dialogue box appears, type –60, and click OK. How did the negative sign affect the rotation?

g) Draw line CD.

h) Double-click on C and rotate CD through an angle of 70°.

i) Double-click on D. Rotate CD through an angle of –32°.

j) Shift-click both rotated lines. From the Construct menu, choose Point at Intersection. Label the point E.

k) What do you know about the measure of ∠CED? Explain.

2. *Rotate a pentagon.*

Follow the instructions in *Investigation 2*, exercise 2 parts a and b, to construct a pentagon. Colour it red.

Click on this tool ...   ... and do this:

a) Double-click on one vertex of the pentagon to mark it as a centre of rotation.

b) Select the pentagon. From the Transform menu, choose Rotate.

c) When the dialog box appears, type 80 and click OK. What happened?

d) Drag a vertex of the original pentagon. Notice what happens to the image pentagon when the original pentagon changes.

e) Double-click on another vertex of the pentagon.

f) Rotate the pentagon through –25° about this point. How does the negative sign affect the rotation?

Communicating the IDEAS

Find an example of a logo or a picture that includes a reflection or a rotation. Write to explain how you know which transformation was used. Find the centre of rotation or the mirror line. Include a sketch in your explanation.

5.2 Using *The Geometer's Sketchpad*: Similar Triangles

An architect often uses the same shape for the main roof and the porch roof when she is designing a house.

Two figures that have the same shape but not necessarily the same size are *similar*.

What do we mean when we say that triangles have the same shape? You will use *The Geometer's Sketchpad* to investigate.

INVESTIGATION 1

Triangles with Corresponding Angles Equal

You will construct two triangles of different sizes but with corresponding angles equal. That is, each triangle will have angles of 37°, 52°, and 91°. You will look for relationships among the measures of corresponding sides.

SET-UP

- From the File menu, choose New Sketch.
- Make sure the Line tool is displayed in the toolbox.
- From the Display menu, choose Preferences. Make sure that Autoshow Labels for Points is checked.

Part A Construct Triangles with the Same Shape.

1. To construct △ABC with ∠CAB = 37° and ∠CBA = 52°:

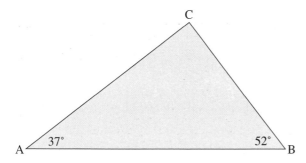

Click on this tool …   … and do this:

a) Draw line AB.

b) Double-click on A. The point flashes to show that it is a centre of rotation.

c) Select line AB. From the Transform menu, choose Rotate. In the dialog box, type 37 then click OK.

d) Double-click on B to make it a centre of rotation.

e) Select line AB. From the Transform menu, choose Rotate. In the dialog box, type –52 then click OK.

f) Shift-click the two rotated lines to select them. From the Construct menu, choose Point at Intersection. This point should be labelled C. If it is not, change the label.

g) Shift-click all three lines. From the Display menu, choose Hide Lines.

h) Shift-click points A, B, and C. From the Construct menu, choose Segment.

2. To construct △DEF with ∠FDE = 37° and ∠FED = 52°:

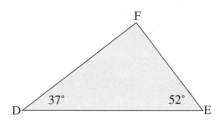

Use the same screen as △ABC.

Click on this tool …	… and do this:
	a) Draw line DE.
	b) Drag point E so that DE is shorter than AB in the first sketch.
	c) Double-click on D to make it a centre of rotation.
	d) Select line DE. From the Transform menu, choose Rotate. In the dialog box, type 37 then click OK.
	e) Double-click on E to make it a centre of rotation.
	f) Select line DE. From the Transform menu, choose Rotate. In the dialog box, type –52 then click OK.
	g) Shift-click the two rotated lines to select them. From the Construct menu, choose Point at Intersection. This point should be labelled F.
	h) Shift-click all three lines. From the Display menu, choose Hide Lines.
	i) Shift-click points D, E, and F. From the Construct menu, choose Segment.

3. To measure ∠ACB:

- Shift-click points A, C, and B in order.
- From the Measure menu, choose Angle.

4. Explain how you could have calculated the measure of ∠ACB.

5. Predict the measure of ∠DFE, then measure to check your prediction.

6. Drag point D, then point A. What do you notice about the measures of ∠ACB and ∠DFE?

Part B Relationships Among the Sides of Triangles with the Same Shape

1. To measure segment AB:

- Select segment AB.
- From the Measure menu, choose Length.

2. Measure segment DE. We say that AB and DE are corresponding sides because they lie between corresponding angles. Which side in △DEF corresponds to side BC?

3. Shift-click AB and DE. From the Measure menu, choose Ratio. What is the result?

4. Measure segments AC and DF.

5. Shift-click AC and DF. From the Measure menu, choose Ratio. What do you notice about the ratios $\frac{AB}{DE}$ and $\frac{AC}{DF}$?

6. Measure segments BC and EF. Find the ratio $\frac{BC}{EF}$. What do you notice about this ratio?

7. Drag point A, then point D. What do you notice about the three ratios?

8. Write a statement to describe the ratios of corresponding sides in triangles with the same shape.

Suppose we know that corresponding sides are in the same ratio. What does this tell us about corresponding angles?

INVESTIGATION 2

Triangles with Corresponding Sides in the Same Ratio

In Section 5.1, you used the Dilate command to make scale diagrams.

In this *Investigation*, you will construct two triangles whose sides are in the same ratio. This means that one is a scale diagram of the other. You will investigate the corresponding angles in the triangles.

SET-UP

- From the File menu, choose New Sketch.
- From the Display menu, choose Preferences. Make sure that Autoshow Labels for Points is checked.

1. To draw △ABC and △AB′C′:

Click on this tool …	*… and do this:*
	a) Draw 3 points A, B, and C.

b) Shift-click points A, B, and C. From the Construct menu, choose Segment.

c) Measure the lengths of AB, AC, and BC.

d) Double-click point A to make it a centre of dilatation.

e) Select △ABC. From the Transform menu, choose Dilate. When the dialog box appears, type 2.00 in the New box and 1.00 in the Old box.

f) Click on each unlabelled vertex.

g) If necessary, click on D. Change D to B′. Click on E. Change E to C′.

h) Segment AB′ is twice as long as AB. Segment AC′ is twice as long as AC. Describe the relationship between BC and B′C′.

2. To compare the measurements of corresponding angles:

   a) Measure ∠BAC and ∠B′AC′, ∠ABC and ∠AB′C′, ∠BCA and ∠B′C′A.

   b) How are the angles in each pair related?

3. **a)** Draw a new △DEF.

   b) Double-click on E to make it a centre of dilatation.

   c) Select △DEF. From the Transform menu, choose Dilate. When the dialog box appears, type 0.5 in the New box and 1.00 in the Old box.

   d) Measure the angles in △DEF and △DE′F′.

   e) From the Edit menu, choose Undo.

   f) Select D as a centre of dilatation.

   g) Repeat parts c and d.

   h) Explain the results.

4. Write a statement about the relationship between the corresponding angles of two triangles that have corresponding sides in the same ratio.

Communicating the IDEAS

The two roofs on page 226 have the same shape although they are not the same size. Look at your results from the *Investigations*. Explain what must be true about the measurements of these two roofs.

5.3 Similar Triangles

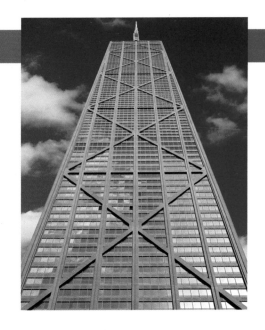

Similar triangles are used in design and construction. In Section 5.2, you investigated some relationships among corresponding sides and angles in similar triangles. Here are the properties.

Take Note

Properties of Similar Triangles

- In similar triangles, corresponding angles are equal.

  Given $\triangle ABC$ is similar to $\triangle PQR$,
  then $\angle ABC = \angle PQR$
  $\qquad \angle BAC = \angle QPR$
  $\qquad \angle ACB = \angle PRQ$

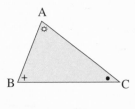
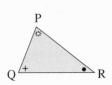

- In similar triangles, the ratios of corresponding sides are equal.

  Given $\triangle ABC$ is similar to $\triangle PQR$, then
  $\dfrac{AB}{PQ} = \dfrac{BC}{QR} = \dfrac{AC}{PR}$

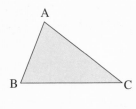
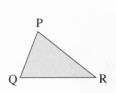

We can use either of the above properties to calculate the measures of angles and sides in similar triangles.

Example 1

Determine the measures of ∠A, ∠B, and ∠C.

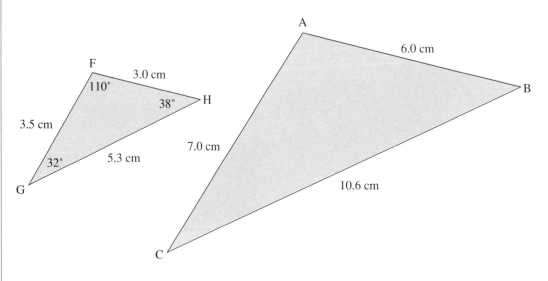

Solution

Check to see if the triangles are similar.

The ratios of corresponding sides are:

$\frac{AB}{FH} = \frac{6.0}{3.0}$, or 2; $\frac{BC}{HG} = \frac{10.6}{5.3}$, or 2; $\frac{AC}{FG} = \frac{7.0}{3.5}$, or 2

Since the ratios of corresponding sides are equal, the triangles are similar.
That is, △FHG is similar to △ABC.

Since the triangles are similar, corresponding angles are equal.

∠A = ∠F = 110°
∠B = ∠H = 38°
∠C = ∠G = 32°

In *Example 1*, the triangles are described so that corresponding angles appear in the same order. That is, we write △**A**B**C** and △**F**H**G** because ∠**A** = ∠**F**;
∠**B** = ∠**H**; and ∠C = ∠G.

We use the symbol ~ to represent the words "is similar to." In *Example 1*, we could write △FHG ~ △ABC.

We can use the ratios written as fractions to calculate the length of a side of a triangle.

Example 2

Triangle ABC is similar to △RPQ. Determine the lengths of RP and AC.

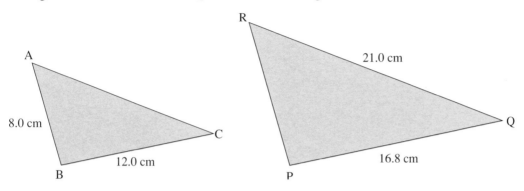

Solution

△RPQ is similar to △ABC, so the ratios of corresponding sides are equal.

$$\frac{RP}{AB} = \frac{PQ}{BC} = \frac{RQ}{AC}$$

Substitute the given lengths.

$$\frac{RP}{8.0} = \frac{16.8}{12.0} = \frac{21.0}{AC}$$

To determine RP, use the first two fractions above.

$$\frac{RP}{8.0} = \frac{16.8}{12.0}$$

Multiply each side by 8.0.

$$RP = \frac{16.8}{12.0} \times 8.0$$

$$RP = 11.2$$

RP is 11.2 cm.

To determine AC, use the second and third fractions above.

$$\frac{21.0}{AC} = \frac{16.8}{12.0}$$

Invert these fractions.

$$\frac{AC}{21.0} = \frac{12.0}{16.8}$$

Multiply each side by 21.0.

$$AC = 21.0 \times \frac{12.0}{16.8}$$

$$AC = 15.0$$

AC is 15.0 cm.

Some similar triangles are found in composite figures, as illustrated by the following example. We can show these triangles are similar if we can find pairs of equal corresponding angles.

Example 3

For each figure below, show that △DEF is similar to △DGH.

a)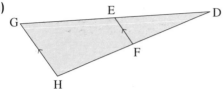

b)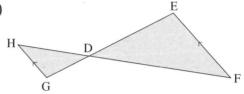

Solution

a) Since EF is parallel to GH, corresponding angles are equal.

$$\angle DFE = \angle DHG$$
$$\angle DEF = \angle DGH$$

Since ∠D is common to both triangles,

$$\angle EDF = \angle GDH$$

Since the corresponding angles in the two triangles are equal, △DEF is similar to △DGH.

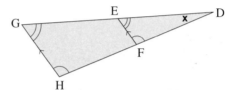

b) Since EF is parallel to HG, alternate angles are equal.

$$\angle DEF = \angle DGH$$
$$\angle DFE = \angle DHG$$

Since opposite angles are equal,

$$\angle EDF = \angle GDH$$

Since the corresponding angles in the two triangles are equal, △DEF is similar to △DGH.

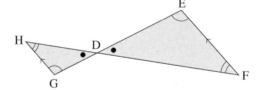

Discussing the IDEAS

1. Explain how you can find out if two triangles are similar.

2. Suppose △ABC is similar to △PQR and corresponding sides are in the ratio 1 : 1. What can you say about the two triangles? Explain.

3. In the solution of *Example 2*, explain why we can invert the second and third fractions to determine AC.

1. Calculate the value of x in each triangle.

a)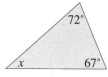

72°

x 67°

b)

26°

x

c)

x

27°   135°

2. Solve each equation.

a) $\frac{a}{3} = \frac{4}{6}$ **b)** $\frac{c}{4} = \frac{5}{10}$ **c)** $\frac{x}{5} = \frac{15}{3}$ **d)** $\frac{m}{10} = \frac{3}{12}$ **e)** $\frac{h}{2} = \frac{25}{5}$ **f)** $\frac{a}{9} = \frac{10}{30}$

5.3 :: Exercises

A **1.** Which triangle is similar to △ABC? Explain.

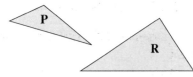

A   B

C

S   Q   P   R

2. For each pair of similar triangles, list the corresponding sides and angles.

a)

A
•
B
D   E
•
x   x
C   F

b)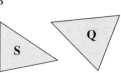

P
o
X
✔
Y
Q ✔
R
o
Z

3. The triangles in each pair are similar. In each case, state the measures of the angles that are not marked.

a)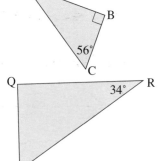

A   B

56°
C
Q   34°   R

P

b)

D
70°   F
E

J

G   40°
H

c)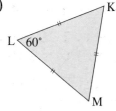

K
L 60°
M
S   T
U

4. These triangles have the same shape, but different orientations.

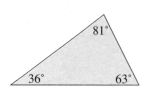
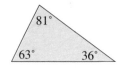

Are the triangles similar? Explain.

5. In each diagram, the triangles are similar. For each pair of triangles, write the ratio of sides that is equal to $\frac{AB}{BC}$.

a)
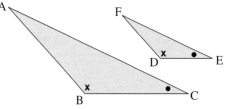

b)
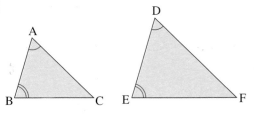

c)
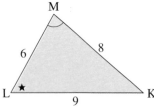

B   **6.** Explain why the triangles in each pair are similar. Determine the values of x and y.

a)
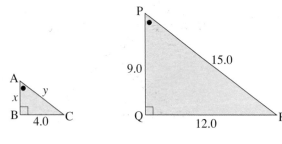

b)

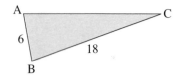

7. State the ratios of the corresponding sides of each pair of similar triangles. Determine each value of x.

a)
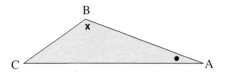

b)

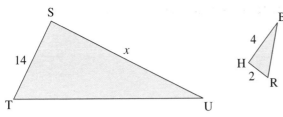

c)

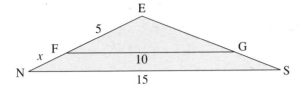

8. State which triangles are similar. Calculate the values of x and y.

a)

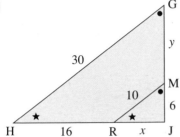

b)

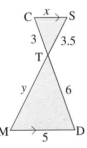

9. For this picture of an ironing board, explain how you know that $\triangle AEC$ is similar to $\triangle BED$.

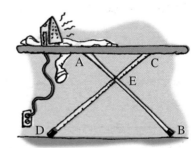

10. Explain why the triangles in each pair are similar. Determine the values of x and y.

a)

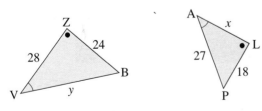

b)

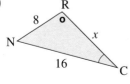 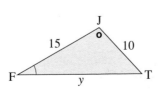

11. The two triangles in each diagram are similar. Determine each length represented by x.

a)

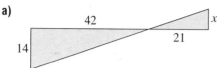

b)

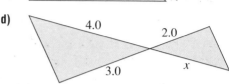

c)

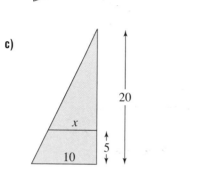

d)

Similar Triangles and Dilatation

Similar triangles are used in many designs.

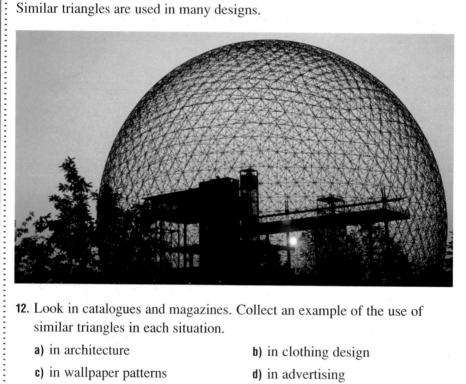

12. Look in catalogues and magazines. Collect an example of the use of similar triangles in each situation.

a) in architecture

b) in clothing design

c) in wallpaper patterns

d) in advertising

You can make a larger or smaller copy of a triangle, or other figure by using a dilatation. A dilatation is a transformation that enlarges or shrinks something by a ratio about a central point.

13. Use the idea described above to enlarge a triangle by the ratio 3 : 1.

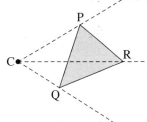

- Use a ruler to draw △PQR on a piece of paper. Do not make it too big since you will be enlarging it.
- Mark a point on the paper and label it C.
- Lightly draw lines CP, CQ, and CR.
- Lay your ruler on segment CP. Mark point K so that CK is three times as long as CP. Lightly draw CK.
- Lay your ruler on CQ and mark point L so that CL is three times as long as CQ. Lightly draw CL.
- Lay your ruler on CR and mark point M so that CM is three times as long as CR. Lightly draw CR.
- Join points K, L, and M. What do you notice about △KLM?

14. Make a design that uses similar triangles.

 15. Name a pair of similar triangles in this figure. Explain how you know the triangles are similar. How many pairs of similar triangles can you find?

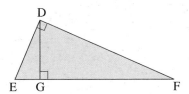

16. In each figure, name a pair of similar triangles. Explain how you know they are similar. Determine the values of *x* and *y*.

a)

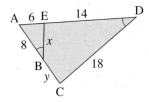

b)

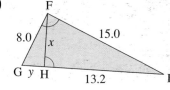

Communicating *the* IDEAS

Write to explain why you only need to measure two pairs of corresponding angles to determine whether two triangles are similar.

Our knowledge of similar triangles helps us to solve problems involving the dimensions of tall or inaccessible objects.

Determining the Height of a Tree

Work with a partner on a sunny day. You will need a metre stick and a measuring tape.

1. In the school grounds, choose a tree whose shadow you can measure. Measure and record the length of the shadow.

2. Hold a metre stick vertically, while your partner measures the length of its shadow. Record this measure.

3. Draw two similar right triangles: one to represent the tree and its shadow; the other to represent the stick and its shadow. Label each triangle with the measures you know.

4. Use the similar triangles to determine the height of the tree.

Example 1

Two students completed the *Investigation* on page 240.

These are the triangles they drew. The triangles are not to scale.

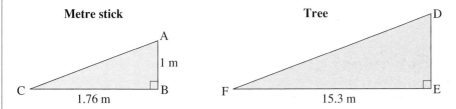

What is the approximate height of the tree?

Solution

The metre stick and the tree are vertical; so, $\angle B = \angle E = 90°$
Since the sun's rays are parallel, the angles between the rays and the ground are equal.

That is, $\angle C = \angle F$

Since two pairs of corresponding angles in the triangles are equal, the angles in the third pair must also be equal.

That is, $\angle A = \angle D$

Since pairs of corresponding angles are equal, $\triangle ABC$ is similar to $\triangle DEF$.
Write the ratios of corresponding sides.
$$\frac{DE}{AB} = \frac{EF}{BC} = \frac{DF}{AC}$$

Substitute the known lengths.
$$\frac{DE}{1} = \frac{15.3}{1.76} = \frac{DF}{AC}$$

Use the first and second fractions.
$$\frac{DE}{1} = \frac{15.3}{1.76}$$
$$DE \doteq 8.69$$

The tree is approximately 9 m high.

Sometimes we wish to know the height of a building or the distance across a river or pond. These heights and distances cannot be measured directly. We can use similar triangles and ratios to calculate the height or distance.

Example 2

Two people, Anne and Bill, are on the same side of a river. They took measurements to calculate the width of the river.

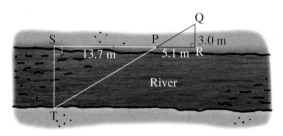

The width to be determined is ST. There is a tree at T.
Anne places a marker at S, directly across the river from T.
Bill places a marker at R then moves to Q, which is 3.0 m from R.
At Q, Bill can see the tree at T.
Anne walks along the river toward R until she reaches P, where she, Bill, and the tree are aligned.
Bill then measures the distances SP and PR.
How wide is the river?

Solution

Triangle STP is similar to △RQP, because the pairs of corresponding angles are equal.

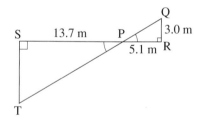

That is, ∠TSP = ∠QRP = 90°
Since opposite angles are equal, ∠STP = ∠RQP

Since two pairs of corresponding angles are equal, the angles in the third pair must also be equal: ∠STP = ∠RQP

Since the triangles are similar, corresponding sides are in the same ratio.

$$\frac{ST}{RQ} = \frac{SP}{RP}$$

Substitute the known measures.

$$\frac{ST}{3.0} = \frac{13.7}{5.1}$$

Multiply each side by 3.0.

$$ST = \frac{3.0 \times 13.7}{5.1}$$

$$ST \doteq 8.06$$

The river is about 8 m wide.

1. In the *Investigation*, how did you know the two triangles you drew were similar?

2. In the solution of *Example 1*:

   a) We show that two pairs of corresponding angles are equal. How do we know that the angles in the third pair are equal?

   b) When we wrote the ratios of corresponding sides, why did we start with DE?

3. Explain how correctly naming two similar triangles helps to identify corresponding sides.

4. What do you need to know about the side lengths of two triangles to know whether the triangles are similar?

5. What do you need to know about the angle measure of two triangles to know whether the triangles are similar?

5.4 Exercises

A **1.** State which triangles are similar. Determine the values of *x* and *y*.

a)

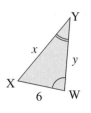
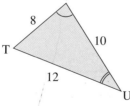

b)

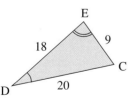
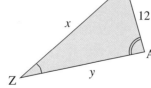

c)

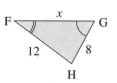
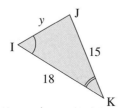

B **2.** Two trees cast shadows as shown.
Determine the height of the
evergreen tree.

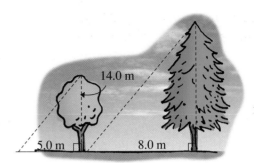

14.0 m

5.0 m   8.0 m

3. How far is it across the river?

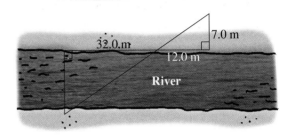

32.0 m   7.0 m

12.0 m

River

4. How high is the support *x* for
the conveyor? The diagram is not
drawn to scale.

12.0 m

x

⟵ 30.0 m ⟶ ⟵15.0 m➤

5. To find the distance AB across
a pond, surveyors measured the
distances shown. Use these
distances to calculate the
distance AB.

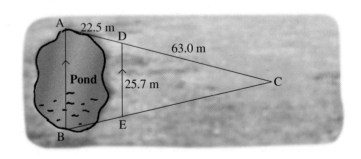

A   22.5 m
D
63.0 m
Pond
25.7 m
C
E
B

6. To determine the distance across
a river, a student uses the sketch
and the measurements shown.
Determine the distance across
the river. The diagram is not
drawn to scale.

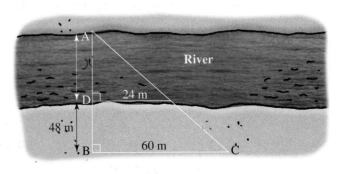

A
River
D   24 m
48 m
B   60 m   C

7. To calculate the height of a building, a student uses the height of a pole and the lengths of the shadows cast by the pole and the building. How tall is the building? The diagram is not drawn to scale.

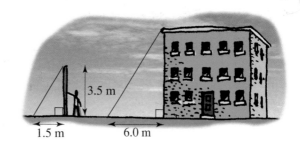

1.5 m   6.0 m   3.5 m

8. A student drew this diagram to determine the length of a bridge over a river. Calculate the length of the bridge.

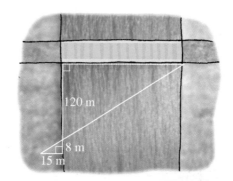

120 m

8 m

15 m

9. A student is 37.5 m from a church. She holds a vertical pencil, 4.8 cm long, 60 mm from her eye. The pencil just blocks the church from her sight. How high is the church?

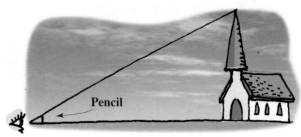

Pencil

10. The shadow of a telephone relay tower is 32.0 m long on level ground. At the same time, a person 1.8 m tall casts a shadow 1.5 m long. What is the height of the tower?

11. A pole 3.8 m high casts a shadow that measures 1.3 m. A nearby tree casts a shadow 7.8 m long. Calculate the height of the tree.

12. A student made a scale drawing of her triangular vegetable garden. Two sides of the garden are 7 m and 9 m. The angle between these sides is 75°. The drawing was a triangle with side lengths 14.0 cm and 18.0 cm, with an angle of 75° between them. The student drew the third side, then measured it. It was 19.7 cm long. What is the length of the third side of the vegetable garden?

Communicating the IDEAS

Write to explain how similar triangles can be used to calculate the height of an object that cannot be measured. Illustrate your explanation with an example.

5.5 Using *The Geometer's Sketchpad*: The Tangent Ratio

In the following *Investigations,* you will learn an important fact about the legs in a right triangle.

Use *The Geometer's Sketchpad*.

The Tangent Ratio in a Right Triangle

SET-UP

- From the File menu, choose New Sketch.
- From the Display menu, choose Preferences. Make sure that Autoshow Labels for Points is checked. Set the angle precision to tenths, and the distance precision to hundredths of a centimetre.

1. To construct right △ABC:

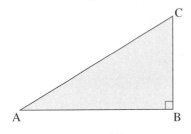

   a) Draw segment AB.

   b) Shift-click point B and segment AB. From the Construct menu, choose Perpendicular Line.

   c) Select the perpendicular line. From the Construct menu, choose Point on Object. The new point should be labelled C. If necessary, move C away from B.

   d) Shift-click points C and A. From the Construct menu, choose Segment.

   e) Select the perpendicular line. From the Display menu, choose Hide Line.

   f) Shift-click points B and C. From the Construct menu, choose Segment.

2. Shift-click points C, A, and B in order. From the Measure menu, choose Angle.

3. Measure the lengths of AB and BC.

4. From the Measure menu, choose Calculate. A calculator appears on the screen.

5. Select the measurement BC by clicking on it. It will appear in the calculator screen.

6. Click on the symbol / for division.

7. Select the measurement AB by clicking on it.

8. Click OK. The ratio $\frac{BC}{AB}$ appears on the screen.

9. Drag point C along the line. What do you notice about the angle measurement? What do you notice about the ratio?

10. The ratio $\frac{BC}{AB}$ is called the tangent of $\angle A$. To see how the ratio changes as the angle changes, make a table:

a) Drag point C until $\angle CAB = 10°$.

b) Shift-click the *measurement* of $\angle CAB$ and the ratio. From the Measure menu, choose Tabulate.

c) Drag point C until $\angle CAB = 20°$. Select the table. From the Measure menu, choose Add Entry.

d) Continue to add entries for $\angle CAB = 30°, 40°, 50°, 60°, 70°$.

e) Describe how the tangent ratio changes as the angle changes.

11. Drag point C until the tangent ratio is 1. What do you notice about $\angle A$? Explain this result.

12. Drag point C until the angle measurement is 25°. To make it exactly 25°, you may need to drag another point. Write the ratio.

13. a) On a scientific calculator, there is a key marked TAN.
Make sure the calculator is in degree mode.
Press DRG until DEG shows in the display.
Press 25 TAN to display the tangent of 25°.

b) On a graphing calculator, there is a key marked TAN.
Make sure the calculator is in degree mode.
Press MODE. Use ▼ ENTER if necessary to highlight **Degree** in the third line. Press CLEAR.
Press TAN 25) ENTER to display the tangent of 25°.

14. Drag point C until $\angle CAB = 43°$. Use a calculator to determine the tangent of 43°.

15. Write a statement to explain the calculator display for the tangent of 43°.

Save the sketch of $\triangle ABC$ for use in *Investigation 2*.

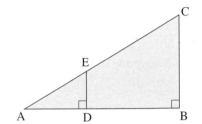
The Tangent Ratio in Similar Triangles

SET-UP

Use △ABC from *Investigation 1*.

1. To construct right △ADE:

   a) Select segment AB. From the Construct menu, choose Point on Object. The new point should be labelled D.

   b) Shift-click point D and segment AB. From the Construct menu, choose Perpendicular Line.

   c) Select the new line and segment AC. From the Construct menu, choose Point at Intersection. The new point should be labelled E.

   d) Select the line through D and E. From the Display menu, choose Hide Line.

   e) Shift-click points D and E. From the Construct menu, choose Segment.

   f) Shift-click points A and D. From the Construct menu, choose Segment.

2. Measure the lengths of DE and AD.

3. Use the Calculate feature to determine the ratio $\frac{DE}{AD}$. What do you notice about the value of this ratio?

4. Explain why $\frac{BC}{AB} = \frac{DE}{AD}$.

5. Drag point D along segment AB. What do you notice about the ratios?

6. Drag point C along segment BC. What do you notice about the ratios?

7. For angle A, the tangent ratio in △ABC is equal to the tangent ratio in △ADE. But the triangles are not the same size. Explain.

Communicating the IDEAS

Write to explain what you learned about the tangent ratio by using *The Geometer's Sketchpad*. Include diagrams with your explanation.

In this house, the main roof has the shape of an isosceles triangle with base angles 20°. The base of the triangle is 22 m.

There is a vertical support every 2 m. How does the builder calculate the length of each support?

If you completed Section 5.5, you do not need to complete the following *Investigation*.

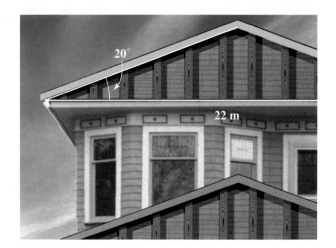

INVESTIGATION

Sides and Angles in Right Triangles: Part I

You will need 1-cm grid paper, a ruler, and a protractor.

1. Along the lines of the grid paper, draw the right triangles shown.

2. Use a protractor to measure ∠A in each triangle. What do you notice?

3. Calculate each ratio:
$$\frac{BC}{AC}, \frac{DE}{AE}, \frac{FG}{AG}, \frac{HI}{AI}.$$
What do you notice? Save your diagram for use in Section 5.9.

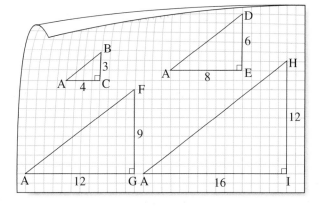

4. Compare your results with those of other students. Did all of you get the same results?

5. How are the triangles related?

In the *Investigation*, you drew four similar right triangles.
The picture at the top of page 249 shows many similar right triangles.
The diagram below shows the four triangles you drew overlapping. Their vertices
are labelled.

To check the triangles are similar, we calculate the ratios of corresponding sides.

In each case, we use the legs: the side
opposite ∠A and the side adjacent to ∠A.

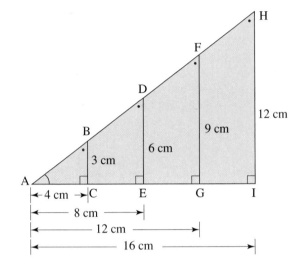

$$\frac{BC}{AC} = \frac{3\ cm}{4\ cm}$$
$$= 0.75$$

$$\frac{DE}{AE} = \frac{6\ cm}{8\ cm}$$
$$= 0.75$$

$$\frac{FG}{AG} = \frac{9\ cm}{12\ cm}$$
$$= 0.75$$

$$\frac{HI}{AI} = \frac{12\ cm}{16\ cm}$$
$$= 0.75$$

All the ratios are equivalent.
The ratios depend only on the measure of
∠A, not on the sizes of the triangles.
We call this ratio the *tangent* of ∠A,
and write it as tan A.

Take Note

The Tangent of an Angle

If ∠A is an acute angle in a right triangle, then

$$\tan A = \frac{\text{length of side } \textit{opposite } \angle A}{\text{length of side } \textit{adjacent to } \angle A}.$$

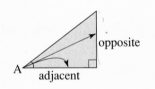

We can use a calculator to determine the tangent of an angle.

Example 1

Determine tan 32°, rounded to 4 decimal places. Draw a diagram to explain
the meaning of the result.

Solution

Using a scientific calculator

Make sure the calculator is in degree mode.
Press 32 [TAN] to display 0.624869351
Rounded to 4 decimal places, tan 32° = 0.6249
We say: "The tangent of 32° is 0.6249."

Using a graphing calculator

Make sure the calculator is in degree mode.
Press [TAN] 32 [ )] [ENTER] to display
.6248693519

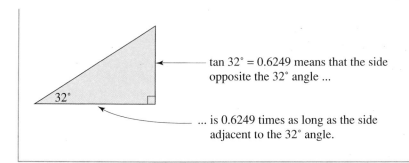

tan 32° = 0.6249 means that the side opposite the 32° angle ...

... is 0.6249 times as long as the side adjacent to the 32° angle.

We can use the tangent ratio to calculate unknown sides in a right triangle.

If we know the angle and the length of one leg in a right triangle, the tangent ratio can be used to determine the length of the other leg.

Example 2

In △ABC, ∠C = 90°, ∠A = 27°, and AC = 5.0 cm
Calculate the length of BC to the nearest tenth of a centimetre.

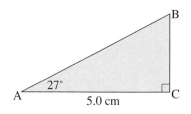

Solution

> **Think ...**
>
> The tangent ratio for ∠A is $\tan A = \dfrac{BC}{AC}$.

Use a scientific calculator.

$\dfrac{BC}{AC} = \tan 27°$ Press 27 [TAN]

Substitute for AC.

$\dfrac{BC}{5.0} \doteq 0.5095$

Multiply each side by 5.0.

BC $\doteq 5.0 \times 0.5095$ Then press [×] 5 [=]

BC $\doteq 2.5476$

The length of BC is 2.5 cm.

Example 3

In △PQR, ∠R = 90°, ∠P = 27°, and QR = 5.0 cm; calculate the length of PR to the nearest tenth of a centimetre.

Solution

Draw a diagram.

> **Think ...**
>
> The tangent ratio for ∠P is $\tan P = \dfrac{QR}{PR}$.

$\dfrac{QR}{PR} = \tan 27°$ Press 27 $\boxed{\text{TAN}}$

Substitute for QR.

$\dfrac{5.0}{PR} \doteq 0.5095$

Multiply each side by PR.

$5.0 \doteq 0.5095 \times PR$

Divide each side by 0.5095.

$\dfrac{5.0}{0.5095} \doteq PR$ Then press $\boxed{1/x}$ $\boxed{\times}$ 5 $\boxed{=}$

$PR \doteq 9.8130$

The length of PR is 9.8 cm.

Compare the results of *Examples 2* and *3*.

In *Example 2*, the length of BC is approximately one-half the length of AC.

In *Example 3*, the length of QR is approximately one-half the length of PR.

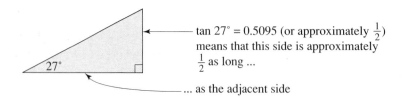

$\tan 27° = 0.5095$ (or approximately $\frac{1}{2}$) means that this side is approximately $\frac{1}{2}$ as long ...

... as the adjacent side

The scientific calculator key sequences are for the *TEXAS INSTRUMENTS TI-34* calculator. The graphing calculator key sequences are for the *TEXAS INSTRUMENTS TI-83* or *TI-83 Plus*. Other calculators may have different keying sequences. Refer to your calculator's manual to find out.

We can use the tangent ratio to solve problems involving right triangles. Sometimes, we also need to use the Pythagorean Theorem.

Example 4

A guy wire supports a tower.
The wire forms an angle of
57° with level ground.
The wire is attached to the
ground 16.5 m from the
base of the tower.

a) At what height is the
guy wire attached
to the tower?

b) How long is the
guy wire?

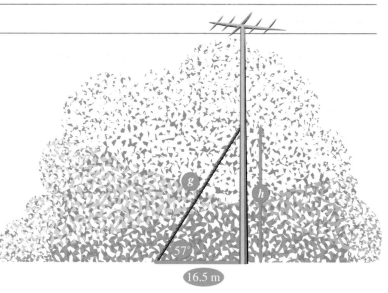

Solution

a) Let *h* metres represent the height at which the guy wire is attached to the tower.

> *Think ...*
> The tangent ratio for 57° is $\tan 57° = \dfrac{h}{16.5}$.

$\dfrac{h}{16.5} = \tan 57°$

$\dfrac{h}{16.5} \doteq 1.5398$

$h \doteq 16.5 \times 1.5398$

$\doteq 25.4077$

The guy wire is attached to the tower at a height of approximately 25.4 m.

b) Let *g* metres represent the length of the guy wire.

> *Think ...*
> We cannot use the tangent ratio because *g* is the hypotenuse of a right triangle. When
> we know 2 legs of a right triangle, we can use the Pythagorean Theorem to calculate
> the hypotenuse.

Use the Pythagorean Theorem.

$g^2 = 16.5^2 + 25.4^2$ Press 16.5 $\boxed{x^2}$ $\boxed{+}$ 25.4 $\boxed{x^2}$ $\boxed{=}$

$= 272.25 + 645.16$

$= 917.41$

$g = \sqrt{917.41}$ Then press $\boxed{\sqrt{x}}$

$\doteq 30.2887$

The guy wire is approximately 30.3 m long.

1. In *Example 1*, suppose your calculator (scientific or graphing) is not in degree mode. Explain how to put it in degree mode.

2. In *Examples 2* and *3*, each right triangle has an angle of 27° and a side of 5 cm. Why are the answers different?

3. In *Example 2*, explain how to use the Pythagorean Theorem to calculate the length of AB.

4. In the solution of *Example 3*, the key $\boxed{1/x}$ was used.
 a) Explain what this key does.
 b) What other key sequence could be used instead?

5. Look at *Examples 2*, *3*, and *4*. In each Think box, the equation is written with the tangent on the left side. In the first line of each solution, the equation is reversed. Explain.

5.6 Exercises

A **1.** In each triangle, name the side:
 a) opposite ∠A **b)** adjacent to ∠A
 i) 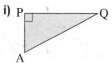 **ii)**

2. For each triangle in exercise 1, what ratio is used to describe tan A?

B **3. a)** Use a calculator to determine the tangent of each angle to 3 decimal places.
 i) tan 35° **ii)** tan 45° **iii)** tan 65°

 b) From part a, why do you think tan 45° is 1?

 c) Will tan 80° be greater or less than 1? Check your prediction with your calculator.

 d) Choose one ratio from part a. Draw a diagram to explain the meaning of the result.

4. *Example 2* was solved using the tangent of ∠A. You can also solve it using the tangent of ∠B.
 a) What is the measure of ∠B?
 b) Solve *Example 2* using tan B.

5. a) Calculate tan A and tan B in each triangle.

i)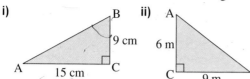

ii)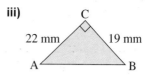

iii)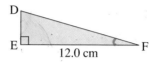

b) Use the Pythagorean Theorem to calculate the length of AB in each triangle in part a. Give the answers to 1 decimal place.

6. Sketch △ABC in which ∠B = 90°, and tan A has each value.

a) $\frac{5}{9}$ b) $\frac{3}{7}$ c) $\frac{4}{11}$ d) $\frac{12}{5}$

7. a) In △DEF, calculate the length of DE for each given angle.

    i) ∠F = 18°   ii) ∠F = 65°
    iii) ∠D = 70°   iv) ∠D = 45°

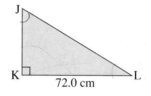

b) Use the Pythagorean Theorem to calculate the length of DF in each triangle in part a.

8. a) In △JKL, calculate the length of JK for each given angle.

    i) ∠L = 24°   ii) ∠L = 75°
    iii) ∠J = 50°   iv) ∠J = 22°

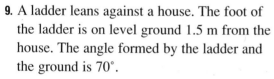

b) Use the Pythagorean Theorem to calculate the length of JL in each triangle in part a.

9. A ladder leans against a house. The foot of the ladder is on level ground 1.5 m from the house. The angle formed by the ladder and the ground is 70°.

a) Calculate how high up the house the ladder reaches.

b) Calculate the length of the ladder.

10. Refer to the roof picture on page 249. The centre of the first support is 2 m from the edge of the roof. Calculate the height of the centre of the first support.

4.121

11. The top of a communications tower is 450 m above sea level. From a ship at sea, its angle of elevation is 4°. The diagram is not to scale.

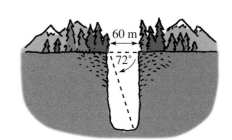

450 m

a) Look at the diagram. Write to explain what is meant by "angle of elevation."

b) How far is the ship from the tower?

c) Suppose the angle of elevation were 8°. How far would the ship be from the tower?

12. A gorge with a rectangular cross section is 60 m wide. The angle of depression of a bottom corner when viewed from the opposite edge is 72°.

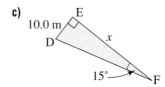

60 m

72°

a) Look at the diagram. Write to explain what is meant by "angle of depression."

b) How deep is the gorge?

c) Suppose the angle of depression were 80°. How deep would the gorge be?

13. Calculate each value of x.

a)

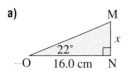

M

x

22°

O   16.0 cm   N

b)

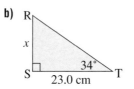

R

x

S   34°

23.0 cm   T

c)

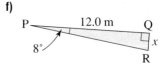

E

10.0 m

D   x

15°   F

d)

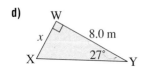

W

x   8.0 m

X   27°   Y

e)

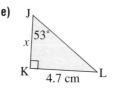

J

53°

x

K   4.7 cm   L

f)

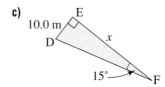

P   12.0 m   Q

8°   x

R

g)

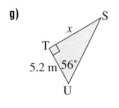

S

x

T

5.2 m   56°

U

h)

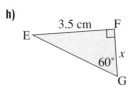

3.5 cm   F

E

60°   x

G

i)

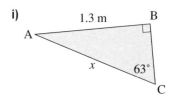

1.3 m   B

A

x   63°

C

Communicating *the* IDEAS

Explain why the tangent of an angle is useful. Make up an example to illustrate your explanation.

Francine is a carpenter. She is installing baseboards in an old house. She must set her saw to cut the correct angle for the section at the bottom of the stairs.

Francine takes measurements and, from these, she knows that tan A is $\frac{23}{25} = 0.92$. She needs to determine the measure of $\angle A$. She can do this in different ways.

Use a scale drawing.

- Use grid paper. Draw right $\triangle ABC$ with legs 23 cm and 25 cm, and $\angle B = 90°$.
- Use a protractor to measure $\angle A$.

Use a scientific calculator.

- Check the calculator is in degree mode.
- Press 23 $\boxed{÷}$ 25 $\boxed{=}$ $\boxed{\text{TAN}^{-1}}$ to display 42.61405597.

Use a graphing calculator.

- Check the calculator is in degree mode.
- Press $\boxed{\text{TAN}^{-1}}$ 23 $\boxed{÷}$ 25 $\boxed{)}$ $\boxed{\text{ENTER}}$ to display 42.61405597.

To the nearest degree, $\angle A = 43°$
Note that $\boxed{\text{TAN}^{-1}}$ is obtained from $\boxed{\text{2nd}}$ $\boxed{\text{TAN}}$.

Example 1

Calculate the measure of $\angle A$.

Solution

> *Think ...*
> The tangent of $\angle A$ is $\frac{3}{4}$.

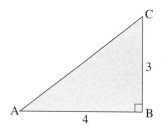

$\tan A = \frac{3}{4}$

Make sure the calculator is in degree mode.
Press 3 $\boxed{÷}$ 4 $\boxed{=}$ $\boxed{\text{TAN}^{-1}}$ to display 36.86989765
To the nearest degree, $\angle A = 37°$

Example 2

In right △ABC:

a) Calculate tan A, and ∠A to the nearest degree.

b) Calculate tan C, and ∠C to the nearest degree.

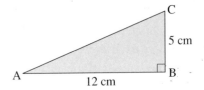

Solution

a) The side opposite ∠A is BC. The adjacent side is AB.

$$\tan A = \frac{BC}{AB}$$
$$= \frac{5}{12}$$ Press 5 [÷] 12 [=]
$$\doteq 0.1466$$ Then press [TAN⁻¹]

$\angle A \doteq 22.6198°$

∠A is 23° to the nearest degree.

b) The side opposite ∠C is AB. The adjacent side is BC.

$$\tan C = \frac{AB}{BC}$$
$$= \frac{12}{5}$$ Press 12 [÷] 5 [=]
$$= 2.4$$ Then press [TAN⁻¹]

$\angle C \doteq 67.3801°$

∠C is 67° to the nearest degree.

We can check the results of *Example 2*.
The sum of the angles in △ABC should be 180°.
$23° + 67° + 90° = 180°$
The sum of the angles is 180°, so the results are correct.

Discussing the IDEAS

1. On page 257, which of Francine's methods to determine the measure of ∠A would result in the more accurate answer? Which method would be quicker to use? Explain.

2. a) In *Example 2*, how are ∠A and ∠C related? Explain.

   b) How are tan A and tan C related? Explain.

   c) Do you think these relationships are true for all right triangles? Experiment with a triangle to test your predictions.

A **1.** Determine ∠A to the nearest degree.

a) tan A = 0.250 b) tan A = 0.709 c) tan A = 1.365 d) tan A = 3.271

e) tan A = 0.549 f) tan A = 0.933 g) tan A = 2.050 h) tan A = 4.556

2. Determine ∠A to the nearest degree.

a) $\tan A = \frac{1}{2}$ b) $\tan A = \frac{2}{5}$ c) $\tan A = \frac{4}{3}$ d) $\tan A = \frac{5}{4}$

e) $\tan A = \frac{3}{4}$ f) $\tan A = \frac{3}{5}$ g) $\tan A = \frac{5}{2}$ h) $\tan A = \frac{10}{7}$

B **3.** In each triangle

a) Calculate tan A, and ∠A to the nearest degree.

b) Calculate tan C, and ∠C to the nearest degree.

c) Check that the sum of the three angles is 180°.

i)
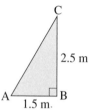

ii)
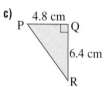

4. In each triangle, calculate tan R and ∠R to the nearest degree.

a)
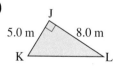

b)
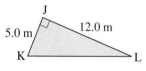

c)
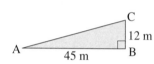

5. In each triangle, calculate tan K and ∠K to the nearest degree.

a)
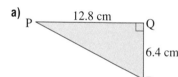

b)

c)
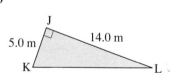

d)
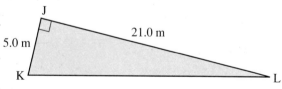

6. An extension ladder is leaning against a wall.
The foot of the ladder is 3.6 m from the wall.
The top of the ladder reaches 8.6 m up the wall.

a) Calculate the measure of the angle formed
by the ladder and the ground.

b) Calculate the length of the ladder.

7. Calculate tan A and ∠A to the nearest degree, then calculate tan B and ∠B
to the nearest degree. Check the results by showing that the sum of the three
angles in each triangle is 180°.

a)
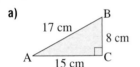

b)
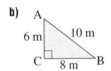

c)
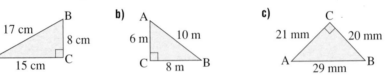

8. You will need three sheets of paper, a ruler, and a protractor.

a) Measure the length and the width of the paper.

b) On one sheet, use your ruler to draw the diagonal (below left). Calculate the
measure of the angle formed by the diagonal and the bottom edge of the paper.

c) Take another sheet and fold it in half. Draw a line from one corner to the
midpoint of the opposite side (below centre). Calculate the measure of the
angle formed by this line and the bottom edge of the paper.

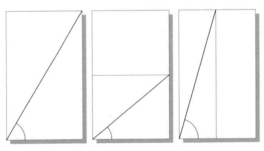

d) Take a third sheet of paper and fold it in half the other way. Draw a line
from one corner to the midpoint of the top edge (above right). Calculate the
measure of the angle formed by this line and the bottom edge of the paper.

e) Use a protractor to check the angles in parts b to d.

9. A person is cutting a piece of trim to edge the stairs. She needs to cut the trim at an angle so that the trim lies flat on the floor. What is the measure of ∠A?

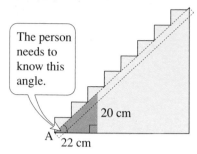

The person needs to know this angle.

20 cm

A 22 cm

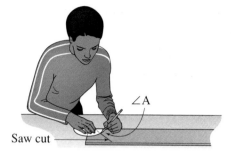

∠A

Saw cut

10. A guy wire supports a tower. The wire is attached to the tower at a height of 25.0 m. The guy wire is attached to the ground 8.4 m from the base of the tower. Calculate the measure of the angle formed by the guy wire and the ground.

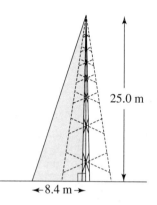

25.0 m

← 8.4 m →

11. You will need grid paper, a ruler, and a protractor.

a) Draw 4 segments on the grid paper as shown.

b) For each segment, complete a right triangle by using the lines on the grid paper. The first one is done for you.

c) Determine the length of each leg by counting the squares on the grid paper.

d) Calculate the measure of the angle formed by each segment and the horizontal leg: ∠A, ∠B, ∠C, and ∠D.

e) Check your answers by using the protractor.

f) Compare your results with those of your classmates. Did all of you draw the same triangles? Explain.

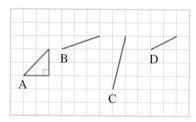

Communicating *the* IDEAS

Suppose you have a photograph of a roof. You do not have a protractor. Explain how you could use the tangent ratio to calculate the angle between the roof line and the horizontal. Draw a diagram to illustrate your explanation.

There are two trigonometric ratios, called the sine ratio and the cosine ratio, that involve the hypotenuse of a right triangle. You will explore these in the following investigations.

Use *The Geometer's Sketchpad*.

INVESTIGATION 1

The Sine Ratio in a Right Triangle

SET-UP

- From the File menu, choose New Sketch.

- From the Display menu, choose Preferences. Make sure Autoshow Labels for Points is checked. Set the angle precision to tenths, and the distance precision to hundredths of a centimetre.

1. To construct right △ABC, see *Investigation 1*, exercise 1, page 246.

2. Measure ∠CAB, BC, and AC.

3. To calculate the ratio $\frac{BC}{AC}$, see *Investigation 1*, exercises 4 to 8, pages 246, 247.

4. Drag point C along the line. What do you notice about the angle measurement? What do you notice about the ratio? This ratio is called the sine of ∠A.

5. Construct points D and E and segment DE as in *Investigation 2*, page 248.

6. Measure DE and AE.

7. Use the Calculate feature to determine the ratio $\frac{DE}{AE}$.

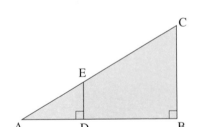

8. Drag point C along the line. What do you notice about the ratios $\frac{BC}{AC}$ and $\frac{DE}{AE}$? Explain.

9. Drag point C until ∠CAB = 37°. Record the ratio.

10. Use a scientific or graphing calculator. Use the ⌊ SIN ⌋ key to calculate the sine of 37°. Record its value.

11. Drag point C along the line until ∠CAB = 62°. Use a calculator to determine the sine of 62°.

12. Write a statement to explain the calculator display for the sine of 62°.

Save the sketch of △ABC for use in *Investigation 2*.

The Cosine Ratio in a Right Triangle

Use △ABC from *Investigation 1*.

1. Measure segments AB and AD.

2. Use the Calculate feature to determine the ratios $\frac{AB}{AC}$ and $\frac{AD}{AE}$.
   What do you notice about the values of these ratios? These ratios are called the cosine of ∠A. Explain why $\frac{AB}{AC} = \frac{AD}{AE}$.

3. Drag point C along the line until ∠CAB = 25°.
   Record the values of the ratios $\frac{AB}{AC}$ and $\frac{AD}{AE}$.

4. Use a scientific or graphing calculator. Use the ⌊ COS ⌋ key to calculate the cosine of 25°. Record its value.

5. Drag point C along the line until ∠CAB = 43°. Use a calculator to determine the cosine of 43°.

6. Write a statement to explain the calculator display for the cosine of 43°.

Communicating *the* IDEAS

Write to explain what you learned about the sine and cosine ratios by using *The Geometer's Sketchpad*. Include diagrams with your explanation.

When firefighters reach the site of a fire, they assess the height of the building. Their truck has an aerial ladder that extends to 30.5 m. Suppose the angle between the ladder and the ground is 77°. What is the highest window the ladder can reach?

The tangent ratio does not apply because we do not know either of the legs in the right triangle. We know the length of the hypotenuse.

If you completed Section 5.8, you do not need to complete the following *Investigation*.

INVESTIGATION

Sides and Angles in Right Triangles: Part II

You will need a ruler and the triangles you drew in the *Investigation*, page 249.

1. For each triangle, measure and record the length of the hypotenuse: AB, AD, AF, AH.

2. Calculate each ratio:
$$\frac{BC}{AB}, \frac{DE}{AD}, \frac{FG}{AF}, \frac{HI}{AH}.$$
What do you notice?

3. Calculate each ratio:
$$\frac{AC}{AB}, \frac{AE}{AD}, \frac{AG}{AF}, \frac{AI}{AH}$$
What do you notice?

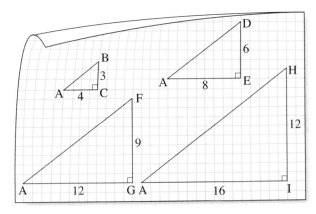

4. Compare your results in exercises 2 and 3 with those of other students. Did all of you get the same results?

5. a) Explain why the ratios in exercise 2 are the same.

b) Explain why the ratios in exercise 3 are the same.

In the *Investigation*, you compared the ratios of corresponding sides in similar triangles. This diagram shows the lengths of the sides for the overlapping triangles.

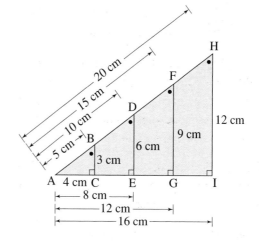

These are the ratios of the sides opposite ∠A and the hypotenuse:

$$\frac{BC}{AB} = \frac{3 \text{ cm}}{5 \text{ cm}} \qquad \frac{DE}{AD} = \frac{6 \text{ cm}}{10 \text{ cm}}$$
$$= 0.6 \qquad\qquad = 0.6$$

$$\frac{FG}{AF} = \frac{9 \text{ cm}}{15 \text{ cm}} \qquad \frac{HI}{AH} = \frac{12 \text{ cm}}{20 \text{ cm}}$$
$$= 0.6 \qquad\qquad = 0.6$$

All the ratios are equivalent. We call this ratio the *sine* of ∠A, and we write it as sin A.

These are the ratios of the sides adjacent to ∠A and the hypotenuse:

$$\frac{AC}{AB} = \frac{4 \text{ cm}}{5 \text{ cm}} \qquad \frac{AE}{AD} = \frac{8 \text{ cm}}{10 \text{ cm}}$$
$$= 0.8 \qquad\qquad = 0.8$$

$$\frac{AG}{AF} = \frac{12 \text{ cm}}{15 \text{ cm}} \qquad \frac{AI}{AH} = \frac{16 \text{ cm}}{20 \text{ cm}}$$
$$= 0.8 \qquad\qquad = 0.8$$

All the ratios are equivalent. We call this ratio the *cosine* of ∠A, and we write it as cos A.

The Sine and Cosine of an Angle

If ∠A is an acute angle in a right triangle, then

$$\sin A = \frac{\text{length of side } \textit{opposite } \angle A}{\text{length of hypotenuse}}$$

$$\cos A = \frac{\text{length of side } \textit{adjacent to } \angle A}{\text{length of hypotenuse}}$$

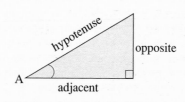

We can use a calculator to determine the sine and cosine of an angle.

Example 1

Determine sin 32° and cos 32°, rounded to 4 decimal places. Draw a diagram to explain the meaning of the results.

Solution

Using a scientific calculator

Make sure the calculator is in degree mode.
Press 32 [SIN] to display 0.529919264
Press 32 [COS] to display 0.848048096

Using a graphing calculator

Make sure the calculator is in degree mode.
Press [SIN] 32 [ )] [ENTER] to display
.5299192642
Press [COS] 32 [ )] [ENTER] to display
.8480480962

Rounded to 4 decimal places,
$\sin 32° = 0.5299$
We say: "The sine of 32°
is 0.5299."

Rounded to 4 decimal places,
$\cos 32° = 0.8480$
We say: "The cosine of 32°
is 0.8480."

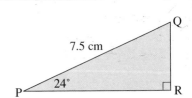

$\sin 32° = 0.5299$ means that
the side opposite the 32° angle
is 0.5299 times as long as
the hypotenuse.

$\cos 32° = 0.8480$ means that the side
adjacent to the 32° angle is 0.8480 times
as long as the hypotenuse.

The sine and cosine ratios can be used to find the lengths of the legs in a right
triangle when the hypotenuse and one angle are known.

Example 2

In right $\triangle PQR$, $\angle R = 90°$, $\angle P = 24°$, and
$PQ = 7.5$ cm; calculate the lengths of RQ and
PR to the nearest tenth of a centimetre.

7.5 cm

24°

P   R   Q

Solution

Think ...

The sine ratio for $\angle P$ is $\sin P = \dfrac{RQ}{PQ}$.
Use the sine ratio to calculate RQ.

The cosine ratio for $\angle P$ is $\cos P = \dfrac{PR}{PQ}$.
Use the cosine ratio to calculate PR.

Use a scientific calculator.
$\dfrac{RQ}{PQ} = \sin 24°$ Press 24 [SIN]

Substitute for PQ.
$\dfrac{RQ}{7.5} \doteq 0.4067$

Multiply each side by 7.5.

$RQ \doteq 7.5 \times 0.4067$ Then press $\boxed{\times}$ 7.5 $\boxed{=}$
$RQ \doteq 3.0505$

$\dfrac{PR}{PQ} = \cos 24°$ Press 24 $\boxed{\text{cos}}$

Substitute for PQ.

$\dfrac{PR}{7.5} \doteq 0.9135$

Multiply each side by 7.5.

$PR \doteq 7.5 \times 0.9135$ Then press $\boxed{\times}$ 7.5 $\boxed{=}$
$PR \doteq 6.8515$

RQ is 3.1 cm and PR is 6.9 cm, to the nearest tenth of a centimetre.

Example 3

A 6.1-m ladder leans against a wall. The angle formed by the ladder and the ground is 71°.

a) How far is the foot of the ladder from the wall?

b) How far up the wall does the ladder reach?

Solution

Let b metres represent the distance from the foot of the ladder to the wall.
Let h metres represent the distance the ladder reaches up the wall.

> **Think ...**
>
> The ladder is the hypotenuse of a right triangle.
>
> The cosine of 71° is $\dfrac{b}{6.1}$.   The sine of 71° is $\dfrac{h}{6.1}$.
>
> Use the cosine ratio to calculate b.   Use the sine ratio to calculate h.

a) $\dfrac{b}{6.1} = \cos 71°$ b) $\dfrac{h}{6.1} = \sin 71°$

$\dfrac{b}{6.1} \doteq 0.3255$ $\dfrac{h}{6.1} \doteq 0.9455$

$\ b \doteq 6.1 \times 0.3255$ $\ h \doteq 6.1 \times 0.9455$

$\ b \doteq 1.9859$ $\ h \doteq 5.7676$

The foot of the ladder is approximately 2.0 m from the wall.   The ladder reaches approximately 5.8 m up the wall.

1. In *Example 2*

a) What is the measure of ∠PQR?

b) Determine the sine of ∠PQR. Compare it to the sine and cosine of ∠QPR. What relationship do you notice?

c) Do you think this relationship will be true for other angles that add to 90°? Check by calculating the sine and cosine of two other angles that add to 90°.

2. In *Example 2*, explain how to use the Pythagorean Theorem to calculate the length of PR.

3. In *Example 3b*, explain how to use the Pythagorean Theorem to calculate how far up the wall the ladder reaches.

5.9 Exercises

A 1. In each triangle, name the side:

a) opposite ∠A   b) adjacent to ∠A   c) that is the hypotenuse

i) R ... G ... A

ii) L ... Q ... A

2. For each triangle in exercise 1, what ratio is used to express each value?

a) sin A   b) cos A

B 3. Use a calculator. Determine the sine and cosine of each angle to 3 decimal places.

a) 76°   b) 43°   c) 19°   d) 81°   e) 62°

f) 14°   g) 47°   h) 71°   i) 9°   j) 28°

4. Determine each value of x. Use the Pythagorean Theorem to determine each value of y.

a)
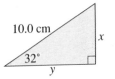

10.0 cm, x, 32°, y

b)
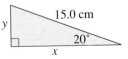

15.0 cm, y, 20°, x

c)
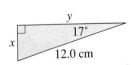

y, 17°, x, 12.0 cm

d)
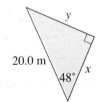

y, 20.0 m, 48°, x

5. a) Sketch △ABC in which ∠B = 90°, and sin A has each value.

    i) $\frac{5}{11}$ **ii)** $\frac{4}{9}$ **iii)** $\frac{3}{5}$ **iv)** $\frac{7}{10}$

  b) Sketch △ABC in which ∠B = 90°, and cos A has each value in part a.

6. In △ABC (below left), ∠B = 90°, ∠A = 53°, and AC = 25 m; determine the lengths of AB and BC to 1 decimal place.

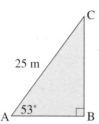
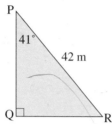

7. In △PQR (above right), ∠Q = 90°, ∠P = 41°, and PR = 42 m; determine the lengths of PQ and QR to 1 decimal place.

8. a) Use a calculator to determine each value to 3 decimal places.

    i) sin 58°   **ii)** cos 40°   **iii)** cos 60°   **iv)** sin 18°

  b) Choose one ratio in part a. Draw a diagram to explain the meaning of the result.

9. A 10.0-m ladder leans against a vertical wall at an angle of 73° (below left).

  a) Calculate the height the ladder reaches up the wall.

  b) Calculate the distance from the foot of the ladder to the wall.

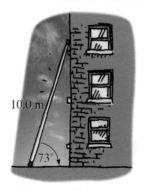
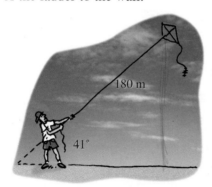

10. A kite has a string 180 m long (above right). The string makes an angle of 41° with the ground. Determine the height of the kite.

11. In △PQR, ∠Q = 90° and PR = 15.0 cm; calculate the lengths of PQ and QR for each given angle.

  a) ∠P = 58°   **b)** ∠R = 42°

12. Recall the situation for the firefighters on page 264.

  a) How high up the building can the ladder reach?

  b) How far will the foot of the ladder be from the building?

13. A snow plow has a 3.2-m blade set at an angle of 30°. How wide a path will the snow plow clear?

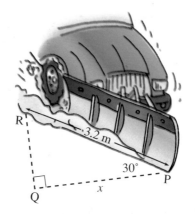

14. A storm causes some 14.0-m hydro poles to lean over. One pole leans at an angle of 72° to the ground. How high is the top of the pole from the ground?

15. When a ship is at T, the navigator observes a lighthouse L due west on the shore. The ship sails 19.5 km north to point S. The navigator measures ∠TSL and finds that it is 33°.

a) How far is the ship from the lighthouse now?

b) Use the Pythagorean Theorem to calculate the distance LT.

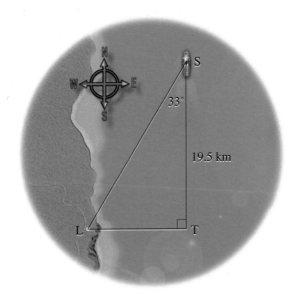

Communicating *the* IDEAS

Sometimes people get the sine and cosine ratios mixed up. Make up a memory aid to help people remember that the sine ratio involves the side opposite the angle and the cosine ratio involves the side adjacent to the angle. Write about your memory aid.

Calculating Angles in a Right Triangle: Part II

In many situations, it is important to know the measure of an angle. It is not always practical to use a protractor. When you know the hypotenuse and one leg in a right triangle, you can use the cosine or sine ratio to determine an angle.

To use the cosine ratio, you need the lengths of:
- the hypotenuse
- the side adjacent to the angle

To use the sine ratio, you need the lengths of:
- the hypotenuse
- the side opposite the angle

Example 1

Two students started a painting company to make money during the summer. They bought a ladder that extends from 7 m to 10 m. For safety, the angle between the ladder and the ground should be between 60° and 78°.

The ladder is set at a length of 8 m. It is placed to reach the top floor window of a house. The foot of the ladder is 2 m from the wall. Is it safe to climb the ladder?

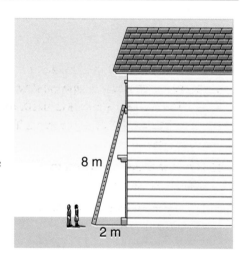

Solution

Draw right △ABC to represent the situation.

> **Think ...**
> We need to determine ∠A.
> We know AB, which is adjacent to ∠A; and AC, the hypotenuse.
> We will use the cosine ratio.

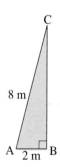

$$\cos A = \frac{AB}{AC}$$
$$\cos A = \frac{2}{8}$$
$$= 0.25$$

Use a scientific calculator to determine ∠A.

Check the calculator is in degree mode.

Press 2 $\div$ 8 $=$ cos⁻¹ to display 75.5224878.

To the nearest degree, ∠A = 76°; this is between 60° and 78°.
The ladder is safe to climb.

Example 2

A sign above a store is 9 m above the ground. Use the safety regulation in
Example 1. Can the sign be painted using a 10-m ladder?

Solution

For safety, the angle between the ladder and the ground must be
between 60° and 78°. Draw a right triangle to represent the situation.

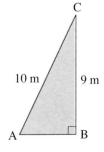

> **Think ...**
>
> We need to determine ∠A.
> We know BC, which is opposite ∠A; and AC, the hypotenuse.
> We will use the sine ratio.

$\sin A = \dfrac{BC}{AC}$

$\sin A = \dfrac{9}{10}$

$\qquad = 0.9$

Use a scientific calculator to determine ∠A.

Check the calculator is in degree mode.

Press 9 $\div$ 10 $=$ SIN⁻¹ to display 64.158067.

To the nearest degree, ∠A = 64°; this is between 60° and 78°.
The ladder is safe to climb. The sign can be painted using a 10-m ladder.

Discussing the IDEAS

1. Explain how to complete *Examples 1* and *2* using the tangent ratio.

2. In *Example 2*, explain how to use the Pythagorean Theorem to calculate the distance
   from the foot of the ladder to the wall.

A **1.** Use a calculator to determine ∠A to the nearest degree.

    a) sin A = 0.602   **b)** cos A = 0.872   **c)** tan A = 1.033

    d) sin A = 0.951   **e)** cos A = 0.201   **f)** tan A = 4.316

2. Determine ∠K to the nearest degree.

    a) cos K = 0.425   **b)** sin K = 0.367   **c)** sin K = 0.519

    d) cos K = 0.785   **e)** sin K = 0.910   **f)** cos K = 0.612

3. Determine ∠A to the nearest degree.

    a) $\sin A = \frac{2}{3}$ **b)** $\cos A = \frac{1}{4}$ **c)** $\cos A = \frac{5}{8}$

    d) $\sin A = \frac{3}{7}$ **e)** $\cos A = \frac{4}{5}$ **f)** $\sin A = \frac{2}{9}$

4. In each diagram, state which ratio should be used to determine ∠A.
Then calculate ∠A to the nearest degree.

a)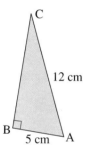

b)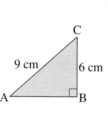

c)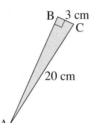

d)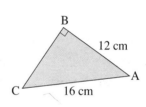

B **5.** In each triangle, calculate sin P then ∠P to 1 decimal place.

a)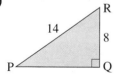

b)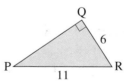

6. For each triangle in exercise 5, determine the measure of ∠R.

7. Calculate ∠A in each triangle to the nearest degree.

a)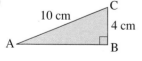

b)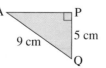

c)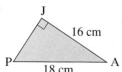

d)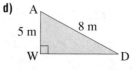

8. In right △PQR, ∠Q = 90°, PR = 4.0 cm, and QR = 3.0 cm

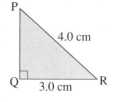

a) Determine the measures of ∠P and ∠R to the nearest degree.

b) Use trigonometry to determine the length of PQ to the nearest tenth of a centimetre.

c) Use the Pythagorean Theorem to check the length of PQ.

9. In △PQR, ∠Q = 90° and PQ = 15.0 cm; determine the measures of ∠P and ∠R in each case, to the nearest degree.

a) when PR = 28.0 cm   b) when QR = 25.0 cm

10. A storm caused some 15.0-m hydro poles to lean over. The top of one pole is 12.0 m above the ground. Calculate the measure of the angle between this pole and the ground.

11. Cars, vans, and trucks have windshields at different angles. Since windshields are rounded, it is difficult to use a protractor to measure the angles at which they are set. You can use your skills with trigonometry to calculate and compare the angles.

You will need a measuring tape and a calculator.

a) Measure the sloping height of the windshield, AC.

b) Measure the vertical height of the windshield, AB.

c) Use the lengths in parts a and b to determine the angle the windshield makes with the horizontal, ∠ACB.

d) Repeat parts a to c for a different vehicle.

e) Compare your results with those of other students.

12. A guy wire is 15.0 m long. It supports a vertical television tower. The wire is fastened to the ground 9.6 m from the base of the tower.

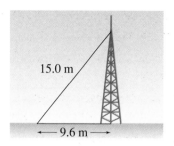

a) Calculate the measure of the angle formed by the guy wire and the ground.

b) Use the Pythagorean Theorem to calculate how far up the tower the guy wire is.

13. A 6.0-m ladder is leaning against a wall. The foot of the ladder is 1.8 m from the wall.

a) Calculate the measure of the angle formed by the ladder and the ground.

b) Use the Pythagorean Theorem to calculate how high up the wall the ladder reaches.

14. A truck travels 6 km up a mountain road. The change in height is 1250 m. Angle CAB is the angle of inclination of the road.

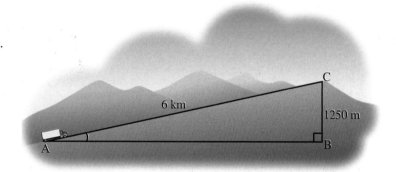

a) Write to explain what is meant by the "angle of inclination."

b) What is the measure of the angle of inclination?

15. We can only use the cosine and sine ratios in a right triangle. Any triangle can be cut to form two right triangles by drawing an altitude. When the triangle is isosceles, the two right triangles are congruent. This roof has the shape of an isosceles triangle. The altitude is marked.

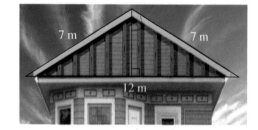

a) Calculate the measure of the angle between the sloping edge of the roof and the horizontal.

b) Determine the measure of the angle at the peak of the roof.

Communicating *the* IDEAS

Choose one trigonometric ratio. Explain what you need to know to be able to use that ratio to determine the measure of an angle. Explain why you might choose that ratio to determine an angle, instead of using a protractor.

5.11 Measuring Inaccessible Heights

Astronomers use trigonometry to measure the distances to the moon and stars. You can use trigonometry to measure inaccessible heights, such as the height of your school. You can use a *clinometer* to measure the angle between a horizontal line and the line of sight to the top of an object.

Work with a partner to make a clinometer.

To make and Use a Clinometer

You will need an enlarged photocopy of a 180° protractor, a rectangular piece of heavy cardboard, a straw, a needle, strong thread, a small weight such as a washer, scissors, glue, and tape.

- Glue the copy of the 180° protractor to the piece of cardboard. Carefully cut around the edge of the protractor.

- Use the needle to pull the thread through the cardboard at the vertex of the 90° angle of the protractor. Tape one end of the thread to the back of the cardboard. Attach the weight to the other end of the thread.

- Tape the straw along the straight edge of the protractor, as shown.

- Hold the protractor with the straight edge on top and the weight hanging down.

- Look through the straw to the top of the object you want to measure.

- Have your partner record the acute angle measurement where the thread lies against the protractor.

Mary and Tanya are measuring the height of their school flagpole. Tanya looks through the straw on the clinometer and Mary reads 40° from the scale on the protractor.

Since the angle between the thread and the straw is 40°, the angle between the horizontal and the line of sight to the flagpole is 50°.

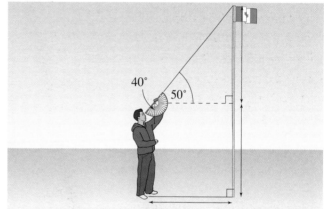

The students measure Tanya's distance from the flagpole. They do this by counting steps from Tanya's position to the flagpole. To get an average step length:

- The students marked a point on the floor of a hallway.
- They measured 4 m, then made another mark.
- Mary walked from one point to the other, counting the number of steps. She recorded this number.
- Mary repeated the walk two more times, recording the number of steps each time.
- The students added the three numbers, then divided by 3. This produces a mean step count for walking 4 m.

- To calculate the step length, they divided 4 m by the mean step count. For example, if the mean step count was 9, the step length is

$$\frac{4\ m}{9} \doteq 0.44\ m$$

The step length is approximately 44 cm.
Mary took 18 steps to walk from Tanya's position to the flagpole.
The distance is $18 \times 0.44\ m = 7.92\ m$.

Tanya and Mary used trigonometry to calculate the height of the flagpole. They drew a diagram, and labelled it with the unknown height, x, and their measurements. When Tanya held the clinometer, the distance from the ground to the centre of the clinometer was 1.60 m.

The students used the tangent ratio to calculate x.

$$\tan 50° = \frac{x}{7.92} \qquad\qquad \text{Press } 50\ \boxed{\text{TAN}}$$

$$1.192 \doteq \frac{x}{7.92}$$

Multiply each side by 7.92.

$$7.92 \times 1.192 \doteq x$$

$$x \doteq 9.439$$

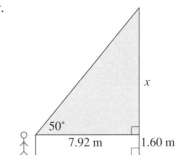

The students added 1.60 m to the length x, to determine the height of the flagpole.

$$9.439 + 1.60 = 11.039$$

The flagpole is approximately 11.0 m tall.

1. Choose a tall object in your area. It could be a tree, a flagpole, an apartment building, or a lamppost.

   a) Determine the distance from you to the object using the step count method, or a measuring tape.

   b) Use a clinometer and trigonometry to determine the height of the object.

Communicating *the* IDEAS

Write a report that summarizes the results of your investigation. Include diagrams and calculations in your report.

Solving Problems Using More than One Triangle

Trigonometry was originally developed to solve navigation problems. Today, it is a powerful tool used to solve many problems. Some of the most interesting problems involve more than one right triangle.

Example

Two students want to determine the heights of two buildings. They stand on the roof of the shorter building.

The students use a clinometer to measure the angle of elevation of the top of the taller building. The angle is 50°.

From the same position, the students measure the angle of depression of the base of the taller building. The angle is 30°.

The students then measure the horizontal distance between the two buildings. The distance is 43.3 m. The students drew this diagram.

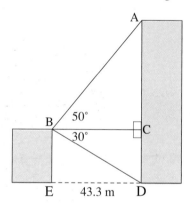

How tall is each building?

Solution

Since the distance between the buildings is measured horizontally,
BC = ED = 43.3 m

> **Think ...**
>
> In right △BCD, we know ∠B and side BC, which is adjacent to ∠B. We need to determine CD, which is opposite ∠B. Use the tangent ratio; the tangent of ∠B.

In △BCD,

$$\tan B = \frac{CD}{BC}$$

Substitute for ∠B and BC.

$$\tan 30° = \frac{CD}{43.3}$$

Multiply each side by 43.3.

$43.3 \times \tan 30° = CD$ Press 43.3 ⊠ × ⊡ 30 ⊡ TAN ⊡ = ⊡

$24.9993 \doteq CD$

$CD \doteq 25.0$

From the diagram, BE = CD = 25.0

The shorter building is about 25 m tall.

Think ...

In right △ABC, we know ∠B and side BC, which is adjacent to ∠B. We need to determine AC, which is opposite ∠B. Use the tangent ratio; the tangent of ∠B.

In △ABC,

$$\tan B = \frac{AC}{BC}$$

Substitute for ∠B and BC.

$$\tan 50° = \frac{AC}{43.3}$$

Multiply each side by 43.3.

$43.3 \times \tan 50° = AC$ Press 43.3 ⊠ × ⊡ 50 ⊡ TAN ⊡ = ⊡

$51.6029 \doteq AC$

$AC \doteq 51.6$

From the diagram, the height of the taller building is CD + AC = 25.0 + 51.6

$= 76.6$

The taller building is about 77 m tall.

Discussing *the* IDEAS

1. Why is it important to mark all right angles when drawing a diagram to model a problem?

2. In the *Example*

  a) Explain what is meant by the "angle of elevation."

  b) Explain what is meant by the "angle of depression."

1. Solve each equation.

a) $\frac{AB}{3} = 5$

b) $6 = \frac{CD}{7}$

c) $0.325 = \frac{EF}{6}$

d) $1.263 = \frac{GH}{6.5}$

e) $\frac{JK}{2.1} = 0.433$

f) $\frac{PQ}{1.7} = 0.524$

2. Solve each equation. Give the answer to 3 decimal places where necessary.

a) $\frac{3}{AB} = 5$

b) $6 = \frac{7}{CD}$

c) $0.324 = \frac{6}{EF}$

d) $1.263 = \frac{6.5}{GH}$

e) $\frac{2.1}{JK} = 0.433$

f) $\frac{1.7}{PQ} = 0.524$

5.12 Exercises

A **1.** Calculate the length of BC in each diagram, to 1 decimal place.

a)

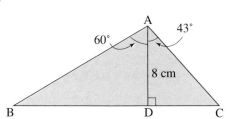

b)

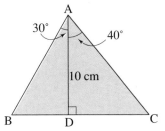

2. Calculate the measure of ∠ABC in each diagram, to the nearest degree.

a)

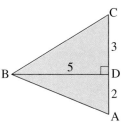

b)

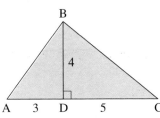

c)

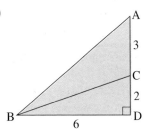

B **3.** Two buildings are 31.7 m apart. From the 12th floor of the shorter building, the angle of elevation to the top of the taller building is 27°. The angle of depression to the base of the taller building is 48°. What is the height of the taller building?

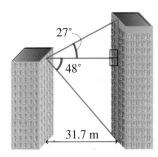

4. A vertical microwave tower is supported by a series of guy wires. The anchor point of one guy wire is 9.8 m from the base of the tower. The angle of inclination of this wire is 37°. The angle of elevation of the top of the tower from this anchor point is 58°.

a) How tall is the tower?

b) How far from the top of the tower is the guy wire attached to the tower?

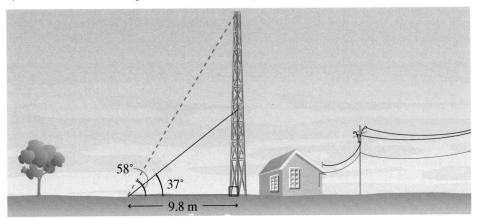

5. One of the world's longest suspension bridges crosses the Humber Estuary in England. The towers of the bridge reach about 135 m above the level of the bridge. The angles of elevation of the towers measured from the centre of the bridge and either end are 10.80° and 18.65°, respectively. How long is the bridge?

6. From the top of a 110-m fire tower, a fire ranger observes two fires, one at an angle of depression of 5° and the other at an angle of depression of 2°. Assume the fires and the tower are in a straight line. Determine the distance between the fires in each case. The diagrams are not drawn to scale.

a) when the fires are on the same side of the tower

b) when the fires are on opposite sides of the tower

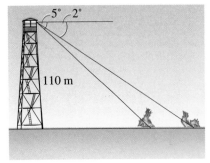

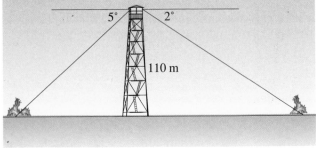

7. Two trees are 100 m apart. From a point midway between them, the angles of elevation to their tops are 12° and 16°. How much taller is one tree than the other? The diagram is not drawn to scale.

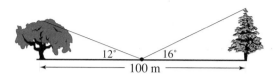

For exercises 8 and 9, you will need an empty carton and either a metre stick or a yardstick.

8. a) Measure the length and height of the carton. Place the stick inside the carton as shown (below left). Use your measurements to determine the angle between the stick and the horizontal.

b) Place the carton on the floor near a wall, with the stick just touching the wall. Use the length of the stick and your answer to part a to calculate how high up the wall the stick reaches. Check your answer by measuring.

c) Give some reasons why your calculated result may differ from your measured result.

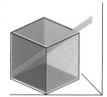

9. a) Place the metre stick along the main diagonal of the carton (above right). Then repeat exercise 8.

b) Explain why the angle of inclination in part a is less than the angle of inclination of the stick in exercise 8a.

C **10.** A cell tower 25 m tall is on top of a hill. To an observer standing on level ground near the foot of the hill, the angle of elevation of the top of the tower is 50° and the angle of elevation of the bottom of the tower is 32°. Calculate the height of the top of the tower from the observer's position on the ground.

Communicating the IDEAS

Write to explain how two right triangles can be used to solve a problem. Make up your own problem and solve it.

Testing Your Knowledge

Geometry Tools

Properties of Similar Triangles

- In similar triangles, corresponding angles are equal.

  Given △ABC is similar to △PQR,

  then, ∠ABC = ∠PQR
  ∠BAC = ∠QPR
  ∠ACB = ∠PRQ

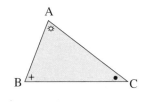
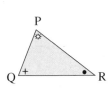

- In similar triangles, the ratios of corresponding sides are equal.

  Given △ABC is similar to △PQR, then

  $$\frac{AB}{PQ} = \frac{BC}{QR} = \frac{AC}{PR}$$

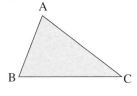
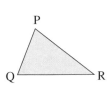

Trigonometry Tools

In right △ABC

$$\sin A = \frac{\text{length of side opposite } \angle A}{\text{length of hypotenuse}} = \frac{BC}{AC}$$

$$\cos A = \frac{\text{length of side adjacent to } \angle A}{\text{length of hypotenuse}} = \frac{AB}{AC}$$

$$\tan A = \frac{\text{length of side opposite } \angle A}{\text{length of side adjacent to } \angle A} = \frac{BC}{AB}$$

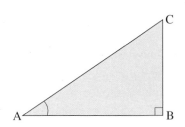

1. Determine each value of *a*. Explain your reasoning.

a)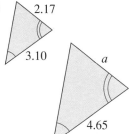

2.17

3.10

a

4.65

b)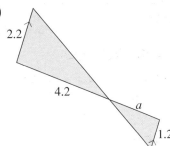

2.2

4.2

a

1.2

2. In the diagram below, DF, EG, and AB are parallel. Determine the values of *w*, *x*, *y*, and *z*. Give the answers to 1 decimal place where necessary.

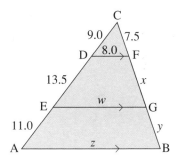

C

9.0   7.5

D 8.0 F

13.5

x

E   *w*   G

11.0

y

A   *z*   B

3. Find a pair of similar triangles and explain why they are similar.

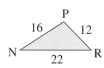

P

16   12

N   22   R

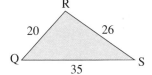

R

20   26

Q   35   S

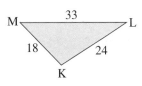

M   33   L

18   24

K

4. Determine the measure of ∠J to the nearest degree.

a) tan J = 0.425   **b)** tan J = 1.534   **c)** tan J = $\frac{5}{3}$ **d)** tan J = $\frac{7}{10}$

5. Determine the sine and cosine of each angle to 4 decimal places.

a) 13°   **b)** 44°   **c)** 23°   **d)** 74°   **e)** 30°

6. Determine the measure of ∠S to the nearest degree.

a) sin S = 0.788   **b)** cos S = 0.731   **c)** cos S = $\frac{3}{5}$

d) sin S = $\frac{5}{8}$ **e)** cos S = 0.899   **f)** sin S = 0.122

7. In △EFG (below left), ∠E = 35° and EF = 9.0 cm

   a) Calculate the length of FG.

   b) Use the Pythagorean Theorem to calculate the length of EG.

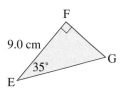

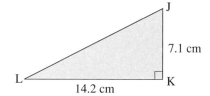

8. In △JKL (above right)

   a) Calculate tan L and ∠L to the nearest degree.

   b) Use the Pythagorean Theorem to calculate the length of JL.

9. In △VWX, determine the measures of ∠W and ∠X to the nearest degree.

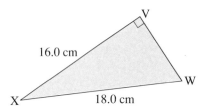

10. Calculate sin F, then ∠F to 1 decimal place.

a)

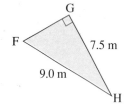

b)

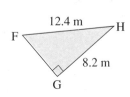

11. Calculate each value of x to 1 decimal place.

a)

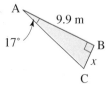

b)

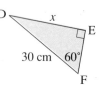

c)

d)

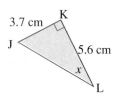

12. Francesco is flying a kite on a string 150 m long. The string makes an angle of 68° with the ground. Francesco is holding the end of the kite string 2 m above the ground. How high is the kite?

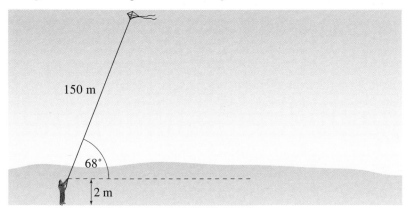

13. The sun's rays are at an angle of 42° to the ground. A hydro pole casts a shadow 18 m long. Calculate the height of the pole.

14. A roller coaster climbs vertically 60 m at an angle of 25° from the lowest to the highest point of the track. It then plunges over the high point to begin the ride. Calculate the length of track that takes the roller coaster from the ground to the highest point.

15. The foot of a 6.0-m ladder is on a level patio 1.5 m from the wall, against which it leans.

a) Calculate the angle formed by the ladder and the ground.

b) Calculate how high up the wall the ladder reaches.

c) Suppose the ladder slips another 0.5 m from the wall. Repeat parts a and b for this situation.

16. Mathematical Modelling A window in an apartment building is 32 m above the ground. From the window, the angle of elevation of the top of an apartment building across the street is 36°. The angle of depression of the bottom of the apartment building is 47°. Determine the height of the taller building.

1. Graph each function. Sketch the graphs. What do the graphs have in common?

   a) $y = x + 2$ **b)** $y = -0.2x + 2$ **c)** $y = 2 - x$

   d) $y = x^2 + 2$ **e)** $y = -x^2 + 2$ **f)** $y = 2x^2 + 2$

2. A group of students worked during the summer delivering flyers. The table shows the number of flyers delivered each week.

Week	1	2	3	4	5	6	7	8	9	10
Number delivered	375	390	390	305	90	555	380	395	390	390

   a) Graph the data.

   b) During one week, it rained almost every day. The students were unable to make many deliveries. During which week did this occur? Explain.

   c) Draw a new table. Calculate the total number of flyers that have been delivered each week. That is, for week 2, 375 + 390 have been delivered; for week 3, 375 + 390 + 390 have been delivered. These are the cumulative totals.

   d) Graph the cumulative totals for each week.

   e) Look at the graphs in part a and d. Do they represent functions? Explain.

3. This table relates the numbers of hours two students worked with the amount of money they earned.

Hours	10	20	30	40
Earnings, student A ($)	65	130	195	260
Earnings, student B ($)	90	140	190	240

   a) Explain how you know that each relationship is linear.

   b) On the same grid, graph *Earnings* against *Hours* for each student.

   c) What is the slope of each line? What does it represent?

   d) Write an equation that relates each student's earnings, E dollars, to the number of hours worked, n.

   e) Describe each situation as representing direct or partial variation.

4. Determine the y-intercept of the graph of each equation.

   a) $y = \frac{5}{2}x + 7$ **b)** $y = -2(x - 3) + 5$ **c)** $y = \frac{1}{5}(x - 10) + 3$

5. Determine the equation of the line passing through each pair of points.

a) A(2, 7) and B(3, 11)   **b)** C(−2, 3) and D(2, 11) **c)** E(2, 3) and F(4, 8)

6. A portrait photographer charges a $40 sitting fee, plus an additional cost per person. The cost, C dollars, for a portrait with x people is given by $C = 40 + 12x$.

a) Graph the function for up to 10 people.

b) Calculate the cost for a portrait with 6 people.

c) How many people are in a portrait that costs $136?

7. Elaine sells long distance telephone service to corporate clients. Her pay is based on the number of hours of time she sells.
For times between 0 and 200 hours, her pay, y dollars, is given by the equation $y = 1.2x$, where x is the number of hours sold.
For times greater than 200 hours, Elaine's pay is given by $y = 0.8x + 80$.

a) Graph Elaine's pay.

b) Why is the function not a straight line?

8. The equation of each line is given in the form $Ax + By + C = 0$. Determine the coordinates of two points on the line, then graph the line.

a) $x - y - 4 = 0$ **b)** $-2x + 3y + 18 = 0$ **c)** $5x - 3y - 15 = 0$

9. The equations and graphs of two linear systems are shown. Estimate the solution of each system.

a)

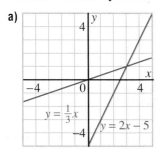

b)

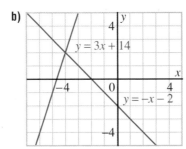

10. Solve each linear system.

a) $y = x + 1$
$\quad y = 2x - 4$

b) $y = 2x + 8$
$\quad y = 5x - 1$

c) $y = 6x + 2$
$\quad y = x - 13$

d) $3x + y = 1$
$\quad y = x - 3$

e) $2x + 3y = 8$
$\quad y = x + 11$

f) $2x + 9y = -2$
$\quad y = 2x + 22$

11. Solve each linear system.

a) $4x - y = 1$
$x + y = 9$

b) $3x + y = 2$
$5x + y = -2$

c) $x - 4y = 0$
$x - 3y = 2$

d) $3x + 2y = 5$
$-x + y = 5$

e) $2x + 3y = 0$
$x + 2y = -1$

f) $-x - 3y = 13$
$2x - 8y = 2$

12. A video game rental is $4.50 and a DVD rental is $6.00. Suppose you pay $55.50 for 11 items rented. How many of each type did you rent?

13. After 6 games, Kathie logged 18 penalty minutes. Suppose the rate continues.

a) There are 53 games in the season. How many penalty minutes will Kathie log by the end of the season?

b) How many games will Kathie play before she logs 30 penalty minutes?

14. Colleen paid $38.95 for 50 L of fuel for her car. Colleen's car tank holds 70 L. How much would a full tank of fuel cost?

15. The height of an image produced on a screen by a slide projector is proportional to the distance from the lens to the screen. When the lens is 2.3 m from the screen, the height of a building in a picture is 14.2 cm. Suppose the lens is 12.8 m from the screen. What will the height of the building be?

16. The aircraft (below left) is a CF-18. It is flown by the Canadian Armed forces. The wingspan is 11.4 m.

a) What is the scale of the drawing?

b) What is the overall length of the aircraft?

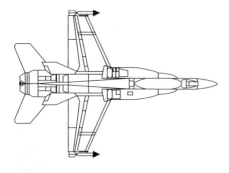

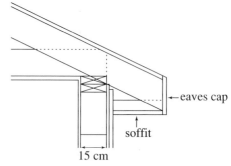

17. The picture (above right) shows a section of the end of a roof. The walls contain insulation 15 cm thick.

a) What is the scale of the drawing?   **b)**   What is the width of the soffit?

c) What is the height of the eaves cap?

18. The maximum voltage produced by a generator is proportional to the rate at which the coils turn. When the coils turn at 30 rev/min, the maximum voltage is 159 V. What will the maximum voltage be when the coils turn at 165 rev/min?

19. a) Explain why the two triangles are similar.

   b) Determine the value of x.

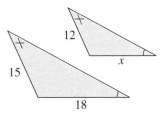

20. Calculate each value of x and y. Give the answers to 2 decimal places if necessary.

a)

b)

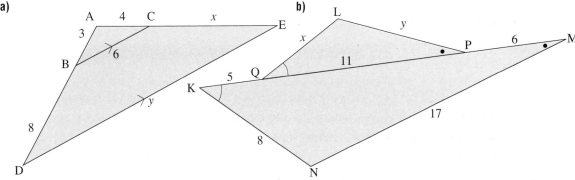

21. Calculate each value of x and y to 1 decimal place.

a)

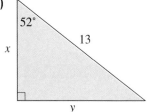

b)

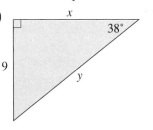

c)

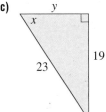

22. Althea wanted to measure the height of a tall building. Althea knows the height of a shorter building. She walked away from the shorter building until the tops of both buildings were in her line of sight. Althea measured the distance to the shorter building. She used a map to determine the distance between the two buildings. Use the information in the diagram. Calculate the height of the taller building.

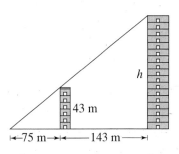

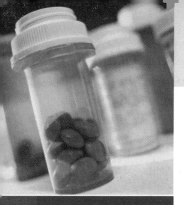

Project

Medicine Doses

Suggested Group Size: 3

Materials: grid paper, various packages for over-the-counter medicines (optional)

This project highlights a real situation. You can apply some of the mathematics you learned in the preceding chapters while working as part of a team. In business, a company's success depends on the strength of its team. Plan together and listen to everyone's ideas on how to approach a problem.

THE TASK

The members of your group will act as public health workers. You will investigate instructions for taking medicine. You will work with proportional reasoning to calculate doses, and analyse dose patterns of over-the-counter medicines.

ASSESSMENT

There are many valid responses to this project. Your teacher may choose to use the rubric on page 297 to assess your work.

BACKGROUND

Everyone gets sick at some time. When we are sick, the doctor may prescribe medicine. For the medicine to be safe and effective, the dose must be carefully calculated.

Before beginning your project, investigate the topic. Use the Internet or visit your local pharmacy to answer these questions.

- Why are some medicines available "over the counter" while others are dispensed by a pharmacist?
- How do people know what dose of medicine to take?
- Why do some medicines need to be taken three or four times a day?

- Manufacturers of over-the-counter medicines, such as aspirin, do not know who will be taking the medicine they package.
  - How do manufacturers help people determine the dose to take?
  - What are some of the warnings on over-the-counter medicines?

GETTING STARTED

When doctors write prescriptions, they use abbreviations to instruct the pharmacists. The symbol, R_X, usually begins the instructions on a prescription. Here are other abbreviations.

SIG	take This abbreviation starts the directions.	tbs	tablespoon
disp	dispense For example, **disp 260** means **give 260 pills in total**	prn	take as needed For example, **prn sleep** means **take as needed for sleep**
q	every	f7	for 7
d	day	tsp	teaspoon
dq	once a day, every day	cap	capsules
bid	twice a day	po	by mouth, orally
tid	three times a day	pc	after meals
qid	four times a day		

The pharmacist prepares the medicine, then types a label that includes the doctor's instructions in plain language.

1. Each of the following patients requires Amoxicillin, an antibiotic. Each prescription includes:
   - the amount of medicine in one capsule (for example, 250 mg)
   - how much medicine to take
   - how long the patient should take the medicine

Project

Read each prescription. Refer to the abbreviation chart. Calculate how many capsules the pharmacist should dispense to each of the patients below.

a)

Dr. George Hanley
7 Bridle Road,
Village Green, Ontario

Name: <u>Fred Partinger</u>

Rx: Disp Amoxicillin 250 mg.
SIG: 1 cap tid f7days.

b)

Dr. Julia Samson
8652 Swan Blvd., #657
Yorktown, Ontario

Name: <u>Madeleine Frawley</u>

Rx: Disp Amoxicillin 250 mg.
SIG: 1 cap tid f10days.

c)

Dr. P. Sealy
40 Nutleigh Cres.
Palmerston, Ontario

Name: <u>Paulina Tan</u>

Rx: Disp Amoxicillin 500 mg.
SIG: 1 cap tid f10days.

d)

Dr. S. Choi
90 Eggerton Road,
Toronto, Ontario

Name: <u>John Selvakumar</u>

Rx: Disp Amoxicillin 500 mg.
SIG: 1 cap tid f7days.

Many over-the-counter medicines state "For children under 12, consult physician." One capsule of the medicine may be too strong for a child. When doctors prescribe medicine for a child, they base the dose on the child's mass.

2. a) A typical daily dose of Amoxicillin for a child is 30 mg/kg/day. This means a child takes 30 mg a day for each kilogram of her mass. The dose is divided into 3 single doses, 8 h apart. The children below were prescribed Amoxicillin. Copy and complete this table. Calculate the daily and single doses.

Name	Mass (kg)	Daily dose (mg)	Single dose (mg)
Mariette	14.0		
Jonas	18.5		
Regan	21.0		
Pierre	28.0		
Sumara	31.0		

b) Use grid paper. Graph the *Single dose* against *Mass*. Describe your graph.

c) For gastroenteritis (an upset stomach), a typical dose of Erythromycin for a child is 40 mg/kg/day. This is given in 3 single doses. Draw a table. Complete the table for the children in part a.

d) Use grid paper. Graph the *Single dose* against the *Mass*. Describe your graph.

3. Suppose you buy an over-the-counter medicine. The label on the medicine has instructions similar to those from the pharmacist, but it includes doses for various types of people.

The label will include how much, how often, and how long to take the medicine.

A label on a fever medicine for children indicates these doses:

Under 2 – as directed by physician

12 to 16 kg	(2–3 years)	5 mL
17 to 21 kg	(4–5 years)	7.5 mL
22 to 26 kg	(6–8 years)	10 mL
27 to 31 kg	(9–10 years)	12.5 mL
over 31 kg	(over 11 years)	15 mL

May be repeated every 6 – 8 hours.

a) Use the data above. Graph the *Single dose* against *Mass* for this medicine. What do you notice? How is this graph different from the graphs in exercise 2?

b) Use the data above. Graph the *Single dose* against *Age* for this medicine. What do you notice? Explain the differences between this graph and the graph in part a.

c) The label indicates that 5 mL of the medicine can be given to any child with a mass between 12 kg and 16 kg. Give two reasons why the label does not give an exact dose for each mass. Justify your reasons.

PROJECT PRESENTATION

Apply what you learned in Getting Started to begin your project. Your finished project should include:

- answers to exercises 1 to 3 above, with full solutions, graphs, and written explanations
- written displays of the major ingredients, and the dose directions for different over-the-counter medicines for 3 illnesses
- written prescriptions as described on page 296
- graphs of dose against mass and dose against age for 3 children's medicines

Project

As public health workers, you are investigating various drugs for an information session at the local health centre. You have been assigned to research medications for:

- hay fever
- coughs
- muscle and joint pain
- sore throats
- flu
- eye infections

Each member of your group should choose **one** of these illnesses. Examine the labels for at least 4 different over-the-counter medicines for that illness. Make sure that one of your chosen medicines includes instructions for children under 12.

1. Prepare a display of the major ingredients and dose directions for these medicines.

2. Use the abbreviation chart as a guide. Choose one of your medicines.

   a) Write a prescription for a 6-year-old.

   b) Write a prescription for an adult.

3. Draw a graph of dose against mass and another graph of dose against age for the children's medicines.

Career Opportunities

Many careers are connected to the dispensing of medicines. Doctors, nurses, pharmacists, and pharmacy assistants work directly with the patient to prescribe and dispense the correct medicine. Behind the scenes, chemists, scientists, and laboratory technicians work to discover new medicines to combat illnesses. The manufacturing sector of the pharmaceutical industry employs people with a wide variety of skills.

Rubric for the Medicine Doses project

Level 1	Level 2	Level 3	Level 4
Knowledge/Understanding			
The student: • determines some doses accurately. • constructs part of the graphs correctly.	• determines most doses accurately. • constructs most parts of the graphs correctly.	• determines almost all doses accurately. • constructs almost all parts of the graphs correctly.	• determines the doses accurately. • constructs the graphs correctly.
Thinking/Inquiry/Problem Solving			
The student: • describes graphs demonstrating some understanding that similarities and differences exist for medical reasons. • makes reasonable comments about labels.	• describes graphs demonstrating some understanding of connections between medicine and the similarities and differences in graphs. • shows some reasoning about labels.	• describes graphs demonstrating an understanding of most connections between medicine and the similarities and differences in graphs. • shows reasoning about labels.	• describes graphs demonstrating an understanding of the medical significance of the similarities and differences in graphs. • shows analytical reasoning about labels.
Communication			
The student: • uses some medical terms correctly. • labels parts of graphs correctly. • presents a project with little organization.	• uses most medical terms correctly. • labels most parts of graphs correctly. • presents a project with some organization.	• uses almost all medical terms correctly. • labels graphs correctly. • presents an organized project.	• uses medical terms correctly. • labels graphs correctly and effectively. • presents a creative, organized project.
Application			
The student: • interprets a few of the given data and a few project instructions correctly. • states a few ingredients or dose directions correctly. • makes a few correct statements about medicine.	• interprets some given data and project instructions correctly. • states some ingredients and dose directions correctly. • makes statements that show some understanding of medicine.	• interprets most given data and project instructions correctly. • states ingredients and dose directions correctly. • makes statements that show an understanding of medicine.	• interprets the given data and project instructions correctly. • states the ingredients and dose directions correctly. • makes statements that show sound understanding of medicine.

MATHEMATICAL MODELLING

Projectile Motion

Suppose you threw a ball straight up in the air. After 2 seconds, it landed back in your hand.

Suppose your friend took one picture of the ball at every 0.25 seconds.

1. Which set of pictures illustrates the positions of the ball? Explain.

a)

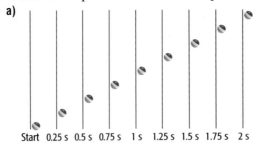

Start 0.25 s 0.5 s 0.75 s 1 s 1.25 s 1.5 s 1.75 s 2 s

b)

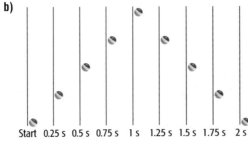

Start 0.25 s 0.5 s 0.75 s 1 s 1.25 s 1.5 s 1.75 s 2 s

c)

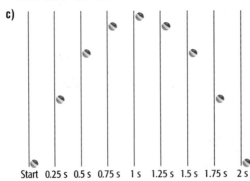

Start 0.25 s 0.5 s 0.75 s 1 s 1.25 s 1.5 s 1.75 s 2 s

2. How could you find the maximum height of the ball?

You will return to this problem in Section 6.4.

FYI Visit www.awl.com/canada/school/connections

For information about the problem on page 298, click on <u>MATHLINKS</u>, followed by Foundations of Mathematics 10. Then select a topic under Projectile Motion.

This section reviews these concepts:

- Graphing linear functions
- Differences in tables of values
- Line of best fit

Patterns with Lines

You will need two sheets of 1-cm grid paper.

1. On one sheet of grid paper, graph each line through the given point with the given slope. Use the entire sheet, with the x-axis near the middle.

   a) $(0, 3)$, slope 3   b) $(0, 2)$, slope 2   c) $(0, 1)$, slope 1

   d) $(0, -1)$, slope -1 e) $(0, -2)$, slope -2 f) $(0, -3)$, slope -3

2. a) Write to describe any patterns in the graphs in exercise 1.

   b) Draw two more lines on your graph that belong to the same pattern.

3. Recall that the equation of the line through $(0, b)$ with slope m is $y = mx + b$.

   a) Choose one line on the grid. Write its equation.

   b) Write the equations of the other lines on the grid.

4. All the points in exercise 1 were on the y-axis. How do you think the pattern would change if you used points on the x-axis?

5. Check your prediction in exercise 4. Graph lines through the given points with the given slope.

   a) $(3, 0)$, slope 3   b) $(2, 0)$, slope 2   c) $(1, 0)$, slope 1

   d) $(-1, 0)$, slope -1 e) $(-2, 0)$, slope -2 f) $(-3, 0)$, slope -3

6. Recall that the equation of the line through $(a, 0)$ with slope m is $y = m(x - a)$.

   a) Choose one line in exercise 5. Write its equation.

   b) Write the equations of the other lines in exercise 5.

7. For each equation, make a table of values for $x = 0, 1, 2, 3, 4$. Draw a graph. If you have a graphing calculator, use it to check your work.

   a) $y = x + 2$ b) $y = 2x - 1$ c) $y = -x + 4$

8. a) Add a third column to the table for the equation in exercise 7a. Calculate the differences in the *y*-values.

b) Complete *Difference* columns for the tables for the equations in exercise 7 parts b and c.

c) What do you notice about the *Difference* columns for all these functions?

d) What do the *Difference* columns tell about the functions?

x	*y*	Difference
0		
1		
2		
3		
4		

9. A function is defined by $y = -3x + 1$. This is the relation between the coordinates of each point on the graph of the function.

a) Which of these points lie on the graph of the function?
A(1, −2), B(0, 0), C(2, 5), D(−1, 4), E(0, 1)

b) Write the coordinates of two other points that lie on the graph.

c) Graph the function.

10. One event in the Human Powered Vehicle competition is the 200-m flying start trial. People push the vehicle to the starting line, then let go. The vehicle is timed as it goes through the course. The table shows some records for the event.

a) Make a scatter plot of the data on grid paper. Remember to include all years from 1974 to 1992 on the horizontal axis.

b) Use a ruler to draw a line of best fit.

c) A record was also set in 1980. Use the line of best fit to estimate this record.

d) The last record was set in 1992. Why do you think no further records have been set since then?

200-m Flying Start Records	
Year	Time (in seconds)
1974	10.4
1975	10.0
1976	9.4
1977	9.1
1979	8.8
1985	7.2
1986	6.8
1992	6.5

6.1 The Shape of a Parabola

Almost all the functions you have worked with in this book are linear functions. Their graphs are straight lines. Recall that for a linear function:

- The equation can be written in the form $y = mx + b$.
- When the values of x in a table of values are consecutive, the values of y have constant differences.

In this chapter, we will study a type of function whose graph is a curve called a *parabola*. When a ball is thrown from one person to another, the path of the ball would be a parabola if there were no air resistance. You see the parabolic shape in satellite dishes, sideline microphones at football games, searchlight mirrors, and suspension bridges.

Constructing a Parabola

You will need a ruler, plain paper, grid paper, and scissors.

1. Fold a sheet of $8\frac{1}{2}$ inches by 11 inches plain paper in half lengthwise. Open the paper.

2. Use a ruler to draw two lines from the bottom centre to the top corners.

3. Place a ruler along one diagonal line. Mark every centimetre along the line. Repeat for the other line.

4. Number the marks on each line from 1 to 27, starting at the bottom.

5. Use a ruler to draw a line to join 27 on the left line to 1 on the right line. Join 26 on the left to 2 on the right. Continue until you have joined 1 on the left line to 27 on the right line.

6. The lines you draw form the outline of a curve. Mark the point on the curve where it crosses the fold. This is the vertex.

7. Carefully cut out the curve.

8. Draw axes on a sheet of grid paper, so the x-axis is toward the bottom of the page, and the y-axis is in the middle of the page.

9. Lay the cut-out curve on the grid paper so the vertex is at (0, 0) and the fold is along the y-axis.

10. Trace the curve as accurately as possible on the grid paper.

Save your graph for use in *Investigation 3*.

In *Investigation 1*, although you only drew straight lines, you created a curve. This curve is a *parabola*.

Properties of a Parabola

- Every parabola has an *axis of symmetry*. If you fold the paper on which a parabola has been drawn along its axis of symmetry, the left and right sides coincide.
- The point where the parabola intersects the axis of symmetry is its *vertex*.
- The width of the parabola increases as you move away from the vertex.

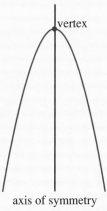

Relating a Number and its Square

You will need a ruler and 1-cm grid paper.

1. Copy and complete this table for integer values of *x* from −4 to 4.

   Square each *x*-value to get the corresponding *y*-value.

2. On grid paper, draw axes so the *x*-axis is close to the bottom of the page, and the *y*-axis is in the middle.

3. The scale on the *x*-axis should go from −4 to 4, and on the *y*-axis from −1 to 20.

4. Plot the data from the table in exercise 1.
   Join the points to form a smooth curve.
   Do not use a ruler!

x	y
−4	
−3	
−2	
−1	
0	
1	
2	
3	
4	

5. Mark the axis of symmetry on your graph. What is its equation?

6. Label the vertex of the curve. What are the coordinates of the vertex?

Save your graph for use in *Investigation 3*.

You have now graphed two curves. They may look different, but they are both parabolas.

The Step Pattern of a Parabola

You will need a ruler and the parabolas from *Investigation 1* and *Investigation 2*.

Use the traced parabola from *Investigation 1*.

1. From the vertex, measure 1 cm to the right. Then turn your ruler and draw a vertical line segment to meet the curve. Measure this segment and write its length on the graph.

2. From the top of the segment you just drew, measure 1 cm to the right. Turn your ruler and draw a new vertical line segment to meet the curve. Measure its length and write it on the graph.

3. Continue measuring 1 cm to the right, then up to the curve, as shown, until you reach the top of the graph.

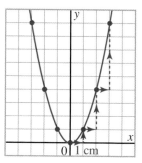

4. Copy this table. Enter the lengths of the vertical line segments in the middle column.

Vertical line segment	Length of vertical line segment (cm)	Length of vertical line segment / Length of first vertical line segment
First		
Second		
Third		
Fourth		
Fifth		

5. Divide the length of each vertical segment in the middle column by the length of the first segment in the middle column (the number in the shaded cell of the table). Write the results in the third column.

6. Repeat exercises 1 to 5 using the parabola you drew in *Investigation 2*.

7. What do you notice about the numbers in the third column in each table?

In *Investigation 3*, you should have discovered the following property.

Take Note

Step Pattern of a Parabola

For a parabola, there is a step pattern of 1, 3, 5, ...

- Starting at the vertex, measure 1 unit right, then measure up to the curve. Call this length *a*.

- Measure 1 unit right, then measure up to the curve. This vertical length is 3 times as long as *a*, or 3*a*.

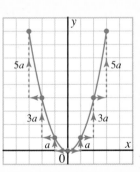

- Measure 1 unit right, then measure up to the curve. This vertical length is 5 times as long as *a*, or 5*a*.

- This pattern of measurements continues.

If we start at the vertex and measure left then up, we get the same results because a parabola is symmetrical.

For a parabola that opens down, a similar pattern holds.

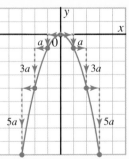

Example 1

Determine whether each curve is a parabola.

a)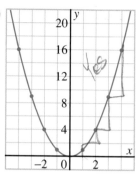

b)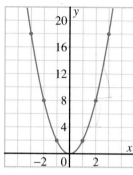

c)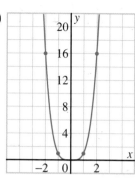

Solution

> **Think ...**
>
> Each graph has a vertex and an axis of symmetry. Check each graph for the step pattern from the vertex: 1 right 1 up, 1 right 3 up, 1 right 5 up, and so on.

Draw broken vertical and horizontal lines to join adjacent points. Count squares to calculate the vertical distance for each horizontal distance of 1 unit.

a) The vertical distances are 1, 3, 5, 7. This is a step pattern for a parabola. The graph is a parabola.

b) The vertical distances are 2, 6, 10. These can be written $2 \times 1, 2 \times 3, 2 \times 5$. This is a step pattern for a parabola. The graph is a parabola.

c) The vertical distances are 1, 15. This is not a step pattern for a parabola. The graph is not a parabola.

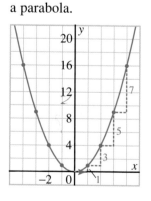

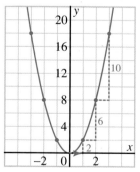

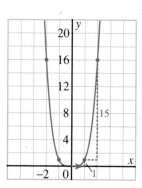

Example 2

This table of values represents a function.

a) Graph the data on a grid. Draw a smooth curve through the points.

b) Is the graph a parabola? Explain.

x	y
−3	−5
−2	0
−1	3
0	4
1	3
2	0
3	−5

Solution

a)

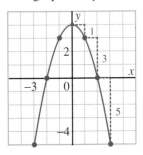

b) Draw broken lines to join adjacent points. Calculate the vertical distances. They are 1, 3, 5. This is a step pattern for a parabola. The graph is a parabola.

In *Example 2*, the coordinates of the vertex are (0, 4). This example illustrates that the step pattern can be used for a parabola whose vertex is not at the origin.

Discussing the IDEAS

In *Investigation 3*, why is the first entry in the middle column of the tables always 1?

A 1. Determine whether each set of points on the grids below could lie on a parabola. Explain.

a)
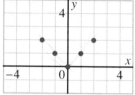

b)
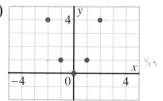

c)
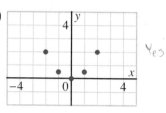

d)
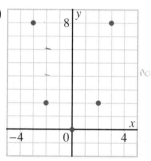

2. Copy each set of points on grid paper. Use the step pattern to add 4 more points to each grid to form a parabola. Draw a smooth curve through the points.

a)
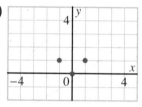

b)
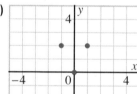

c)
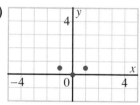

d)
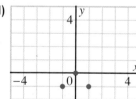

B 3. Use grid paper and the symmetry of a parabola. For each parabola, find the coordinates of four other points on the parabola.

a) The vertex of a parabola is $(0, 0)$ and the point $(-1, 1)$ lies on the curve.

b) The vertex of a parabola is $(0, 5)$ and the point $(1, 6)$ lies on the curve.

c) The vertex of a parabola is $(0, 1)$ and the point $(3, 10)$ lies on the curve.

4. Each incomplete table contains data for a parabola. The coordinates of the vertex of each parabola are in heavy type.

Copy each table. Use the step pattern to complete the tables.

a)

x	y
−3	9
−2	4
−1	1
0	**0**
1	1
2	4
3	9

b)

x	y
−3	10
−2	5
−1	2
0	**1**
1	2
2	5
3	10

c)

x	y
−3	12
−2	7
−1	4
0	**3**
1	4
2	7
3	12

d)

x	y
−3	9
−2	3
−1	0
0	**−1**
1	0
2	3
3	8

e)

x	y
−3	13
−2	8
−1	5
0	**4**
1	5
2	8
3	13

5. Plot the data from each table in exercise 4 on a grid.

a) What is similar about the parabolas?

b) What is different about the parabolas?

6. a) Plot these data on a grid. Join the points with a smooth curve.

b) Use the table and graph from part a. Explain why the curve is a parabola.

x	y
−3	18
−2	8
−1	2
0	0
1	2
2	8
3	18

7. Each incomplete table contains data for a parabola. Copy each table. Use the step pattern to complete the tables.

a)

x	y
−3	
−2	8
−1	2
0	0
1	2
2	
3	

b)

x	y
−3	
−2	
−1	3
0	1
1	3
2	9
3	

c)

x	y
−3	
−2	11
−1	5
0	3
1	
2	
3	

d)

x	y
−3	
−2	7
−1	1
0	−1
1	
2	
3	

8. Plot the data from each table in exercise 7 on a grid.

a) Identify the coordinates of the vertex of each parabola.

b) What is similar about these parabolas?

c) How are these parabolas different from those in exercises 4 and 5?

9. You can construct a parabola by paper folding. You will need grid paper and a ruler.

a) Draw a horizontal line close to the bottom of the grid paper.

b) On the line, mark 13 points equally spaced. Number the points 1, 2, 3… from left to right. (Make sure the points are dark enough to be seen through the paper.)

c) A few centimetres above the point labelled 7, mark a point A.

d) Fold the paper so that point 1 coincides with point A. Crease the paper, then unfold it.

e) Use a ruler to draw a (light) vertical line from point 1 up to the crease. Mark a point where the line meets the crease.

f) Repeat parts d and e for each of points 2 to 13.

g) Draw a smooth curve through the points.

h) Check that this curve has the properties of a parabola, as described on pages 304 and 306.

i) Examine the parabolas your classmates drew. How did the position of the point A affect the shape of the parabola?

10. Use your parabola from exercise 9. Measure the distance from one point to A, then from that point to the horizontal line. Repeat this for two other points. What do you notice?

C **11.** In *Investigation 1*, you constructed a pattern inside lines at an angle of approximately 42°. How do you think the shape of the curve would be affected if you used a larger angle? a smaller angle? Investigate to find out.

Communicating *the* IDEAS

Write to explain why a semicircle is not a parabola.

Parabolas and Quadratic Functions

In earlier chapters, you used an equation to graph a function on a graphing calculator. The function $y = x^2$ is an example of a quadratic function. The graph of a quadratic function is always a parabola.

Other quadratic functions have equations such as:

$$y = 3x^2 \qquad\qquad A = \pi r^2 \qquad\qquad y = 2x^2 - 3x + 1$$

- The equation of a quadratic function has one term with an exponent of 2.
- The equation of a quadratic function can be written using letters other than x and y.

Take Note

Definition of a Quadratic Function

A *quadratic function* has an equation that can be written in the form $y = ax^2 + bx + c$, where a, b, and c are constants, and $a \neq 0$.

INVESTIGATION 1

Graphing a Quadratic Function from Its Equation

1. Graphing $y = x^2$

   a) Press ⌈ Y= ⌋. If necessary, use ▲ to place the cursor beside Y1=. If necessary, press ⌈CLEAR⌋ to remove any equations.

   b) Press ⌈X,T,θ,n⌋ ⌈ x^2 ⌋ to enter the equation $y = x^2$.

   c) Press ⌈ 2nd ⌋ ⌈WINDOW⌋ for ⌈TBLSET⌋. Change the settings to those shown below left. This sets the table so the value of x begins at 0 and increases by 1.

```
TABLE SETUP
 TblStart=0■
 ΔTbl=1
Indent: Auto Ask
Depend: Auto Ask
```

```
WINDOW
 Xmin=-9.4
 Xmax=9.4
 Xscl=1
 Ymin=-6.2
 Ymax=6.2
 Yscl=1
 Xres=1
```

d) Press WINDOW. Change the settings to those shown on the screen on page 312. Remember to use the (-) key to input −9.4 and −6.2.

e) Press GRAPH. What features of a parabola do you see on the screen? What are the coordinates of the vertex?

f) Press 2nd GRAPH for TABLE. The screen shows the table of values for the function $y = x^2$. The values of x are in the first column and the values of y are in the second column. Use ▲ and ▼ to scroll up and down to view the numbers in the table. What patterns do you see in the numbers?

g) Sketch the graph and label it with its equation.

2. Graphing $y = 3x^2$

Leave $Y1 = x^2$ from exercise 1.

a) Press Y=. Use ▼ to put the cursor beside Y2 =.

b) Press 3 X,T,θ,n x^2 to enter the equation $y = 3x^2$.

c) Press 2nd GRAPH to view the table of values. The x- and y-values for the graph of $Y1 = x^2$ are in the first two columns. The y-values for the graph of $Y2 = 3x^2$ are in the third column. How do the numbers in the third column relate to those in the second column? Is there a step pattern in the third column? Explain.

d) Press TRACE to see the graphs. How does the graph of $y = 3x^2$ compare to the graph of $y = x^2$?

e) Sketch the graph of $y = 3x^2$ and label it with its equation.

3. Graphing $y = -x^2$

a) Press Y=. Leave $Y1 = x^2$. Press ▼ CLEAR to clear the equation from Y2 =.

b) Press (-) X,T,θ,n x^2 to enter the equation $y = -x^2$ as Y2.

c) Press 2nd GRAPH to view the table of values. How are the numbers in the second and third columns related?

d) Press TRACE to see the graphs. How are the two graphs related?

e) Sketch the graph of $y = -x^2$ and label it with its equation.

4. Graphing $h = 4t^2 + 2$

This equation has no x or y. To graph it with a calculator, we change the letters. Since the equation is written $h =$, we use y in place of h, and x in place of t.

a) Press [Y=]. Clear any equations in the list.

b) Put the cursor beside Y1 =. Press 4 [X,T,θ,n] [x²] [+] 2 to enter the equation $y = 4x^2 + 2$.

c) Press [WINDOW] and set the values: Xmin = −4.7, Xmax = 4.7, Xscl = 1, Ymin = −2, Ymax = 20, Yscl = 5.

d) Press [TRACE] to see the graph.

e) Use [◀] or [▶] to move the cursor to find the value of y when $x = 1.2$.

f) What is the vertex of this parabola? Use the equation to explain why the vertex is not at (0, 0).

g) Check your understanding. Change the equation in Y1 = to produce a congruent parabola with vertex (0, 3).

In this *Investigation*, you should have discovered these properties of the quadratic function $y = ax^2 + c$.

Take Note

Properties of a Quadratic Function $y = ax^2 + c$

The graph of $y = ax^2 + c$ is a parabola with vertex (0, c).

• When a is positive, the parabola opens up.

• When a is negative, the parabola opens down.

A hanging cord or chain has a curve that approximates a parabola. In *Investigation 2*, you will collect data then use a graphing calculator to determine the equation of a quadratic function whose graph best fits the data.

Estimating the Equation of the Parabola of Best Fit

You will need:

- a very long shoelace or a cord (preferably round rather than flat)
- a large piece of grid paper (or several sheets taped together).

1. Draw coordinate axes on the grid paper so the *x*-axis is close to the bottom of the paper and the *y*-axis is in the middle.

2. Tape the grid paper to a bulletin board.

3. Use two pins at the same height to suspend the shoelace against the grid paper. The lowest point of the shoelace should be at the origin (0, 0), as shown.

4. Write the coordinates of 7 points through which the shoelace passes.

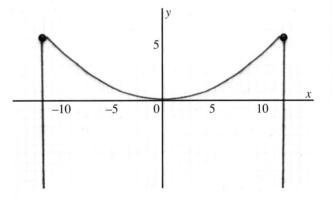

5. Make a scatter plot of the data. Follow these steps.

Step 1. Enter the data.

- Press [STAT] **1.** If there are data in lists L1 and L2, use [▲] to move the cursor to the heading of each column, then press [CLEAR] [ENTER] to clear the data.

- Enter the *x*-coordinates of the points in list L1 and the *y*-coordinates in list L2. Remember to press [ENTER] after each number is input.

Step 2. Set up the plot.

- Press [2nd] [Y=] **1** to select Plot 1. Press [ENTER].

- Select the first plot type, and make sure that L1 and L2 are beside Xlist and Ylist, respectively. Your screen should look like the first screen on page 316.

Step 3. Set up the window.

- Press [WINDOW]. Enter appropriate values for your data.

Step 4. Graph the data.

- Press [GRAPH]. Your scatter plot should look similar to that below right.

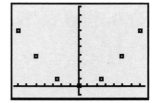

6. The graph suggests there may be a quadratic relationship between the horizontal distances and heights.

   If a parabola of best fit were drawn, its vertex would lie at the origin. Recall that the equation of a parabola with vertex at the origin and opening up has the form $y = ax^2$, where a is positive.

   When you graph equations of the form $y = ax^2$ for different values of a, you can find a curve that appears to pass through the points on the screen. For the screen below left, the value of a was 0.2. This is too large. For the screen below right, the value of a was correct.

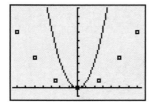
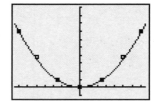

   Use the calculator to plot an equation for the function.

   a) Press [Y=] and clear any equations. Put the cursor beside Y1 =.

   b) Enter an equation of the form $y = ax^2$. Press [＿＿] [X,T,θ,n] [x^2] [ENTER] [GRAPH], where [＿＿] represents a positive number; for example, 0.1.

   c) The calculator plots both the scatter plot and the graph of the equation. Decide how you should change the value of a in the equation so the graph will pass through as many points as possible.

   d) Repeat part b, changing the value of a until you have an equation whose graph is a good fit for the data.

   e) Sketch the graph and write the equation of the parabola of best fit from part d.

1. Look at the sketches you drew in *Investigation 1*.

   a) How is the step pattern for the parabola $y = x^2$ related to the step pattern for the parabola $y = 3x^2$?

   b) How is the step pattern for the parabola $y = x^2$ related to the step pattern for the parabola $y = -x^2$?

2. In *Investigation 2*, how did you use the curve to choose a value for a in the equation $y = ax^2$?

3. Share your results from *Investigation 2* with other students. Discuss what factors may have led to different results.

4. Suppose a parabola has vertex $(0, 7)$. How do you know that its equation cannot have the form $y = ax^2$?

5. What are the advantages of using a graphing calculator to estimate the equation of a parabola of best fit? Are there any disadvantages? Explain.

6.2 Exercises

A 1. Which equations represent quadratic functions? Explain.

   a) $y = 3x^2$
   b) $y = 2x + 5$
   c) $y = \frac{2}{7}x + 1$
   d) $h = 2t^2 + 1$
   e) $A = s^2$
   f) $P = 2(w + l)$

2. For each equation, state whether its graph opens up or down and state the coordinates of the vertex. If you have a graphing calculator, check your answers by graphing each equation.

   a) $y = 3x^2$
   b) $y = -2x^2$
   c) $y = 0.5x^2$
   d) $y = x^2 + 1$
   e) $y = -4x^2 + 2$
   f) $y = -4x^2 - 2$

B 3. This graph was drawn by starting at $(0, 0)$ and applying the step pattern:

   2 right 1 up, 2 right 3 up, 2 right 5 up…

   2 left 1 up, 2 left 3 up, 2 left 5 up….

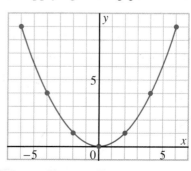

a) Copy the graph on grid paper.

b) Enter the coordinates of the points into a graphing calculator. Make a scatter plot of the data.

c) Enter an equation of the form $y = ax^2$ into the calculator. Graph the equation.

d) Examine the scatter plot and the graph. Change the value of a in the equation until the graph passes through the points on the scatter plot.

e) Write the equation for the parabola of best fit for these data.

4. Repeat exercise 3, using each set of step patterns.

    a) 1 right 2 up, 1 right 6 up, 1 right 10 up…
        1 left 2 up, 1 left 6 up, 1 left 10 up…

    b) 1 right 1 down, 1 right 3 down, 1 right 5 down…
        1 left 1 down, 1 left 3 down, 1 left 5 down…

    c) 3 right 1 up, 3 right 3 up, 3 right 5 up…
        3 left 1 up, 3 left 3 up, 3 left 5 up…

5. Plot the data from each table. Use the method of *Investigation 2* to find an equation of a parabola of best fit. Remember to set the window to view all points.

a)

x	y
−6	36
−4	16
−2	4
0	0
2	4
4	16
6	36

b)

x	y
−3	10
−2	5
−1	2
0	1
1	2
2	5
3	10

c)

x	y
0	0
2	−8
4	−32
6	−72
8	−128
10	−200
12	−288

d)

x	y
0	−1
2	−3
4	−9
6	−19
8	−33
10	−51
12	−73

6. Relating the Edge Length and Surface Area of a Cube

Consider a cube with edge length 1 cm. Suppose you cover each face with construction paper. Will you need twice as much paper to cover a cube with edge length 2 cm? Will you need 10 times as much paper to cover a cube with edge length 10 cm?

To find out, investigate the relationship between the edge length and the surface area.

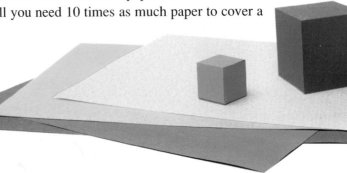

a) Copy and complete this table for cubes.

Edge length (cm)	Surface area (cm²)
0	0
1	6
2	
3	
4	
5	

b) Plot the points on a grid. Join the points with a smooth curve.

c) Is the graph a parabola? Explain.

d) How is this graph different from other parabolas? Why is it not sensible to extend the graph to the left?

e) Use the process of *Investigation 2* to find a parabola of best fit for the data. See pages 315, 316 for details. Enter the data into a graphing calculator and display a scatter plot.

- Enter the edge lengths in L1 and the surface areas in L2. You may need to clear the lists first.
- Set up the plot.
- Clear any equations in the Y= list.
- Use these WINDOW settings:
  Xmin = 0, Xmax = 5, Xscl = 1, Ymin = 0, Ymax = 160, Yscl = 5
- View the graph.
- Press Y= and enter an equation of the form $y = ax^2$ for a suitable value of a.
- Graph your equation and the scatter plot.
- If necessary, change the value of a until the curve is a parabola of best fit for the data. Sketch the graph and write the equation of the parabola.

f) Use the equation from part e to determine the area of paper needed to cover a cube with edge length 10 cm.

7. Relating the Speed of a Car and the Braking Distance

An important statistic for car performance is braking distance. This is the distance a car travels after the brakes are applied. The following data, for a 1999 Chevrolet Monte Carlo SS, was published in *Car and Driver* magazine.

Speed (km/h)	Braking distance (m)
0	0
20	1.9
40	7.5
60	16.8
80	29.8
100	46.6
120	67.1

a) Draw coordinate axes on grid paper so the origin (0, 0) is close to the lower left corner of the paper. For the *x*-axis, use a scale of 1 cm for 10 km/h. For the *y*-axis, use a scale of 1 cm for 5 m.

b) Plot the points. What type of relationship does the graph suggest? How is it different from the graphs in previous exercises? How is it similar to the graph in exercise 6? Explain.

c) Enter the data into a graphing calculator and display a scatter plot. Use the process of *Investigation 2* to find the equation of a parabola of best fit.

d) Use the equation from part c to predict the braking distance when the car is travelling at 140 km/h.

Communicating *the* IDEAS

Write to explain the relationship between a quadratic function and a parabola. Use sketches and equations in your explanation.

In Section 6.2, you collected data and estimated the equations of parabolas of best fit. In this section, you will investigate data that can be modelled by a parabola with its vertex not at the origin.

If you have a CBR, complete *Investigation 1* and *Investigation 2*. If you do not have a CBR, complete *Investigation 2* only.

Rolling Down the Ramp

You will use a motion detector to record the distance to a cylinder as it rolls down a ramp.

You will need this equipment:

- a long surface (such as a desk or a board) that is raised a little on one end to form a ramp
- a full can or plastic bottle to roll down the ramp
- a CBR unit and a TI-83 graphing calculator

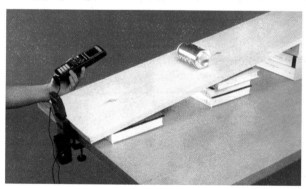

Step 1. Physical Set-Up

- Set up the equipment as shown. Clamp the CBR to the lower end of the ramp. Open the CBR so the sensor points up the ramp.
- Hold the can at the top of the ramp and let it go. Make sure you catch the can before it hits the CBR. Practise rolling the can several times until you can do this:
  - the can rolls straight down the ramp
  - you stop the can so that it does not hit the CBR

Step 2. CBR Set-Up

- Use the connecting cable to connect the CBR to the calculator. Make sure the connections are secure.

- Follow these steps to run the RANGER program.

For the TI-83 Plus

- Turn on the calculator, and press [APPS].
- Press **2** to choose [CBL/CBR].
- Press any key.
- Press **3** to choose RANGER.
- Press [ENTER] to display the main menu.
- Press **1** to choose SETUP/SAMPLE.

For the TI-83

- Turn on the calculator, and press [PRGM].
- Choose RANGER. If this program is not present, see page 36 for instructions.
- Press [ENTER] [ENTER] to display the main menu.
- Press **1** to choose SETUP/SAMPLE.

By default, the calculator is set to collect data for 15 s. This is too much time for this activity. Follow these steps to change the time to 3 s.

For both calculators

- On the REALTIME line, there is a small triangular cursor beside YES. Press [ENTER] to change this to NO.

- Press [▼] to move the cursor down to the next line. Press **3** to change the time to 3 s.

- Move the cursor to the top of the screen, beside START NOW. The screen should look like the one shown. Press [ENTER] .

```
MAIN MENU    ▶START NOW

REALTIME:  NO
  TIME(S):  3
 DISPLAY:  DIST
BEGIN ON:  [ENTER]
SMOOTHING:  NONE
   UNITS:  METERS
```

Step 3. Collect and Graph the Data.

- When you are ready to start, press [ENTER] on the calculator. Let the can go as you practised earlier.

- While the calculator is collecting the data, you will hear a clicking sound and the word TRANSFERRING… appears. A graph of distance against time will be displayed.

- Part of the graph on the screen should be a parabola. To try again, press [ENTER] **5** to select REPEAT SAMPLE.

- Leave the graph on the screen while you complete the following steps. Since the data for your graph are now in your calculator, you can disconnect the cable so that others can use the CBR.

Step 4. Select the Relevant Data.

You are interested only in the data collected while the can was moving down the ramp. Change the display to show the points corresponding to these times only:

- Press ENTER to get to the Plot menu. Choose **4** to select PLOT TOOLS. Choose **1** for SELECT DOMAIN. Your plot will reappear with the words LEFT BOUND?. Move the cursor to the point where the parabola begins, and press ENTER. The words RIGHT BOUND? will appear. Move the cursor to the point where the parabola ends, and press ENTER. After a short time, a screen showing your selected graph will appear.
- Press ENTER **7** to quit the RANGER program.

Step 5. Graph the Parabola of Best Fit.

The curve looks like half a parabola, opening down, with its vertex on the y-axis but not at $(0, 0)$. Therefore, the parabola of best fit has an equation of the form $y = ax^2 + c$, where a is negative and $(0, c)$ is the vertex.

- To find the vertex, press TRACE. Use ◄ or ► to move the cursor along the curve to the highest point. This point is the vertex of the parabola of best fit. Record the coordinates of this point.

In the calculator, enter an equation for the function:

- Press Y= .
- Enter an equation of the form $y = ax^2 + c$. Remember to choose a negative number for a and use the second coordinate of the vertex as c.

  For example, if the vertex is $(0, 5)$, enter $Y1 = -2x^2 + 5$, or $Y1 = -0.5x^2 + 5$.
- Press GRAPH to see the parabola on the graph with the scatter plot.
- Change the value of a in the equation until the parabola passes through, or close to, as many of the points as possible. This will be the parabola of best fit. Record its equation.
- Do you think a parabola is a good model for these data? Explain.

Step 6. Repeat Steps 3 to 5, with a Steeper Ramp.

1. Compare the two parabolas of best fit.

2. How did the height of the ramp affect the equation?

The Gateway Arch

The photograph on the facing page shows the Gateway Arch in St. Louis, Missouri. Both the upper and lower curves formed by the arch appear to be parabolic.

Assume the axis of symmetry is the y-axis and the point at the top of the upper arch is the vertex of the upper curve.

1. Use the grid to write the coordinates of the vertex and 4 other points on the upper curve. Use the scale to express these coordinates in metres. For example, suppose a point has coordinates (4, 11). Each square is 1 cm. The scale is 1 cm to 15 m. In metres, the coordinates of the point are (4 × 15, 11 × 15), or (60, 165).

2. On grid paper, make a scatter plot of the data. Draw a smooth curve through the points.

3. Use the method of Section 6.2, *Investigation 2*, to estimate the equation of a parabola of best fit for the data.

   The parabola opens down and its vertex is on the y-axis. Therefore, the equation has the form $y = ax^2 + c$, where:
   • a is negative
   • c is the y-coordinate of the vertex

   Experiment to determine the value of a to make the parabola pass through, or close to, as many points as possible.

4. Write the equation of the parabola of best fit on the scatter plot.

5. Do you think a parabola is a suitable model for the upper curve? Explain.

6. Repeat exercises 1 to 5 for the lower curve.

7. Compare the equations of the parabolas of best fit for both curves. Explain the differences.

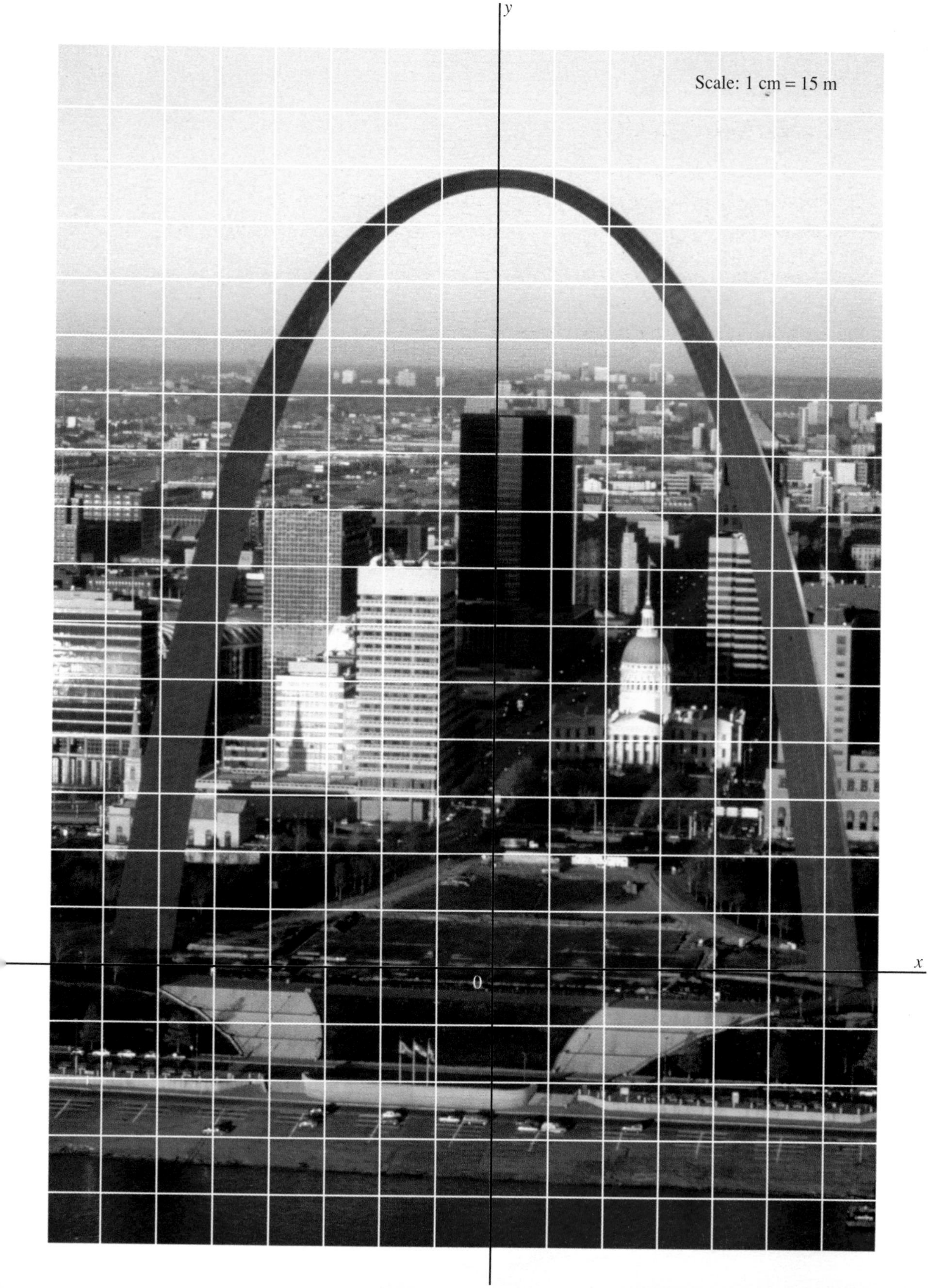
y
Scale: 1 cm = 15 m
x
0

Discussing the IDEAS

1. In *Investigation 1*, what do the coordinates of the vertex tell about the motion of the cylinder?

2. In *Investigation 1*, how do you think the equation would change if you used a longer ramp?

3. In *Investigation 2*, why is *a* negative?

4. Share the results of *Investigation 2* with your classmates. Did all of you agree on the equations of the parabolas of best fit for both curves? Explain any differences.

6.3 Exercises

A 1. Each parabola has an equation of the form $y = ax^2 + c$. For each parabola:
   i) State the coordinates of the vertex.
   ii) State whether *a* is positive or negative.

a)
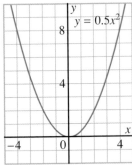

b)
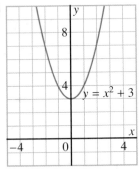

c)
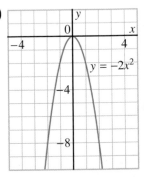

d)
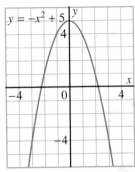

e)
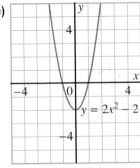

f)
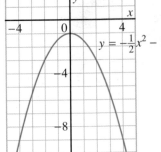

2. Each incomplete table contains data for a parabola. Each vertex is on the y-axis. Copy each table. Use the step pattern to complete it.

a)

x	y
-4	
-3	
-2	
-1	
0	0
1	1
2	4
3	
4	

b)

x	y
-8	
-6	
-4	
-2	
0	1
2	2
4	5
6	10
8	

c)

x	y
-8	
-6	
-4	
-2	
0	0
2	-1
4	-4
6	
8	

d)

x	y
-4	
-3	
-2	
-1	
0	3
1	2
2	-1
3	
4	

3. For each parabola in exercise 2, write the coordinates of the vertex, then graph the parabola.

B 4. Relating the Areas and the Perimeters of Squares

a) Draw five different squares on grid paper. Calculate their perimeters and areas.

b) Copy and complete this table. The first two rows have been completed.

Side length (cm)	Perimeter (cm)	Area (cm^2)
0	0	0
1	4	1
2	8	

c) Enter the *Perimeter* data in list L1 and the *Area* data in list L2.

d) Construct a scatter plot.

e) Assume the curve is a parabola. What are the coordinates of its vertex? What is the form of the quadratic equation?

f) Estimate, then enter, an equation for a parabola of best fit. Graph the data and your equation on the same screen. Change the value of a until the parabola is a parabola of best fit.

g) Write the equation of a parabola of best fit.

h) Use the equation or the graph to complete each exercise.
   i) A square has a perimeter of 100 cm. What is its area?
   ii) A square has a perimeter of 5 cm. What is its area?
   iii) A square has an area of 20 cm^2. What is its perimeter?

iv) Suppose you have 60 m of fence to enclose a square field for horses. What is the area of the field?

v) Suppose you want a field with an area twice as large as that in part iv. Will you need twice as much fence? Explain.

5. a) Look at the values of *x* and *y* in each table. Do the data represent a quadratic function? Explain.

b) Copy and complete each table.

i)

x	y	Difference
0	0	
1	1	
2	4	
3	9	
4	16	
5	25	

ii)

x	y	Difference
0	0	
2	1	
4	4	
6	9	
8	16	
10	25	

iii)

x	y	Difference
0	0	
1	2	
2	8	
3	18	
4	32	
5	50	

iv)

x	y .	Difference
0	0	
2	2	
4	8	
6	18	
8	32	
10	50	

v)

x	y	Difference
0	0	
1	3	
2	12	
3	27	
4	48	

vi)

x	y	Difference
0	0	
2	3	
4	12	
6	27	
8	48	

c) What do you notice about the *Difference* columns? What does this tell you about the relationship between *x* and *y*?

6. For each table in exercise 5:
- Plot the data in the first two columns as a scatter plot.
- Enter a quadratic equation to find a parabola of best fit. Since each vertex is (0, 0), each equation has the form $y = ax^2$.
- Sketch each graph and write its equation.

a) What do you notice about the values of a in the equations for the tables whose x-values increase by 1?

b) What do you notice about the values of a in the equations for the tables whose x-values increase by 2?

7. Predict which equation could model each situation. Explain.
- **i)** $y = 40x$
- **ii)** $y = 40x^2$
- **iii)** $y = -2x^2 + 40$
- **iv)** $y = x + 40$
- **v)** $y = -x + 40$

a) Jeans are on sale for $40.00 a pair. You graph the costs for 1, 2, 3… pairs.

b) An arch opening down has a maximum height of 40 m. You graph the shape of the arch.

c) Bob has 40 m of ribbon. Each present he wraps requires 1 m of ribbon. You graph the length left after wrapping each present.

d) Each square metre of cloth costs $40.00. You graph the costs for several different squares.

e) A company is to pave a square patio. It charges an initial fee of $40.00 to examine the location. It charges $1 per square metre for the paving stones and installation.

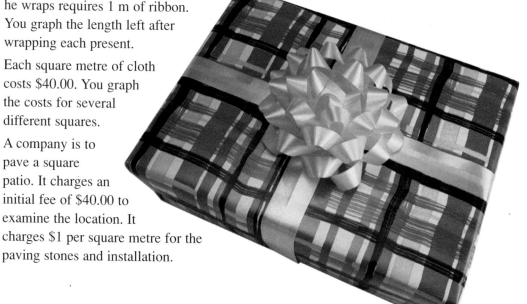

Communicating *the* **IDEAS**

Choose either *Investigation 1* or *Investigation 2*. Write a report of your findings so that someone who did not complete the investigation could understand. Describe the problem you investigated, the conclusions you reached, and how you arrived at your conclusions. Use written explanations, tables, graphs, equations, or calculations.

On page 298, you considered the motion of a ball thrown upward. You were shown three possible scenarios:

- A ball that kept moving upward.
- A ball that rose the same distance each quarter second, reached the top, then fell the same distance each quarter second.
- A ball that rose a shorter distance each quarter second, reached the top, then fell a greater distance each quarter second.

Since we know the ball landed after 2 seconds, the first model is not reasonable.

In the second model, the distance travelled each quarter second is the same. The relationship between the height of the ball and the elapsed time is *linear*.

In the third model, the distance travelled each quarter second appears to follow the step pattern of a parabola. The relationship between the height of the ball and the elapsed time is *quadratic*.

Is either model true for the motion of a ball?

A stroboscopic picture of a ball is shown at the right. By taking measurements from this picture, you can determine how the height of the ball is related to the elapsed time.

You need this information.

- The scale of the picture is 1 cm = 0.2 m.
- The time between flashes of the stroboscope is 0.1 s.

Step 1. Collect the Data.

1. Copy the table below.

2. There are 17 images of the ball in the picture. Number the images from the bottom left to the top then to the bottom right. Write these numbers in the first column of the table.

3. Measure the height of each image above the white line, in centimetres. Use the red dots at the centres of the images. Record the heights in the second column.

4. Use the time between flashes to calculate the elapsed time for each image. The time for the ball on the white line will be 0 seconds. Record these times in the third column.

5. Use the scale of the picture to calculate the actual heights of the ball in metres. Record the heights in the fourth column.

6. Make a scatter plot. Plot *Time* horizontally and *Actual height* vertically.

Image number	Height of ball on picture (cm)	Time (s)	Actual height of ball (m)

Step 2. Examine Your Graph.

7. Which of the models on page 298 looks most like your graph?

8. Add a fifth column to the table. Label it *Differences*. Calculate the differences in the *Actual heights*. Does a linear or a quadratic function describe the motion of the ball? Explain.

9. What are the coordinates of the vertex of the graph? How is this graph different from the parabolas you studied earlier?

Since the vertex of this parabola does not lie on the *y*-axis, it cannot be modelled by an equation of the form $y = ax^2$, or $y = ax^2 + c$. It requires an equation of the form $y = ax^2 + bx + c$. You will learn more about this in the next section.

Communicating *the* IDEAS

Write a report of your findings so that someone who did not complete this section could understand. Describe the problem you investigated, the conclusions you reached, and how you arrived at your conclusions. Use written explanations, tables, graphs, or calculations.

In Section 6.2, a quadratic function was defined as one whose equation can be written in the form $y = ax^2 + bx + c$, where $a \neq 0$. You examined many parabolas with equations of the form $y = ax^2$, and $y = ax^2 + c$ in Sections 6.2 and 6.3. All the parabolas had the y-axis as their axis of symmetry.

In this section, you will examine the graphs of quadratic functions with equations of the form $y = ax^2 + bx + c$. These graphs have an axis of symmetry that is not the y-axis.

Example 1

a) Graph the function $y = x^2 - 4x + 3$.

b) Determine the coordinates of the vertex and the equation of the axis of symmetry.

c) Determine the y- and x-intercepts.

Solution

- Press ⟨ Y= ⟩ and clear any equations or plots.
- Make sure the cursor is beside Y1=. Enter the equation $y = x^2 - 4x + 3$.
- Press ⟨WINDOW⟩. Change the settings to Xmin = −9.4, Xmax = 9.4, Xscl = 1, Ymin = −10, Ymax = 10, Yscl = 1.

a) Press ⟨GRAPH⟩ to obtain the graph below left.

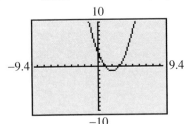
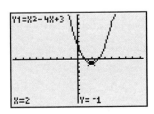

b) To determine the coordinates of the vertex:

- Press ⟨TRACE⟩, and move the cursor along the curve until the least value of y is seen. The graph above right shows that the coordinates of the vertex are $(2, -1)$.

Since this is a parabola, the axis of symmetry passes through the vertex. The equation of the axis of symmetry is $x = 2$.

c) Press [GRAPH]. The *y*-intercept is the value of *y* when $x = 0$.

- Press [TRACE]. Use [◄] or [►] to move the cursor to the point where $x = 0$. The *y*-intercept is 3.

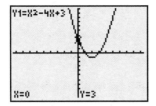

The *x*-intercept is the value of *x* when $y = 0$. The graph shows there are two *x*-intercepts.

- Press [TRACE]. Move the cursor to the leftmost point on the graph where $y = 0$ (below left). This *x*-intercept is 1.

- Move the cursor along the curve to the other point where $y = 0$ (below right). This *x*-intercept is 3.

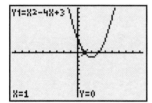
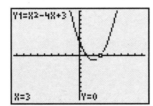

The *y*-intercept is 3, and the *x*-intercepts are 1 and 3.

In *Example 1*, the *x*-intercepts of the graph are 1 and 3.

When $x = 1$, the function $y = x^2 - 4x + 3$

$$\begin{align}\text{becomes } y &= (1)^2 - 4(1) + 3\\ y &= 1 - 4 + 3\\ y &= 0\end{align}$$

When $x = 3$, the function $y = x^2 - 4x + 3$

$$\begin{align}\text{becomes } y &= (3)^2 - 4(3) + 3\\ y &= 9 - 12 + 3\\ y &= 0\end{align}$$

Since $y = 0$ when $x = 1$ and $x = 3$, we say that 1 and 3 are the *zeros* of the function $y = x^2 - 4x + 3$. The zeros are the values of *x* for which $y = 0$; that is, the values of *x* when $x^2 - 4x + 3 = 0$.

The equation, $x^2 - 4x + 3 = 0$, is called a *quadratic equation*. It has two solutions, $x = 1$ and $x = 3$. These solutions are called the *roots* of the equation.

Take Note

The Zeros of a Function

The *x*-intercepts of the graph of a function are the *zeros* of the function. The zeros are the values of *x* for which $y = 0$.

A parabola always has a vertex. It is an important point. The *y*-coordinate of the vertex represents the *maximum value* of the function when its graph opens down. The *y*-coordinate of the vertex represents the *minimum value* of the function when the graph opens up.

Example 2

Graph the function $y = -x^2 + 4x - 3$. Use the graph to determine:

a) the maximum value of the function

b) the zeros of the function

Solution

- Press ⬚ Y= and clear any equations or plots.
- Make sure the cursor is beside Y1=. Enter the equation $y = -x^2 + 4x - 3$. Use (-), and not −, to enter the negative sign in front of x^2.
- Press WINDOW. Change the settings to Xmin = −9.4, Xmax = 9.4, Xscl = 1, Ymin = −10, Ymax = 10, Yscl = 1.
- Press GRAPH to obtain the graph below left.

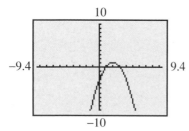
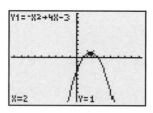

a) From the graph, the function $y = -x^2 + 4x - 3$ has a maximum value. This maximum value is the *y*-coordinate of the vertex of the parabola.
- Press TRACE, and move the cursor to the point with the greatest *y*-coordinate, above right.

The coordinates of the vertex are (2, 1), so the maximum value of the function is 1.

b) The zeros of the function are the *x*-intercepts of the graph. The screen shows there are two zeros.

- Press TRACE, and move the cursor to the left point where $y = 0$ (below left). This zero of the function is 1.
- Move the cursor to the right point where $y = 0$ (below right). This zero is 3.

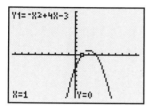

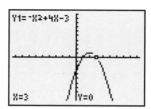

The zeros of the function are 1 and 3.

Discussing the IDEAS

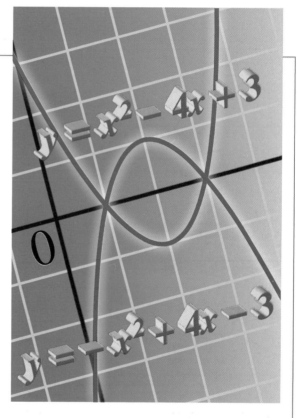

1. How are the parabolas in *Example 1* and *Example 2* different from the parabolas you graphed in earlier sections? How are they similar?

2. In *Example 1*, the *y*-intercept is (0, 3). Since parabolas are symmetrical, the reflection of this point in the axis of symmetry will also be on the parabola. What are the coordinates of the reflected point?

3. How is the parabola in *Example 2* different from the parabola in *Example 1*? Explain.

4. How many different *x*-intercepts could the graph of a quadratic function have? Explain.

5. How many different *y*-intercepts could the graph of a quadratic function have? Explain.

6. How can you tell from the equation of a quadratic function whether it has a maximum value or a minimum value?

A 1. Identify these properties of each quadratic function below.
 i) the coordinates of the vertex
 ii) the y-intercept
 iii) the x-intercepts
 iv) the coordinates of the image of the point corresponding to the
 y-intercept reflected in the axis of symmetry

a)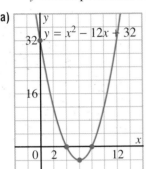

b)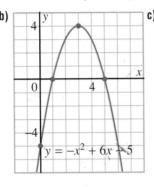

c)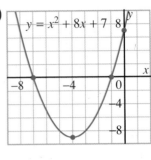

2. Write to explain how the vertex of a parabola is related to its axis of symmetry.

B 3. Graph each quadratic function. Identify each property, where possible.
 i) the coordinates of the vertex
 ii) the y-intercept
 iii) the x-intercepts

a) $y = 2x^2 - 8$
b) $y = x^2 - 6x + 8$
c) $y = -x^2 - 6x$
d) $y = 2x^2 + 6x$
e) $y = 0.5x^2 - 2x + 7$
f) $y = 2x^2 - 6x + 5$

4. Graph each function, then use Trace to find its zeros.

a) $y = x^2 - 3x + 2$
b) $y = -x^2 + 8x - 7$
c) $y = 2x^2 - 8x$
d) $y = 2x^2 - 6x - 8$
e) $y = -2x^2 + 6x + 8$
f) $y = 3x^2 + 6x$

5. Draw as many rectangles as you can that have a perimeter of 28 cm. Calculate
 the area of each rectangle. Copy and complete this table.

Length (cm)	Width (cm)	Area (cm²)
14	0	0
13	1	
12	2	

a) Graph the data. Plot *Width* horizontally and *Area* vertically. Join the
 points with a smooth curve.

b) Is this curve a parabola? Explain.

c) Write the coordinates of the vertex.

d) Write the equation of the axis of symmetry.

e) Write the coordinates of the *x*-intercepts of the graph.

f) What is the maximum value of the function?

g) What are the length and width of the rectangle when the function is a maximum?

h) Find two points on the graph with *y*-coordinate 40. Explain why there are two rectangles with the same area.

i) How many rectangles did you draw with an area greater than 40 cm²?

6. A hockey arena seats 1600 people. The cost of a ticket is $10.00. At this price, every ticket is sold.

To increase revenue, the manager decides to increase the ticket price. But if she increases the price, fewer people will buy tickets. The table shows how many people will buy tickets at each price.

a) Copy the table. Use the information to complete the third column.

Ticket price ($)	Number of people	Revenue ($)
10	1600	16 000
15	1300	
20	1015	
25	760	
30	517	

b) Graph the data. Plot *Ticket price* horizontally and *Revenue* vertically.

c) Draw a smooth curve through the plotted points.

d) What price should the manager set to earn the highest revenue?

e) Draw a horizontal line at a revenue of $18 000.

f) Between what two ticket prices (approximately) will the revenue be greater than $18 000?

g) Explain why the manager might decide to charge less than the price that produces the maximum revenue.

7. Private and military aircraft often carry a flare pistol to attract rescuers. The height of the flare above the ground is a quadratic function of the elapsed time since firing. If the flare is fired upward at 92 m/s, its height, *h* metres, is given by the equation $h = -4.9t^2 + 92t$, where *t* seconds is the time since firing.

a) Graph the equation.

b) From the graph, determine:
   i) the maximum height of the flare
   ii) the time it takes the flare to reach maximum height
   iii) the time when the flare returns to the ground
   iv) the two different times when the height of the flare is 50 m

c) From your answer in part iv above, determine the time for which the flare is higher than 50 m.

8. In a science experiment, a ball was thrown vertically upward. Stroboscopic camera equipment recorded its height every 0.2 s.

   a) Graph the data. Choose suitable scales. Plot *Time* horizontally and *Height* vertically. Sketch a parabola of best fit through the points.

   b) What are the coordinates of the vertex of the graph? How does this point relate to the maximum height reached by the ball?

   c) Estimate the time it took the ball to reach one-half of its maximum height, on its way up.

   d) Determine the horizontal intercepts of the graph. Explain how the horizontal intercepts relate to the time the ball was in the air.

   e) What is the relation between the coordinates of the vertex and the total time the ball was in the air?

Time (s)	Height above the ground (m)
0.0	0.0
0.2	1.8
0.4	3.2
0.6	4.2
0.8	4.8
1.0	5.0
1.2	4.8
1.4	4.2
1.6	3.2
1.8	1.8
2.0	0.0

FOCUS
TECHNOLOGY

Using the Calculate Menu

You can use the Calculate menu to determine the maximum or minimum value, and the zeros of a quadratic function.

9. Determine the minimum value and the zeros of the quadratic function $y = x^2 - 9x + 6$.

   • Press [Y=] and clear any equations or plots.

   • Make sure the cursor is beside Y1=. Enter the equation $y = x^2 - 9x + 6$.

   • Press [WINDOW]. Change the settings to Xmin = −9.4, Xmax = 9.4, Xscl = 1, Ymin = −20, Ymax = 20, Yscl = 5.

   • Press [GRAPH] to obtain the graph below left.

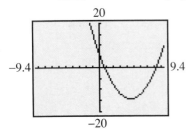

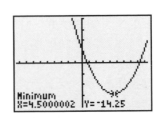

To determine the minimum value:

- Press [2nd] [TRACE] to obtain the Calculate menu. Press **3** to choose **minimum**. The words "Left Bound?" appear.
  Move the cursor close to the minimum point, and on the left side of this point. Press [ENTER]. The words "Right Bound?" appear. Move the cursor close to the right of the minimum point. Press [ENTER]. The word "Guess?" appears.

  Press [ENTER]. The calculator displays the screen, bottom right on page 338. It shows the coordinates of the minimum point. The coordinates are rounded to (4.5, –14.25).

To determine the zeros:

- Press [2nd] [TRACE] to obtain the Calculate menu. Press **2** to choose **zero**. The words "Left Bound?" appear. Move the cursor toward one of the points where the graph intersects the x-axis, and to the left side of this point. Press [ENTER]. The words "Right Bound?" appear. Move the cursor to the right side of the point where the graph intersects the x-axis. Press [ENTER]. The word "Guess?" appears. Press [ENTER]. The calculator displays the screen below left, to show the coordinates of this point to many decimal places.

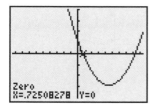
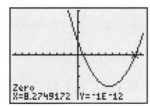

Repeat the process with the cursor close to the other point where the graph intersects the x-axis. The calculator displays the screen above right. To three decimal places, the zeros are 0.725 and 8.275.

10. Use the Calculate menu. For each function, determine:

  i) its maximum or minimum value

  ii) its zeros

  a) $y = 2x^2 + 7x - 12$ **b)** $y = -x^2 - 7x + 10$ **c)** $y = x^2 - 7x + 4$

Communicating *the* IDEAS

The vertex of a parabola is an important point. Write to explain its importance.

Testing Your Knowledge

Algebra Tools

Quadratic Functions

- The equation of every quadratic function has the form $y = ax^2 + bx + c$, where $a \neq 0$.
- The graph of every quadratic function is a parabola.
- Every parabola has an axis of symmetry and a vertex.
- An equation of the form $y = ax^2$ has vertex $(0, 0)$ and axis of symmetry the y-axis, or $x = 0$.
- An equation of the form $y = ax^2 + c$ has vertex $(0, c)$ and axis of symmetry $x = 0$.
- When a is positive, the graph of $y = ax^2 + bx + c$ opens up. The function $y = ax^2 + bx + c$ has a minimum value.
- When a is negative, the graph of $y = ax^2 + bx + c$ opens down. The function $y = ax^2 + bx + c$ has a maximum value.
- The zeros of the function $y = ax^2 + bx + c$ are the x-intercepts of the graph of this function; that is, the values of x for which $y = 0$.
- Step pattern
  When x increases by 1, there is a pattern in the differences of the y-values.
  Starting at $(0, 0)$, go 1 right 1 up, then 1 right 3 up, then 1 right 5 up, ...
  and go 1 left 1 up, 1 left 3 up, 1 left 5 up, ...

For example:

$y = x^2$

x	y
−4	16
−3	9
−2	4
−1	1
0	0
1	1
2	4
3	9
4	16

Start
1
3
5
7

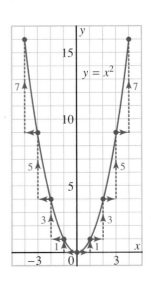

Getting in Shape

With cheerleading tryouts just 12 days away, Marsha decided to get in shape. Here is her fitness schedule.

Day / Exercise	1	2	3	4	5	6	7	8	9	10	11
Run 1 km	1	1	1	1	1	1	1	1	1	1	1
Sit-ups		2	2	2	2	2	2	2	2	2	2
Chin-ups			3	3	3	3	3	3	3	3	3
Knee bends				4	4	4	4	4	4	4	4
Jumping jacks					5	5	5	5	5	5	5
Push-ups						6	6	6	6	6	6
Toe touches							7	7	7	7	7
Leg lifts								8	8	8	8
Somersaults									9	9	9
Back flips										10	10
Cartwheels											11

1. Copy and complete this table. Count how many of each exercise Marsha did over the course of her training. Use 1 for the first exercise (Run 1 km), use 2 for the second exercise (Sit-ups), and so on. For example, Marsha ran 1 km every day for 11 days, so the total for exercise 1 is 11.

Exercise	Total performed
1	11
2	
3	

2. Plot the data on a grid. What shape do the points appear to have?

Testing Your Knowledge

3. Use the graph.

    a) Is this shape parabolic? Check whether the data satisfy the step pattern of a parabola.

    b) What is the maximum value of this function? What does this value tell you about Marsha's exercises?

    c) Suppose you wrote the equation of the parabola of best fit. Would the value of a be positive or negative? Explain.

4. Suppose Marsha added 4 more types of exercises and prepared for 15 days instead of 11. What would the maximum value of the function be?

5. Suppose Marsha prepared for only 10 days. How would this affect the maximum value of the function? Explain.

6. A parabola passes through the points A(1, 4), B(2, 7), and C(3, 12). Plot these points on a grid. Use the step pattern to determine the coordinates of the vertex of the parabola.

7. Make a table of values, then draw a graph of each quadratic function.

    a) $y = 2x^2$ **b)** $y = -3x^2$ **c)** $y = x^2 + 2$ **d)** $y = -x^2 - 4$ **e)** $y = 0.5x^2$

8. For each parabola in exercise 7, write the coordinates of the vertex and state whether it is a maximum or a minimum point.

9. Graph each quadratic function. Identify each property, where possible:

    i) the coordinates of the vertex

    ii) the y-intercept

    iii) the x-intercepts

    a) $y = 3x^2 - 5$ **b)** $y = -x^2 + 2$ **c)** $y = 2x^2 + 1$

10. Graph each quadratic function. Identify each property, where possible:

    i) the coordinates of the vertex

    ii) the y-intercept

    iii) the x-intercepts

    iv) the coordinates of the image of the point corresponding to the y-intercept reflected in the axis of symmetry

    a) $y = x^2 + 6x + 8$ **b)** $y = -2x^2 + 14x - 20$

11. Determine the zeros of each quadratic function.

    a) $y = x^2 + 5x + 4$ **b)** $y = x^2 - 3x$ **c)** $y = -x^2 + 3x + 10$

12. On a suspension bridge, the roadway is hung from huge cables passing through the tops of high towers. In the photograph of Vancouver's Lions Gate Bridge below, the main cables appear to have the shape of a parabola.

   a) Use the grid to write the coordinates of five points on the cable. The side length of one square represents 27 m. Use this scale to express the coordinates in metres.

   b) Graph the data on grid paper. Draw a smooth curve through the points.

   c) Enter the data into a graphing calculator and display a scatter plot.

   d) Press ⌷Y=⌷. Enter an equation of the form $y = ax^2$.

   e) Change a until the graph passes through as many points as possible.

   f) Write the equation of the parabola of best fit.

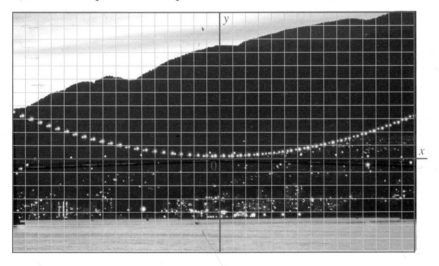

13. **The Relationship between Perimeter and Area**

   Draw as many rectangles as you can that have a perimeter of 20 cm and side length a whole number of centimetres.

   a) Calculate the area of each rectangle. Copy and complete this table.

   b) Graph the data. Plot *Width* horizontally and *Area* vertically. Join the points with a smooth curve.

Length (cm)	Width (cm)	Area (cm²)
0	0	0
9	1	9
	2	
	3	
	4	
	5	

c) What shape does this graph appear to have?

d) Check whether the data satisfy a step pattern.

e) Mark the vertex. Write its coordinates.

f) Draw the axis of symmetry. Write its equation.

g) Write the x-intercepts of the graph. What do they represent?

h) What information does the vertex tell about the area of a rectangle with perimeter 20 cm?

i) Find two points on the graph with y-coordinate 21. Why are there two rectangles with the same area?

j) How many rectangles in the scatter plot have an area greater than 21 cm^2?

14. When a soccer ball is kicked with a vertical speed of 20 m/s, its height, h metres, after t seconds is given by the equation $h = -4.9t^2 + 20t$.

a) Graph the function.

b) What is the maximum height of the soccer ball?

c) For how many seconds is the soccer ball in the air?

d) For how many seconds is the soccer ball higher than 3 m?

Testing Your Knowledge

15. **Mathematical Modelling** In a science experiment, water dripped steadily from a burette. The falling drops appear to be stopped in this stroboscopic picture.

The scale of the picture is 1 cm = 0.1 m.

The time between successive flashes of the stroboscope is 0.1 s.

a) Five drops are visible in the picture. Suppose these are numbered in order, starting at the top. Measure and record the distance from the white line down to the yellow centre of each drop.

b) Use the scale of the picture to calculate the distance each drop has fallen in centimetres.

c) Use the time between successive flashes to calculate the time each drop has been falling.

d) Enter the data into a graphing calculator; the times in list L1 and the distances in list L2.

e) Press ⬚ Y= . Enter an equation of the form $y = ax^2$. Graph the equation.

f) Examine the scatter plot and the graph. Change the value of a until the parabola passes through as many points as possible.

g) Write the equation of the parabola of best fit.

16. **Mathematical Modelling** Suppose a slingshot is used to throw a stone vertically upward at 30 m/s.
The equation for the stone's motion is $h = -4.9t^2 + 30t$, where h is the height of the stone in metres and t is the time in seconds.

Use a graphing calculator to graph the slingshot function.
- Press ⬚ Y= and clear any equations or plots.
- Make sure the cursor is beside Y1=. Enter the equation $y = -4.9x^2 + 30x$.
- Press WINDOW, and change the settings to Xmin = 0, Xmax = 9.4, Xscl = 1, Ymin = -15, Ymax = 55, Yscl = 10.

a) Press 2nd TRACE for Calculate to determine the maximum height of the stone.

b) Find the time it takes for the stone to reach its maximum height.

c) Use Calculate to determine the time it takes for the stone to return to the ground.

Round Robin Scheduling

Have you heard the term "round robin"? This is a way to schedule games in sports. Each player or team plays all the other players or teams once.

Suppose you are organizing a round robin tournament.

Suppose you know the number of players or teams. How can you determine the number of games to be scheduled?

How can you make sure that each player or team plays all the others once?

Until you develop a model to solve this problem, here are some things to consider.

1. In which sports do players or teams play round robin tournaments?

2. **a)** Are there any disadvantages to round robin tournaments?

   b) How might the disadvantages be overcome?

You will return to this problem in Section 7.3.

FYI Visit **www.awl.com/canada/school/connections**

For information related to the above problem, click on <u>MATHLINKS</u>, followed by Foundations of Mathematics 10. Then select a topic under Round Robin Scheduling.

This section reviews these concepts:

• The distributive law

• Simplifying polynomials

• Solving equations

The Distributive Law

During the summer, Samantha worked at the local hardware store. She earned $7 an hour. Here are the hours she worked in one week:

Mon.	Tues.	Wed.	Thurs.	Fri.
6 h	4 h	5 h	5 h	7 h

There are two ways to calculate Samantha's weekly earnings.

1. **Method 1**

   a) Calculate how much Samantha earned each day.

   b) Add the daily amounts.

2. **Method 2**

   a) Calculate how many hours Samantha worked that week.

   b) Multiply the total number of hours by the hourly pay.

3. The results of exercises 1 and 2 illustrate the distributive law. That is, $7(6) + 7(4) + 7(5) + 7(5) + 7(7) = 7(6 + 4 + 5 + 5 + 7)$

4. Use the expression in exercise 3 to explain the distributive law.

5. The area of this figure can be calculated in two ways.

   Method 1: Calculate the area of the large rectangle.

   Method 2: Calculate the area of each small rectangle, then add the areas.

   a) Calculate the area in the two ways described above.

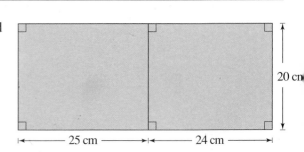

b) Write two expressions for the total area of each diagram.

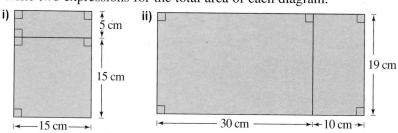

6. Use the distributive law to expand each expression.

a) $3(x + 2)$ **b)** $3(x - 2)$ **c)** $4(x + 5)$ **d)** $4(x - 5)$

e) $-3(x + 2)$ **f)** $-3(x - 2)$ **g)** $-4(x + 5)$ **h)** $-4(x - 5)$

7. Use the distributive law to write each sum as a product.

a) $3(x) + 3(2)$ **b)** $3(x) + 3(-2)$ **c)** $(-3)(x) + (-3)(2)$ **d)** $(-3)(x) + (-3)(-2)$

8. Simplify.

a) $7a - 12 - 5a + 3$ **b)** $6a - 4b + 3c - 5a - 2b + c$

c) $5x + 3y - 7 + x - y + 12$ **d)** $2x^2 + 5x - 3 + 4x^2 + 1$

e) $2m^2 - 7m + 5 + 3m^2 - 4m - 2$ **f)** $3t - 4w - t - w$

9. Add the polynomials.

a) $(7x + 4) + (3x + 2)$ **b)** $(7x - 4) + (3x - 2)$

c) $(-7x + 4) + (-3x + 2)$ **d)** $(-7x - 4) + (-3x - 2)$

e) $(2x^2 + 6x + 4) + (x^2 - 5x + 1)$ **f)** $(x^2 - 2x + 2) + (3x^2 + x - 1)$

g) $(-x^2 - 5x - 2) + (x^2 - 2x + 1)$ **h)** $(x^2 + 3x + 1) + (-2x^2 - x - 1)$

10. Subtract the polynomials.

a) $(7x + 4) - (3x + 2)$ **b)** $(7x - 4) - (3x - 2)$

c) $(-7x + 4) - (-3x + 2)$ **d)** $(-7x - 4) - (-3x - 2)$

e) $(2x^2 + 6x + 4) - (x^2 - 5x + 1)$ **f)** $(x^2 - 2x + 2) - (3x^2 + x - 1)$

g) $(-x^2 - 5x - 2) - (x^2 - 2x + 1)$ **h)** $(x^2 + 3x + 1) - (-2x^2 - x - 1)$

11. Simplify.

a) $(8a)(2a)$ **b)** $(9x)(-3x)$ **c)** $(4a^2)(-7a)$ **d)** $(-3x)(-5x)$

12. Solve.

a) $5x = 30$ **b)** $3x + 2 = 14$ **c)** $3a + 7 = 2a - 4$ **d)** $4x - 6 = x - 15$

Recall that a polynomial with one term is a monomial, and a polynomial with two terms is a binomial. In grade 9, you multiplied a monomial and a binomial, such as $3(x + 2)$, using the distributive law. You represented this product by combining 3 sets of algebra tiles:

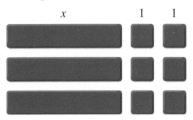

We can use the algebra tiles to form a rectangle.

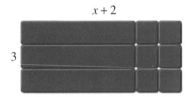

The area of the rectangle is $3(x + 2)$.
By counting tiles, the area is $3x + 6$.
We write $3(x + 2) = 3x + 6$.

Instead of writing the length and width as algebraic terms, we can use algebra tiles.

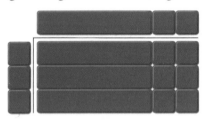

To represent the product $3x(x + 2)$, we make a rectangle $3x$ units long and $(x + 2)$ units wide. We place tiles to represent the length and the width, then fill in the rectangle with tiles. We need 3 x^2-tiles and 6 x-tiles.

The area of the rectangle is $3x(x + 2)$.
By counting tiles, the area is $3x^2 + 6x$.
We write $3x(x + 2) = 3x^2 + 6x$.

When we multiply a polynomial by a monomial using the distributive law, we say we are *expanding* the product.

Multiplying in Algebra

Use the distributive law. $3x(x + 2) = 3x(x) + 3x(2)$
$$= 3x^2 + 6x$$

Example

Expand.

a) $5x(x - 4)$

b) $(-3)(2x^2 - x + 1)$

Solution

Use the distributive law.

a) $5x(x - 4) = 5x(x) + 5x(-4)$
$$= 5x^2 - 20x$$

b) $(-3)(2x^2 - x + 1) = (-3)(2x^2) + (-3)(-x) + (-3)(1)$
$$= -6x^2 + 3x - 3$$

Discussing the IDEAS

1. Use algebra tiles to explain why $3x(x + 2)$ can be written as $(x + 2)3x$.

2. Some polynomials cannot be represented with algebra tiles. Can the polynomials in the *Example* be represented with algebra tiles? Explain your answer.

Practise Your Skills

1. Simplify.

a) 4×7
b) $3(-2)$
c) $(-3) \times 5$
d) $(-6)(-8)$
e) $(-4) \times 6$
f) $5(-7)$
g) $(-6)(-7)$
h) $(-0.5)(-1)$

2. Simplify.

a) $4(7x)$
b) $3(-2x)$
c) $(-3)(5x)$
d) $(-6)(-8x)$
e) $(-4)(6x)$
f) $5(-7x)$
g) $(-6)(-7x)$
h) $(-0.5)(-x)$

3. Simplify.

a) $(4x)(7x)$
b) $(3x)(-2x)$
c) $(-3x)(5x)$
d) $(-6x)(-8x)$
e) $(-4x)(6x)$
f) $(5x)(-7x)$
g) $(-6x)(-7x)$
h) $(-0.5x)(-x)$

A **1.** What product does each diagram represent?

a)

b)

c)

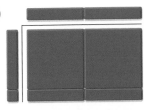

d)

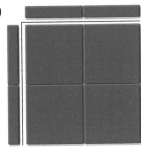

e)

f)

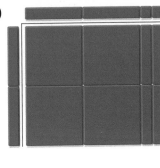

2. Use algebra tiles to expand each product.

a) $2(x + 3)$ **b)** $3(2x + 1)$ **c)** $2x(x + 1)$ **d)** $3x(x + 2)$

3. Match each product with the appropriate set of algebra tiles.

a) $3x(x + 2)$ **b)** $2x(x + 2)$ **c)** $2x(x + 4)$

d) $x(x + 4)$ **e)** $2x(2x + 4)$ **f)** $x(x + 2)$

A

B

C

D

E

F

B **4. Expand.**

a) $4(3x + 2)$ **b)** $3(5b - 6)$ **c)** $(-5)(x + 2)$ **d)** $(-1)(3x - 6)$

e) $2(2a + 5)$ **f)** $(-2)(n + 7)$ **g)** $(-1)(y - 1)$ **h)** $(-2)(3x - 5)$

5. Expand.

a) $x(x + 7)$ **b)** $x(x + 3)$ **c)** $a(3a - 4)$ **d)** $(-x)(4x + 1)$

e) $(-a)(4a + 2)$ **f)** $y(1 - y)$ **g)** $b(3b + 5)$ **h)** $(-n)(2n - 1)$

6. Expand.

a) $3x(x + 4)$ **b)** $2x(3x - 2)$ **c)** $5n(-2n + 3)$ **d)** $3a(-7a - 2)$

e) $(-7y)(y + 3)$ **f)** $(-5x)(3x + 4)$ **g)** $(-2x)(4x - 3)$ **h)** $4x(3 - 6x)$

7. Expand.

a) $2(x^2 + 3x - 4)$ **b)** $-2(x^2 + 3x - 4)$ **c)** $x(x^2 + 3x - 4)$

d) $(-x)(x^2 + 3x - 4)$ **e)** $(-3a)(a^2 - 3a - 2)$ **f)** $(-4m)(-m^2 - m + 1)$

g) $(-x)(2x^2 - 3x + 4)$ **h)** $(-2x)(-x^2 - 2x - 1)$ **i)** $4x(x^2 + x - 3)$

8. a) Use the distributive law to multiply 5×34.

     b) Use the distributive law to multiply $5(3x + 4)$. Evaluate this polynomial for $x = 10$.

     c) Compare your answers to parts a and b. Explain any relationship you discover.

9. Evaluate each binomial for $x = 1$.

a) $x + 3$ **b)** $x - 3$ **c)** $-x + 3$ **d)** $-x - 3$

10. Evaluate each polynomial in exercise 9 for $x = -1$.

11. Evaluate each polynomial for $x = 2$.

a) $x^2 + 3x + 4$ **b)** $-x^2 + 3x - 4$ **c)** $x^2 - 3x + 4$

d) $-x^2 - 3x - 4$ **e)** $x^2 + 3x - 4$ **f)** $-x^2 + 3x + 4$

12. Evaluate each polynomial in exercise 11 for $x = -2$.

Communicating
the IDEAS

Draw a rectangle with length $2x$ and width $(x + 3)$. Write the area of this rectangle as a product. Use the distributive law to write the area as a sum of the areas of two smaller rectangles. Divide the large rectangle into two smaller rectangles and show their lengths and widths.

In Section 7.1, we knew the length and width of a rectangle, and had to decide which tiles completed the rectangle. In this section, we reverse this procedure.

To *factor* a natural number means to write it as a product of other natural numbers (but not 1). For example,

$$12 = 6 \times 2 \qquad\qquad 12 = 4 \times 3 \qquad\qquad 12 = 2 \times 2 \times 3$$

When we write $12 = 2 \times 2 \times 3$, we say that 12 has been *factored fully* because none of the factors can be factored further.

INVESTIGATION

Using Algebra Tiles to Factor

1. Use 2 x-tiles and 6 1-tiles. What polynomial do they represent?

2. Arrange these tiles to form a rectangle.

   a) Write the length of the rectangle as a polynomial.

   b) Write the width of the rectangle as a polynomial.

   c) Write the area of the polynomial as a product of its length and width.

3. Check the rectangles of other students. Did all of you arrange the tiles the same way?

4. Repeat exercises 2 and 3 for these algebra tiles: 4 x-tiles and 4 1-tiles.

5. Explain how the rectangles from the two sets of tiles are different.

In the *Investigation*, you may have found 2 different rectangles for the algebra tiles that represent $2x + 6$:

• Length $2x + 6$, width 1, area $1(2x + 6)$

• Length $x + 3$, width 2, area $2(x + 3)$

We say that 1 and $2x + 6$ are factors of $2x + 6$.
Similarly, 2 and $x + 3$ are factors of $2x + 6$.

The first way to factor is incomplete because the second factor, $2x + 6$, can be factored again.

The second way to factor is complete.
We say that $2x + 6$ is *factored fully* when we write $2x + 6 = 2(x + 3)$.

This chart compares factoring and expanding in arithmetic and algebra.

In Arithmetic	In Algebra
We *multiply* factors to form a product. $$(2)(3) = 6$$ $$(\text{factor})(\text{factor}) = \text{product}$$ We *factor* a number by expressing it as a product of factors. $$6 = (2)(3)$$	We *expand* an expression to form a product. $$2(x + 3) = 2x + 6$$ $$(\text{factor})(\text{factor}) = \text{product}$$ We *factor* a polynomial by expressing it as a product of factors. $$2x + 6 = 2(x + 3)$$

The operations of expanding and factoring are inverses: that is, each operation reverses the other.

Example 1

Use algebra tiles to factor fully. $6x^2 + 9x$

Solution

a) Use 6 x^2-tiles and 9 x-tiles to represent $6x^2 + 9x$.
Arrange the tiles to form a rectangle.

The width and length of this rectangle are x and $6x + 9$.
But $6x + 9$ can be factored again.

Arrange the tiles in a different rectangle.

The length and width are $3x$ and $2x + 3$.

From the diagram, $6x^2 + 9x = 3x(2x + 3)$

Some polynomials cannot be represented by algebra tiles. They have to be factored algebraically.

Example 2

Factor fully.

a) $4x^3 + 6x^2$ **b)** $6x^2 - 9x + 15$

Solution

a) $4x^3 + 6x^2$

Factor each term of the polynomial.

$4x^3 = 2 \cdot 2 \cdot x \cdot x \cdot x$
$6x^2 = 2 \cdot 3 \cdot x \cdot x$

Identify the factors that are common to each term.
Each term has the factors 2 and x and x.
So, $2(x)(x)$, or $2x^2$ is the greatest common factor.
Write each term as a product of the greatest common factor and another monomial.

$4x^3 + 6x^2 = 2x^2(2x) + 2x^2(3)$

Use the distributive law to write the sum as a product.
$4x^3 + 6x^2 = 2x^2(2x + 3)$

b) $6x^2 - 9x + 15$

Factor each term of the polynomial.

$6x^2 = 2 \cdot 3 \cdot x \cdot x$
$9x = 3 \cdot 3 \cdot x$
$15 = 3 \cdot 5$

Identify the factors that are common to each term.
Each term has the factor 3.
3 is the greatest common factor.

Write each term as a product of 3 and another monomial.
$6x^2 - 9x + 15 = 3(2x^2) - 3(3x) + 3(5)$

Use the distributive law to write the sum as a product.
$6x^2 - 9x + 15 = 3(2x^2 - 3x + 15)$

We can check a factored polynomial by expanding.

For example, in *Example 2a*, $4x^3 + 6x^2 = 2x^2(2x + 3)$

To check, expand the right side of this equation.

That is, $2x^2(2x + 3) = 2x^2(2x) + 2x^2(3)$

$$= 4x^3 + 6x^2$$

Since this is the original polynomial, the factoring is correct.

Discussing *the* IDEAS

1. In *Example 1*:

    a) How do we know that $6x + 9$ can be factored further?

    b) How do we know that $3x(2x + 3)$ cannot be factored further?

2. In *Example 2a*, explain how each term in the binomial, $4x^3 + 6x^2$, was written as a product.

Practise Your Skills

1. Factor fully.

    a) 48   **b)** 81   **c)** 45   **d)** 108   **e)** 800   **f)** 120

2. Each number is written in factored form. State the greatest common factor of each pair of numbers.

    a) $8 = 2 \cdot 2 \cdot 2$ **b)** $9 = 3 \cdot 3$ **c)** $10 = 2 \cdot 5$
            $6 = 2 \cdot 3$ $12 = 2 \cdot 2 \cdot 3$ $25 = 5 \cdot 5$

    d) $16 = 2 \cdot 2 \cdot 2 \cdot 2$ **e)** $20 = 2 \cdot 2 \cdot 5$ **f)** $28 = 2 \cdot 2 \cdot 7$
            $32 = 2 \cdot 2 \cdot 2 \cdot 2 \cdot 2$ $36 = 2 \cdot 2 \cdot 3 \cdot 3$ $21 = 3 \cdot 7$

    g) $30 = 2 \cdot 3 \cdot 5$ **h)** $39 = 3 \cdot 13$ **i)** $40 = 2 \cdot 2 \cdot 2 \cdot 5$
            $50 = 2 \cdot 5 \cdot 5$ $26 = 2 \cdot 13$ $60 = 2 \cdot 2 \cdot 3 \cdot 5$

3. Use the lists of factors in exercise 2. State the greatest common factor of each pair of numbers.

    a) 8, 12   **b)** 6, 9   **c)** 16, 36   **d)** 28, 36   **e)** 25, 40

4. Use the lists of factors in exercise 2. State the greatest common factor of each set of numbers.

    a) 6, 8, 12   **b)** 9, 12, 30   **c)** 10, 20, 40   **d)** 12, 16, 36

A **1.** State the greatest common factor of each pair of numbers.

  a) 9, 15   **b)** 16, 28   **c)** 50, 75   **d)** 12, 24   **e)** 12, 4

2. Write each monomial as a product of 2 and another monomial.

  a) $2x$ **b)** $-4x$ **c)** $2x^2$ **d)** $-4x^2$ **e)** $8x^3$

3. Write each monomial as a product of $3x$ and another monomial.

  a) $-6x$ **b)** $3x^2$ **c)** $6x^2$ **d)** $-12x^3$ **e)** $3x^3$

4. Write each monomial as a product of $2a^2$ and another monomial.

  a) $2a^2$ **b)** $-2a^2$ **c)** $4a^2$ **d)** $-8a^2$ **e)** $-8a^3$

5. Find the greatest common factor.

  a) $2x, 4$ **b)** $2x^2, 4x$ **c)** $2x^3, 4x^2$ **d)** $-2x, 4$

  e) $3x, 6$ **f)** $3x^2, 9x$ **g)** $-6x^2, 6x$ **h)** $3x^2, -9$

6. Write each sum as a product.

  a) $3(x) + 3(1)$ **b)** $4(a) + 4(2)$ **c)** $2(3) + 2(2x)$

  d) $3(5) + 3(7x)$ **e)** $(-2)(x) + (-2)(5)$ **f)** $(-1)(4) + (-1)(3x)$

7. Write each difference as a product.

  a) $3(x) - 3(1)$ **b)** $4(a) - 4(2)$ **c)** $2(3) - 2(2x)$

  d) $3(5) - 3(7x)$ **e)** $2(x) - 2(5)$ **f)** $2(x^2) - 2(3)$

8. Find the greatest common factor.

  a) $3x^2, 2x$ **b)** x, x^2 **c)** $a, -a^2$

  d) $-4x, 16$ **e)** $2x^2, 6x$ **f)** $12x^2, 15x$

9. Use algebra tiles to factor each binomial.

  a) $3x + 3$ **b)** $2y + 4$ **c)** $6a + 3$

  d) $5x + 15$ **e)** $4a + 12$ **f)** $4n + 10$

10. Factor.

  a) $3x - 3$ **b)** $-2y + 4$ **c)** $-6a - 3$

  d) $5x - 15$ **e)** $-4a + 12$ **f)** $4n - 10$

B **11.** Factor and check.

  a) $x + 3x^2$ **b)** $4y - y^2$ **c)** $2a - 6$

  d) $9b^2 - 6b$ **e)** $-5b - 15b^2$ **f)** $-4x + 3x^2$

  g) $10b + 12b^2$ **h)** $-4x + 9x^2$ **i)** $-6x^2 - 8x^3$

12. Factor each binomial.

a) $5y + 10$ b) $8m^2 + 24$ c) $6x + 7x^2$

d) $35a + 10a^2$ e) $45d^3 + 36d$ f) $49b^2 + 7b$

g) $3x^2 + 6x$ h) $8y^2 + 4y$ i) $5p^3 + 15p^2$

j) $24m^2 + 16m$ k) $12a + 18a^2$ l) $-28y^2 - 35y$

13. Check each answer in exercise 12 by expanding.

14. Factor.

a) $36x^2 + 12x - 6$ b) $8x^2 + 5x^3 + x$ c) $a^3 + 12a^2 - 4a$

d) $9x^2 + 6x^3 - 12x$ e) $8y^2 - 32y + 24y^3$ f) $16x^2 - 32x + 16x^3$

15. Factor and check.

a) $9b^2 - 3b + 12$ b) $13y^3 + 6y^2 + 3y$ c) $4x^6 + 32x^5 + 48x^4$

d) $12y^4 - 12y^3 + 24y^2$ e) $10a^3 + 17a^2 + 18a$ f) $10z^4 - 15z^2 + 30z$

16. Recall the formula for the total surface area, A, of a cylinder.

$A = 2\pi r^2 + 2\pi rh$

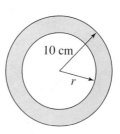

r is the radius and h is the height.

a) Factor the formula.

b) Use the result of part a. Sketch a rectangle whose area is equal to that of the total surface area of a cylinder. Label the dimensions of the rectangle.

c) For one cylindrical can, r is 5 cm and h is 12 cm.
   i) What is the total surface area of the can?
   ii) What are the approximate dimensions of a rectangle that has the same area as the can in part i?

17. In this diagram, the radius of the large circle is 10 cm. The radius of the small circle is r centimetres.

a) What is the area of the large circle?

b) What is the area of the small circle?

c) What is the shaded area?

d) Factor the expression in part c.

e) Suppose $r = 5$ cm. What is the shaded area?

Communicating *the* IDEAS

Write to explain how factoring and expanding are related. Use examples to illustrate your explanation.

7.3 Round Robin Scheduling

On page 346, you considered organizing a round robin tournament. Suppose you know the number of teams. How can you determine the number of games to be scheduled? How can you match up the teams so each team plays every other team once?

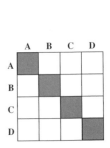

Use a Numerical Model

Suppose there are 4 teams in a round robin tournament.

The teams are A, B, C, and D.

You can count the games each team plays.

Team A plays teams B, C, D. There are 3 games.

This is true for each team.

Since there are 4 teams, the number of games is $4 \times 3 = 12$.

But this counts every game twice (B against A is the same game as A against B).

So, there must be $\frac{12}{2} = 6$ games.

1. Use the numerical method. Determine the number of games in a round robin tournament with each number of teams.

    a) 5 teams   b) 6 teams   c) 7 teams

2. Suppose there are n teams in a round robin tournament. Use the numerical method. Write a formula for the number of games, g, in terms of n.

Use a Grid

Visualize the schedule for 4 teams. The schedule has 16 spaces, but 4 of these cannot be used because a team cannot play itself. There are $16 - 4 = 12$ spaces for games. But this counts each game twice. So, there must be 6 games.

3. Use the grid method. Determine the number of games in a round robin tournament with each number of teams.

    a) 5 teams   b) 6 teams   c) 7 teams

4. Suppose there are *n* teams in a round robin tournament. Use the grid method. Write a formula for the number of games, *g*, in terms of *n*.

5. Compare your formulas in exercise 2 and exercise 4. Use algebra to show that the two formulas are the same.

If you have a graphing calculator, complete exercise 6; otherwise, complete exercise 7.

6. a) Graph the formulas in exercises 2 and 4. How do the graphs show that the formulas are the same?

b) Use the Trace or Tables feature. Determine the numbers of games required for 8, 9, 10, 11, and 12 teams. Show the results in a table.

7. Determine the numbers of games required for 8, 9, 10, 11, and 12 teams. Show the results in a table.

Use a Diagram

8. Explain how the diagram at the right can be used to schedule a round robin tournament with 4 teams.

9. Suppose the diagrams below are used to schedule a tournament. Answer these questions for each diagram.
   i) How many teams are in the tournament?
   ii) How many games are scheduled?

a)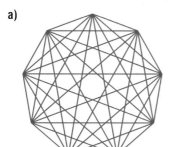

b)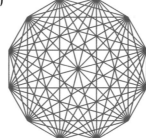

Comparing the Models

10. You used three different models to schedule a tournament. Which model do you think is best? Explain.

Communicating the IDEAS

Write a set of instructions to schedule a round robin tournament. Your instructions should include steps for determining the number of games to be played, and for matching the teams so each team plays all the others once.

In Section 7.1, we learned to multiply a monomial and a polynomial using the distributive law. Recall that a monomial has one term, a binomial has two terms, and a polynomial has one or more terms.

Area Models of Binomial Products

1. **a)** What is the length of this rectangle?

   b) What is the width of this rectangle?

   c) Write the area of the rectangle as the product of its length and its width.

   d) Write the area of the rectangle as the sum of the areas of the algebra tiles.

2. In the rectangle above

   a) How many x^2-tiles are there?

   b) How many x-tiles are there?

   c) How many 1-tiles are there?

3. Explain how the area model shows that $(x + 3)(x + 2) = x^2 + 5x + 6$

4. Use algebra tiles to make an area model to illustrate each product. Sketch each area model. Write an equation similar to that in exercise 3 under each sketch.

   a) $(x + 1)(x + 2)$ **b)** $(x + 4)(x + 2)$

   c) $(x + 3)(x + 5)$ **d)** $(x + 3)(x + 1)$

5. What patterns do you see in the area equations?

We can illustrate the product of two binomials with a diagram.

	x	7
x	x^2	$7x$
2	$2x$	14

The length of the large rectangle is $x + 7$ and its width is $x + 2$.

The area of the rectangle is $(x + 7)(x + 2)$.

The area is also the sum of the areas of the four small rectangles, $x^2 + 7x + 2x + 14$.

Therefore, $(x + 7)(x + 2) = x^2 + 7x + 2x + 14$
$$= x^2 + 9x + 14$$

This method shows that there are four terms in the product.

Example 1

Use algebra tiles to expand $(3x + 1)(x + 2)$.

Solution

Use algebra tiles to make a rectangle with length $3x + 1$ and width $x + 2$.

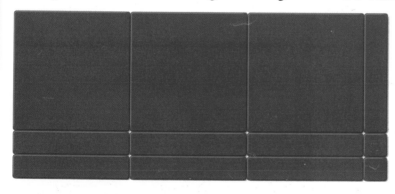

From the diagram, $(3x + 1)(x + 2) = 3x^2 + 7x + 2$

We can multiply two binomials without using algebra tiles.
We use arrows to help remember each pair of terms in the product.

A **1.** i) State the area of each rectangle as the product of its length and width.

ii) State the area as the sum of the areas of the tiles.

a)

b)

c)

2. i) State the area of each rectangle as the product of its length and width.

ii) State the area as the sum of the areas of the smaller rectangles.

a)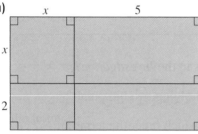

b)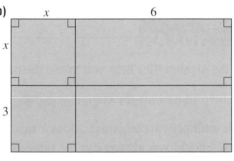

3. a) Write the area of each rectangle as a product.

i)

ii)

iii)

iv)

v)

vi)

b) Write each area in part a as a sum.

4. a) Write the area of each rectangle as a product.

i)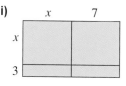

ii)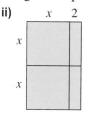

iii)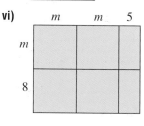

iv)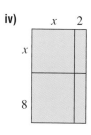

v)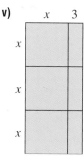

vi)

b) Write each area in part a as a sum.

B **5.** Expand. Use algebra tiles.

   a) $(x + 1)(x + 4)$ **b)** $(n + 2)(n + 5)$ **c)** $(a + 3)(a + 3)$ **d)** $(y + 1)(y + 5)$

6. Expand.

   a) $(a + 1)(a - 2)$ **b)** $(b + 3)(b - 5)$ **c)** $(n - 3)(n - 2)$ **d)** $(y - 4)(y + 5)$
   e) $(b - 6)(b + 3)$ **f)** $(a - 10)(a - 6)$ **g)** $(z - 5)(z - 6)$ **h)** $(b + 10)(b + 5)$

7. State the number that belongs in each square.

   a) $(x + 3)(x - 2) = x^2 + \square\, x - 6$ **b)** $(a + 5)(a + 7) = a^2 + \square\, a + 35$
   c) $(n - 4)(n - 6) = n^2 + \square\, n + 24$ **d)** $(x - 7)(x + 1) = x^2 + \square\, x - 7$

8. Expand.

   a) $(2a + 1)(a - 2)$ **b)** $(3b + 1)(3b + 1)$ **c)** $(2x - 3)(x - 2)$
   d) $(4x + 5)(x + 1)$ **e)** $(2a - 3)(3a + 2)$ **f)** $(5x - 2)(3x + 4)$

9. Expand.

   a) $(x + 3)(x + 7)$ **b)** $(x - 3)(x - 7)$ **c)** $(x - 3)(x + 7)$ **d)** $(x + 3)(x - 7)$

10. Expand.

   a) $(2a + 1)(3a + 2)$ **b)** $(2a - 1)(3a - 2)$
   c) $(2a - 1)(3a + 2)$ **d)** $(2a + 1)(3a - 2)$

11. Expand.

   a) $(a + 2)^2$ **b)** $(3n + 1)^2$ **c)** $(x - 6)^2$ **d)** $(a - 3)^2$ **e)** $(2y - 5)^2$ **f)** $(3b + 5)^2$

12. Patterns in Binomial Products: Part I

a) Expand the products in each list.

i) $(x + 1)(x + 1)$ **ii)** $(x + 1)(x - 2)$ **iii)** $(x + 2)(x + 1)$
$(x + 1)(x + 2)$ $(x + 2)(x - 2)$ $(x + 2)(x + 2)$
$(x + 1)(x + 3)$ $(x + 3)(x - 2)$ $(x + 2)(x + 3)$
$\vdots$ $\vdots$ $\vdots$

b) Describe any patterns you found in part a.

c) Predict the next three products for each list in part a.

d) Suppose you extend the lists in part a upward. Predict the preceding three products in each list.

13. Expand.

a) $(x - 3)(x + 3)$ **b)** $(2a + 1)(2a - 1)$ **c)** $(8n - 3)(8n + 3)$

d) $(4a + 3)(4a - 3)$ **e)** $(3x - 2)(3x + 2)$ **f)** $(5x + 1)(5x - 1)$

14. Describe the pattern you see in the products in exercise 13.

15. Expand.

a) $(6x - 3)(2x - 5)$ **b)** $(3b + 2)(3b - 2)$ **c)** $(5a + 1)(4a - 7)$

d) $(a + 8)(8a + 1)$ **e)** $(2a - 3)(2a - 3)$ **f)** $(3a + 4)(2a - 3)$

16. Expand.

a) $(2 - x)(3 - x)$ **b)** $(5 + a)(3 + a)$ **c)** $(4 - m)(3 + m)$

d) $(6 + t)(3 - t)$ **e)** $(7 - x)(7 - x)$ **f)** $(7 - x)(7 + x)$

C 17. Find the binomial to complete each equality.

a) $(n + 2)(\qquad) = n^2 + 7n + 10$

b) $(x - 3)(\qquad) = x^2 - 7x + 12$

c) $(\qquad)(x + 6) = x^2 + 4x - 12$

d) $(\qquad)(a - 5) = a^2 - 3a - 10$

e) $(x + 2)(\qquad) = x^2 + 5x + 6$

f) $(t - 4)(\qquad) = t^2 + t - 20$

g) $(\qquad)(\qquad) = x^2 + 9x + 20$

h) $(\qquad)(\qquad) = a^2 - 9a + 14$

Communicating the IDEAS

A student in your class says that $(x + 5)^2$ equals $x^2 + 25$. Use an area model to explain why this is not correct.

7.5 Expanding and Simplifying Polynomial Expressions

Like terms have exactly the same variables, so $2x$, $-5x$, and $0.5x$ are like terms, while $3x$ and x^2 are unlike terms. In an earlier grade, you learned to simplify algebraic expressions by collecting like terms. In Section 7.4, after multiplying binomials you collected like terms to write the product in simplest form.

In this section, we will combine these skills to expand then simplify polynomial expressions.

Example 1

Expand. $4(3x + 1)(x - 2)$

Solution

> **Think ...**
>
> $4(3x + 1)(x - 2)$ is the product of a monomial and two binomials. To reduce this to the product of two binomials, multiply the monomial and the first binomial.

$$4(3x + 1)(x - 2) = (4(3x) + 4(1))(x - 2)$$
$$= (12x + 4)(x - 2)$$

Multiply the two binomials.

$$4(3x + 1)(x - 2) = 12x(x) + 12x(-2) + 4(x) + 4(-2)$$
$$= 12x^2 - 24x + 4x - 8$$
$$= 12x^2 - 20x - 8$$

Example 2

Expand, then simplify. $(2x - 1)(3x + 4) + (2x - 3)^2$

Solution

$$\begin{aligned}
(2x - 1)(3x + 4) + (2x - 3)^2 &= (2x - 1)(3x + 4) + (2x - 3)(2x - 3) \\
&= (6x^2 + 8x - 3x - 4) + (4x^2 - 6x - 6x + 9) \\
&= (6x^2 + 5x - 4) + (4x^2 - 12x + 9) \\
&= 6x^2 + 5x - 4 + 4x^2 - 12x + 9
\end{aligned}$$

Collect like terms.

$$\begin{aligned}
(2x - 1)(3x + 4) + (2x - 3)^2 &= 6x^2 + 4x^2 + 5x - 12x - 4 + 9 \\
&= 10x^2 - 7x + 5
\end{aligned}$$

Example 3

Expand, then simplify. $(2x - 1)(3x + 4) - (2x - 3)^2$

Solution

$$\begin{aligned}
(2x - 1)(3x + 4) - (2x - 3)^2 &= (2x - 1)(3x + 4) - (2x - 3)(2x - 3) \\
&= (6x^2 + 5x - 4) - (4x^2 - 12x + 9)
\end{aligned}$$

To subtract a polynomial, add the opposite.

$$\begin{aligned}
(2x - 1)(3x + 4) - (2x - 3)^2 &= (6x^2 + 5x - 4) + (-4x^2 + 12x - 9) \\
&= 6x^2 + 5x - 4 - 4x^2 + 12x - 9
\end{aligned}$$

Collect like terms.

$$\begin{aligned}
(2x - 1)(3x + 4) - (2x - 3)^2 &= 6x^2 - 4x^2 + 5x + 12x - 4 - 9 \\
&= 2x^2 + 17x - 13
\end{aligned}$$

Example 4

Write an expression for the area of the shaded region.

Solution

Divide the shaded region into two rectangles. Label the rectangles with their dimensions.

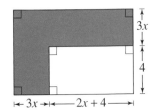

Rectangle A has length $3x + (2x + 4) = 5x + 4$

Rectangle A has width $3x$

Area of Rectangle A $= 3x(5x + 4)$
$= 15x^2 + 12x$

Rectangle B has length $3x$
Rectangle B has width 4
Area of Rectangle B is $3x(4) = 12x$
Area of shaded region $= (15x^2 + 12x) + 12x$
$= 15x^2 + 12x + 12x$
$= 15x^2 + 24x$

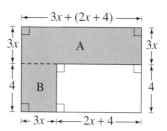

Discussing the IDEAS

1. Describe how you could check the solution to *Example 1*.

2. *Example 4* could be solved by subtracting the areas of two rectangles.
   a) Identify the two rectangles to be subtracted.
   b) Which polynomial products result when using these two rectangles? Solve *Example 4* using these rectangles.

7.5 Exercises

A

1. Simplify.
   a) $x^2 + 6x + 2x + 10$
   b) $x^2 + 6x - 2x + 10$
   c) $x^2 - 2x + 6x + 10$
   d) $x^2 - 6x - 2x + 10$

2. Simplify.
   a) $2(x^2 + 3x)$
   b) $(-2)(x^2 + 3x)$
   c) $2(x^2 - 3x)$
   d) $(-2)(x^2 - 3x)$

3. Simplify.
   a) $2x^2 + 3x + 4 + 3x^2 + 2x + 5$
   b) $2x^2 + 3x - 4 + 3x^2 - 2x - 5$
   c) $2x^2 - 3x - 4 + 3x^2 - 2x + 5$
   d) $2x^2 - 3x + 4 + 3x^2 + 2x - 5$

4. Expand and simplify.
   a) $(x + 2y)(x + 5y)$
   b) $(a - 3b)(a + 2b)$
   c) $(3m - n)(2m - n)$
   d) $(5x + 3y)(4x - y)$
   e) $(6r + s)(r - 3s)$
   f) $(8a + 7b)(7a + 8b)$

B

5. Simplify.
   a) $3(x - 2)(x + 4)$
   b) $2(y - 2)(y - 5)$
   c) $5(-m + 4)(m - 3)$
   d) $-(t - 5)(t + 1)$

6. Expand and simplify.

a) $2(x + 2)(x + 5)$ **b)** $3(m - 1)(m + 4)$ **c)** $5(x + 6)(x - 2)$

d) $7(x - 5)(x - 5)$ **e)** $-3(x + 4)(x - 2)$ **f)** $-2(x + 6)(x - 10)$

7. Simplify.

a) $2(x - 1)^2$ **b)** $3(y + 4)^2$ **c)** $-(3k + 1)^2$ **d)** $2(t - 3)^2$

8. Expand and simplify.

a) $(x + 3)(x - 2) + (5x + 2)$ **b)** $(x - 4)(x - 1) + (3x - 4)$

c) $(x - 5)(x + 1) + (x + 3)(x - 4)$ **d)** $(x + 3)(x - 8) + (2x + 1)(x - 3)$

e) $(x - 4)(x - 6) - (x + 2)(x + 6)$ **f)** $(x - 5)(x + 1) - (x + 2)(x + 4)$

9. Expand and simplify.

a) $(x - 3)(x + 1) + (x - 4)^2$ **b)** $(y - 2)^2 + (y + 1)(y - 7)$ **c)** $(y + 1)^2 + (y - 3)^2$

d) $(x - 3)^2 + (x - 4)^2$ **e)** $(2y - 3)^2 - (5y + 1)^2$ **f)** $(2x + 1)^2 - (4x - 3)^2$

10. What is the area of each shaded region?

a)

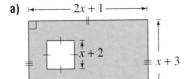
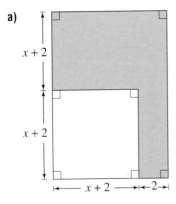

b)

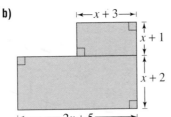
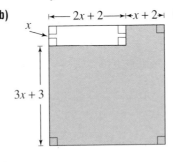

11. What is the area of each shaded region? In each case, write to explain how to calculate the area in two different ways.

a) **b)** **c)**

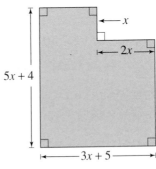

Explain the difference between "expanding" and "simplifying" as they relate to polynomials. Use examples to illustrate your explanation.

Recall that a polynomial with three terms is a trinomial.

In Section 7.4, we learned that the product of two binomials is often a trinomial.
For example, $(x + 3)(x + 4) = x^2 + 7x + 12$.
Factoring is the reverse process.
To factor a trinomial, such as $x^2 + 7x + 12$, is to write it as the product $(x + 3)(x + 4)$.
Factoring trinomials is one of the tools we need to solve quadratic equations algebraically.

INVESTIGATION

Using Algebra Tiles to Factor a Trinomial

1. Use one x^2-tile, five x-tiles, and six 1-tiles. What polynomial do they represent?

2. Arrange the tiles to form a rectangle.

   a) Write the length of the rectangle as a binomial.

   b) Write the width of the rectangle as a binomial.

   c) Write the area of the rectangle as a product of the length and width.

3. Check the rectangles of other students. Did all of you arrange the tiles the same way?

4. a) What are the factors of the trinomial?

   b) How are the numbers that appear in the factors related to the numbers that appear in the trinomial?

5. Write the trinomial as a product of factors.

6. Use algebra tiles to factor each trinomial.

   a) $x^2 + 3x + 2$ **b)** $x^2 + 6x + 8$ **c)** $x^2 + 4x + 4$

7. In each part of exercise 6, how are the numbers that appear in the factors related to the numbers that appear in the trinomial?

Consider this expansion.

$$x^2 + 7x + 12 = (x + 3)(x + 4)$$

7 is the sum of 3 and 4

12 is the product of 3 and 4

We use these relationships to factor a trinomial.

Example 1

Factor. $x^2 + 7x + 6$

Solution

$x^2 + 7x + 6$

Think ...

We want to find two numbers whose sum is 7 and whose product is 6.

Factors of 6	Sum
1, 6	$1 + 6 = 7$
−1, −6	$−1 − 6 = −7$
2, 3	$2 + 3 = 5$
−2, −3	$−2 − 3 = −5$

List pairs of factors of 6. Add each pair of factors.

Look for the factors that have a sum of 7. They are 1 and 6. Write these numbers as the second terms in the binomials.

$x^2 + 7x + 6 = (x + 1)(x + 6)$

Example 2

Factor. $a^2 − 8a + 12$

Solution

$a^2 − 8a + 12$

Think ...

We want two numbers whose sum is −8 and whose product is 12.

Factors of 12	Sum
1, 12	$1 + 12 = 13$
−1, −12	$−1 − 12 = −13$
2, 6	$2 + 6 = 8$
−2, −6	$−2 − 6 = −8$
3, 4	$3 + 4 = 7$
−3, −4	$−3 − 4 = −7$

The factors that have a sum of −8 are −2 and −6.

$a^2 − 8a + 12 = (a − 6)(a − 2)$

When you factor, you can check your work by expanding.

Example 3

Factor $m^2 - 5m - 14$, then check.

Solution

$m^2 - 5m - 14$

Think ...

We want two numbers whose sum is -5 and whose product is -14.

Factors of –14	Sum
$-1, 14$	$-1 + 14 = 13$
$1, -14$	$1 - 14 = -13$
$-2, 7$	$-2 + 7 = 5$
$2, -7$	$2 - 7 = -5$

The factors that have a sum of -5 are 2 and -7.
$m^2 - 5m - 14 = (m + 2)(m - 7)$

Check.
Expand $(m + 2)(m - 7)$.

$(m + 2)(m - 7) = m^2 - 7m + 2m - 14$
$\qquad\qquad\quad = m^2 - 5m - 14$

Since this is the trinomial we started with, the factors are correct.

Example 4

Factor. $x^2 + 3x + 5$

Solution

$x^2 + 3x + 5$

Think ...

We want two numbers whose sum is 3 and whose product is 5.

Factors of 5	Sum
$1, 5$	$1 + 5 = 6$
$-1, -5$	$-1 - 5 = -6$

No pairs of factors add to 3, so there is no solution.
The trinomial $x^2 + 3x + 5$ cannot be factored.

Example 4 illustrates that not all trinomials can be factored as binomial pairs.

1. Explain what factoring a trinomial means.

2. In *Example 1*, the trinomial $x^2 + 7x + 6$ was factored. What is another trinomial that begins with x^2, ends with +6, and can be factored? Explain how you found this trinomial.

3. In *Example 2*, only factors of 12 with the same sign were tested. Explain.

4. In *Example 3*, only factors of −14 with opposite signs were tested. Explain.

5. Can every trinomial be factored? Explain.

6. Is the product of two binomials always a trinomial? Explain.

7.6   Exercises

A 1. Determine two numbers with each given sum and product.

	Product	Sum	Numbers
a)	8	6	
b)	8	−6	
c)	12	8	
d)	12	−8	
e)	−12	4	
f)	−12	−4	

2. For each diagram
   i) Write the trinomial represented by the algebra tiles.
   ii) Rearrange the tiles into a rectangle.
   iii) Use the result to factor the trinomial.

   a)

b)

c)

B **3. a)** Factor each trinomial.

    i) $x^2 + 2x + 1$ **ii)** $x^2 + 3x + 2$ **iii)** $x^2 + 4x + 3$ **iv)** $x^2 + 5x + 4$

b) Describe a pattern in the terms of the trinomials in part a.

c) Describe a pattern in the algebra tile rectangles for the trinomials.

d) Extend the pattern by writing three more trinomials.

4. Factor. Use algebra tiles or an area diagram to model each solution.

    a) $x^2 + 7x + 10$ **b)** $x^2 + 11x + 10$ **c)** $x^2 + 7x + 6$ **d)** $t^2 + 9t + 8$

5. a) Factor.

    i) $x^2 + 7x + 12$ **ii)** $x^2 - 7x + 12$ **iii)** $x^2 + 10x + 25$ **iv)** $x^2 - 10x + 25$

b) For each product of two binomials in part a, why are the signs in the binomials the same?

6. a) Factor.

    i) $x^2 + x - 20$ **ii)** $x^2 - x - 20$ **iii)** $x^2 + 2x - 24$ **iv)** $x^2 - 2x - 24$

b) For each product of two binomials in part a, why are the signs in the binomials different?

7. Factor.

    a) $x^2 - 4x + 4$ **b)** $x^2 - 11x + 24$ **c)** $x^2 - 9x + 18$ **d)** $x^2 - 11x + 18$

    e) $x^2 - 19x + 18$ **f)** $x^2 - 14x + 45$ **g)** $x^2 - 8x + 16$ **h)** $x^2 - 4x + 3$

8. Factor. Check by expanding.

    a) $x^2 + 5x - 6$ **b)** $x^2 - 5x - 6$ **c)** $q^2 + 6q - 7$ **d)** $q^2 - 6q - 7$

    e) $x^2 + 8x - 20$ **f)** $n^2 - 8n - 20$ **g)** $a^2 + 10a - 24$ **h)** $a^2 - 10a - 24$

9. Factor. Check by expanding.

a) $x^2 - 6x + 8$

b) $x^2 + 9x + 18$

c) $a^2 - 11a + 18$

d) $m^2 + 8m + 16$

e) $n^2 - 10n + 25$

f) $n^2 - 13n + 30$

g) $p^2 + 16p + 64$

h) $y^2 - 13y + 42$

i) $x^2 + 15x + 56$

j) $x^2 - 10x - 56$

k) $x^2 + 6x + 5$

l) $a^2 + 8a + 12$

m) $m^2 + 6m + 9$

n) $x^2 - x - 2$

o) $x^2 - 5x + 6$

10. Factor if possible.

a) $r^2 - 9r + 14$

b) $a^2 - 8a - 10$

c) $n^2 - 8n + 9$

d) $m^2 - 9m + 20$

e) $k^2 - 8k + 15$

f) $x^2 + 10x + 12$

g) $a^2 - 2a - 15$

h) $m^2 + 9m + 20$

i) $n^2 - 5n - 12$

11. Factor. Check by expanding.

a) $x^2 + 7x - 8$

b) $a^2 + 5a - 14$

c) $r^2 - 5r - 36$

d) $a^2 - 4a - 45$

e) $n^2 - 3n - 54$

f) $m^2 - 2m - 48$

g) $k^2 - 2k - 63$

h) $x^2 - 7x - 30$

i) $81 - 18a + a^2$

j) $121 + 22m + m^2$

k) $t^2 - 2t - 3$

l) $n^2 + 13n + 42$

m) $x^2 - 17x + 30$

n) $c^2 - 11c + 30$

o) $m^2 + 6m - 55$

12. Consider the trinomial $x^2 - 13x + 36$.

a) Determine its value when $x = 59$.

b) Factor the trinomial.

c) Evaluate the factored expression when $x = 59$.

d) Compare your work for parts a and c. Explain why it might be useful to factor an expression before you evaluate it.

13. a) Factor.

    i) $x^2 + 2x + 1$

        $x^2 + 4x + 4$

        $x^2 + 6x + 9$

    ii) $x^2 - 2x + 1$

         $x^2 - 4x + 4$

         $x^2 - 6x + 9$

b) Describe any patterns you found in part a.

c) For each list in part a, write the next two trinomials.

d) Each trinomial in part a can be written as a *binomial square*. Write to explain the term "binomial square."

14. Factor if possible.

a) $x^2 + 16x + 63$

b) $a^2 + 12a + 30$

c) $x^2 - 4x + 32$

d) $t^2 + 11t + 24$

e) $n^2 - 12n + 35$

f) $k^2 - 5k - 21$

Collecting Data for Binomial Squares

In the previous exercises, you factored many trinomials of the form $x^2 + bx + c$. The values of b and c in these trinomials form data that you can graph and analyse using a graphing calculator. In this exercise, you will collect data from the trinomials that are binomial squares. Recall that each trinomial in exercise 13 is a binomial square. For the binomial square $x^2 - 2x + 1$, $b = -2$ and $c = 1$.

15. a) Find at least ten binomial squares among the exercises of this section. For each binomial square, record the values of b and c.

b) Enter the data in the list editor of a calculator.
   - Press [STAT] **1**. Clear any entries in list L1 by moving the cursor to the heading of L1, then press [CLEAR] [ENTER]. Similarly, clear any entries in list L2.
   - Enter the values of b in list L1 and corresponding values of c in list L2.

c) Graph the data.
   - Press [2nd] [Y=] for Stat Plot. Make sure that Plot 1 is turned on. If it is off, press **1** [ENTER] to turn it on.
   - Press [ZOOM] **6** to set the standard window and graph the data.

d) Use the methods of Chapter 6 to estimate the equation of the parabola that best fits the data.

e) On grid paper, sketch the parabola and write its equation on the graph.

16. Find an integer to replace each square so that each trinomial can be factored. There may be more than one answer for some trinomials.

a) $x^2 + \square\, x + 12$ **b)** $x^2 - \square\, x + 20$ **c)** $x^2 + \square\, x - 18$

d) $x^2 + 5x + \square$ **e)** $x^2 + 4x + \square$ **f)** $x^2 - 2x + \square$

17. Find as many trinomials as you can that begin with x^2, end with $+30$, and can be factored.

Communicating
the IDEAS

Write an example of a trinomial that cannot be factored as the product of two binomials. Explain how you can tell that it cannot be factored as a product of two binomials.

Removing a Square from a Square

You will need 1-cm grid paper and scissors.

1. From the 1-cm grid paper, cut the largest possible square; for example, 17 cm by 17 cm.

2. From the 17-cm square, cut a smaller square from one corner; for example, 12 cm by 12 cm.

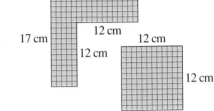

3. The side of the original square is 17 cm. The side of the cut-out square is 12 cm.

   a) What is the area of the original square?

   b) What is the area of the cut-out square?

   c) Use the results of parts a and b to write the area of the L-shaped piece.

4. Cut the L-shaped piece into two congruent parts. Arrange the pieces to form a rectangle.

   a) What is the length of the rectangle?

   b) What is the width of the rectangle?

   c) What is the area of the rectangle?

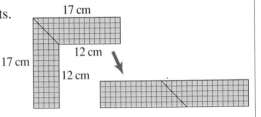

5. Compare the results of exercise 3c and exercise 4c. What do you notice? Explain.

A polynomial that can be expressed in the form $x^2 - y^2$ is called a *difference of squares*. A difference of squares can be factored as the product of the sum and difference of the same quantities.

In the *Investigation*, in exercise 3c, the area of the L-shaped piece is $17^2 - 12^2 = 145$.

In exercise 4c, the area of the L-shaped piece is $(17 + 12)(17 - 12) = (29)(5)$
$$= 145$$

Since the areas are equal, we write $17^2 - 12^2 = (17 + 12)(17 - 12)$
This pattern applies to squares of any size.

In general, if the side of the original square is x, and the side of the cut-out square is y, we write $x^2 - y^2 = (x + y)(x - y)$.

Take Note

Difference of Squares Property

$$x^2 - y^2 = (x + y)(x - y)$$

Difference of the squares of x and y

Product of the sum and the difference of x and y

You can always use this pattern to factor a difference of squares.

Example 1

Factor.

a) $x^2 - 36$

b) $49 - x^2$

Solution

a) $x^2 - 36 = x^2 - 6^2$
$= (x + 6)(x - 6)$

b) $49 - x^2 = 7^2 - x^2$
$= (7 + x)(7 - x)$

Example 2

Factor.

a) $4x^2 - 81$

b) $16m^2 - 121$

Solution

Use mental math to write each monomial term as a square.

a) $4x^2 - 81 = (2x)^2 - (9)^2$
$= (2x + 9)(2x - 9)$

b) $16m^2 - 121 = (4m)^2 - (11)^2$
$= (4m + 11)(4m - 11)$

Sometimes, an expression can be factored as the difference of two squares after removing a common factor.

Example 3

Factor. $8x^2 - 72$

Solution

$8x^2 - 72$

> **Think ...**
>
> $8x^2$ and 72 cannot be written as squares. But $8x^2$ and 72 have a common factor of 8.

Remove 8 as a common factor. $8x^2 - 72 = 8(x^2 - 9)$

Apply the difference of squares. $8x^2 - 72 = 8(x - 3)(x + 3)$

Discussing the IDEAS

1. When you expand $(x + y)(x - y)$, there are only two terms. Why?

2. In *Example 1a*, the expression $x^2 - 36$ was factored as the difference of squares. Suppose the expression had been $x^2 - 6$. Could this expression be factored as a difference of squares? Explain.

3. In *Examples 1* and *2*, the factors were written with the sum of the terms first; for example, $(x + 6)(x - 6)$. Could the factors have been written with the difference of the terms first; for example, $(x - 6)(x + 6)$? Explain.

4. Does the *Investigation* help you to factor the sum of two squares? Explain.

Practise Your Skills

1. Write each number as the square of a number.

   a) 1 b) 4 c) 9 d) 16 e) 25 f) 36

   g) 49 h) 64 i) 81 j) 100 k) 121 l) 144

   m) 169 n) 196 o) 225 p) 400 q) 625 r) 900

2. Write each monomial as a square of a monomial.

   a) $9x^2$ b) $64m^2$ c) $121y^2$ d) $100x^2$ e) $81a^2$

A 1. Factor.

a) $x^2 - 49$ b) $b^2 - 121$ c) $m^2 - 64$ d) $81 - x^2$

e) $y^2 - 100$ f) $x^2 - 36$ g) $4 - y^2$ h) $16 - t^2$

B 2. Factor.

a) $100m^2 - 49$ b) $64b^2 - 1$ c) $121a^2 - 100$ d) $36b^2 - 25$

e) $25p^2 - 81$ f) $144m^2 - 1$ g) $9x^2 - 121$ h) $25q^2 - 1$

3. Factor.

a) $121m^2 - 4$ b) $4x^2 - 169$ c) $9x^2 - 400$ d) $64 - 9b^2$

e) $1 - 100m^2$ f) $49 - 4p^2$ g) $64s^2 - 9$ h) $100 - 49y^2$

4. a) Factor.

i) $16s^2 - 9$ ii) $4x^2 - 49$ iii) $81a^2 - 25$ iv) $100 - 121d^2$

v) $25 - 36q^2$ vi) $625 - 81z^2$ vii) $49 - 400n^2$ viii) $4 - 225f^2$

b) Choose one binomial from part a. Write to explain how you factored the binomial.

5. A circular fountain has a radius of 150 cm. It is surrounded by a circular flower bed with radius 320 cm.

a) Write an expression for the area of a circle, radius 320 cm. Do not expand.

b) Write an expression for the area of a circle, radius 150 cm. Do not expand.

c) Use the results of parts a and b to write an expression for the area of the flower bed.

d) Use the difference of squares to calculate the area of the flower bed.

6. Use difference of squares factoring to evaluate each expression. Use a calculator to check your answers.

a) $27^2 - 26^2$ b) $99^2 - 98^2$ c) $33^2 - 31^2$ d) $65^2 - 62^2$

7. Use difference of squares factoring to calculate the shaded area in each diagram.

a)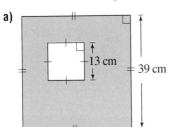

13 cm

39 cm

b)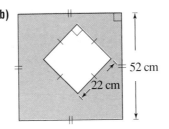

52 cm

22 cm

c)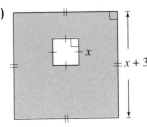

x

$x + 3$

8. A square with side x centimetres is cut from a corner of a square piece of cardboard with side 10 cm.

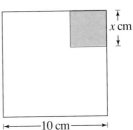

x cm

10 cm

a) Suppose the L-shaped piece of cardboard that remains is cut into 2 parts (below left). Write an expression for the area of each part. Then add the two expressions and simplify the result.

b) Suppose the L-shaped piece of cardboard is cut into 3 parts (below middle). Write an expression for the area of each part. Then add the three expressions and simplify the result.

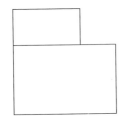
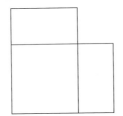
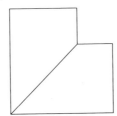

c) Suppose the L-shaped piece of cardboard is cut into two parts (above right), and these parts are rearranged to form a rectangle. Write an expression for:
i) the length and width of the rectangle
ii) the area of the rectangle

d) In parts a, b, and c, you should have obtained the same expression for the area of the L-shaped piece of cardboard. What is another way to explain why this expression represents the area of this piece?

9. Factor fully.

a) $2x^2 - 18$ **b)** $12x^2 - 3$ **c)** $5 - 45x^2$ **d)** $100 - 4x^2$

e) $24 - 150x^2$ **f)** $50 - 18x^2$ **g)** $48x^2 - 3$ **h)** $63x^2 - 48$

Communicating the IDEAS

How would you explain to someone, over the telephone, how to recognize a difference of squares, and how to factor a difference of squares? Write to explain your answer.

7.8 Solving Quadratic Equations by Factoring

In earlier grades, you solved many linear equations, such as $3x + 5 = 17$. You have now developed the tools you need to solve certain quadratic equations algebraically. From Chapter 6, recall that a quadratic equation can be written in the form $ax^2 + bx + c = 0$. Examples of quadratic equations are:

$$x^2 = 25 \qquad x^2 + 3x - 28 = 0 \qquad 3x^2 + 9x = 12$$

The *root* of a quadratic equation is a value of the variable that satisfies the equation.

In quadratic equations, such as $x^2 = 25$, there is no x-term. Such equations can be solved by isolating the variable, then taking the square root of each side. You can use mental math to find the square root of a perfect square.

Example 1

Solve.

a) $x^2 = 25$ **b)** $3x^2 = 27$ **c)** $a^2 - 3 = 13$

Solution

a) $x^2 = 25$
Take the square root
of each side.
$x = \pm\sqrt{25}$
$x = \pm 5$

b) $3x^2 = 27$
Divide each side by 3.
$x^2 = 9$
Take the square root
of each side.
$x = \pm 3$

c) $a^2 - 3 = 13$
Add 3 to each side.
$a^2 = 16$
Take the square root
of each side.
$a = \pm 4$

We can solve some quadratic equations by factoring. The solution by factoring depends on the following important property:

Take Note

Zero-Product Property

If two numbers have a product of 0, then one or both of them must be 0.
That is, if $m \times n = 0$, then either $m = 0$, or $n = 0$, or both.
This is also true for algebraic expressions.
If $(a + b)(c + d) = 0$, then either $a + b = 0$, or $c + d = 0$, or both.

Example 2 illustrates the solution of quadratic equations that have x-terms.

Example 2

Solve and check.

a) $x^2 - x - 6 = 0$

b) $x^2 + 10x + 25 = 0$

Solution

a) $x^2 - x - 6 = 0$

Use mental math to factor.

$(x - 3)(x + 2) = 0$

Either $x - 3 = 0$ or $x + 2 = 0$

$\qquad\qquad x = 3 \qquad\quad x = -2$

b) $x^2 + 10x + 25 = 0$

Use mental math to factor.

$(x + 5)(x + 5) = 0$

That is, $x + 5 = 0$

$\qquad\qquad\qquad x = -5$

Check. Always check by substituting in the original equation.

a) If $x = 3$,

$x^2 - x - 6$

$= 3^2 - 3 - 6$

$= 9 - 3 - 6$

$= 0$

If $x = -2$,

$x^2 - x - 6$

$= (-2)^2 - (-2) - 6$

$= 4 + 2 - 6$

$= 0$

Both solutions are correct.

b) If $x = -5$,

$x^2 + 10x + 25$

$= (-5)^2 + 10(-5) + 25$

$= 25 - 50 + 25$

$= 0$

The solution is correct.

The solutions of the equations in *Example 2* can be illustrated graphically. The roots of these equations are the zeros of the corresponding quadratic functions, $y = x^2 - x - 6$ and $y = x^2 + 10x + 25$. The graphs of these functions intersect the x-axis at the points whose x-coordinates are the roots of the equations.

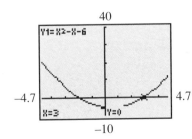
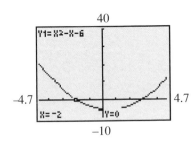
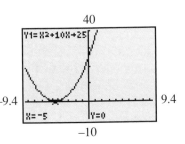

The equation $x^2 - x - 6 = 0$ can be written in factored form as $(x - 3)(x + 2) = 0$. This equation has two roots, $x = 3$ and $x = -2$, where each root corresponds to one factor.

The equation $x^2 + 10x + 25 = 0$ can be written as $(x + 5)(x + 5) = 0$. This equation has only one root, $x = -5$. Since this root corresponds to two factors, we say that this equation has two equal roots.

Sometimes, we must collect like terms to simplify an equation before we can solve it.

Example 3

Solve and check. $x^2 - 5x + 4 = x - 1$

Solution

$x^2 - 5x + 4 = x - 1$

Collect all the terms on the left side, to obtain 0 on the right side of the equation.

$x^2 - 5x + 4 - x + 1 = 0$

Simplify.

$$x^2 - 6x + 5 = 0$$

Factor.

$$(x - 5)(x - 1) = 0$$

$x - 5 = 0$ or $x - 1 = 0$

    $x = 5$ $x = 1$

Check.

If $x = 5$,   left side $= x^2 - 5x + 4$ right side $= x - 1$

                       $= 5^2 - 5(5) + 4$ $= 5 - 1$

                       $= 25 - 25 + 4$ $= 4$

                       $= 4$

If $x = 1$,   left side $= x^2 - 5x + 4$ right side $= x - 1$

                       $= 1^2 - 5(1) + 4$ $= 1 - 1$

                       $= 1 - 5 + 4$ $= 0$

                       $= 0$

Both roots are correct.

In Chapter 6, you found that quadratic functions arise in problems involving objects moving under the influence of gravity.

Example 4

When a soccer ball is kicked with a vertical speed of 20 m/s, its height, h metres, after t seconds is given by the formula:

$h = -5t^2 + 20t$

a) How long after it is kicked is the soccer ball at a height of 15 m?

b) For how long is the soccer ball above 15 m?

Solution

a) When the soccer ball is 15 m high, $h = 15$

Substitute 15 for h in the formula $h = 5t^2 + 20t$.

$15 = -5t^2 + 20t$

This is a quadratic equation. The solution of the equation is the time in seconds when the height is 15 m.

Collect all the terms on one side of the equation.

$5t^2 - 20t + 15 = 0$

Remove 5 as a common factor.

$5(t^2 - 4t + 3) = 0$

Divide each side by 5.

$t^2 - 4t + 3 = 0$

Factor.

$(t - 1)(t - 3) = 0$

Use the Zero-Product Property.

Either $t - 1 = 0$ or $t - 3 = 0$

$\qquad t = 1 \qquad\qquad t = 3$

The soccer ball is at a height of 15 m twice: first on the way up, 1 s after the kick; then on the way down, 3 s after the kick.

b) The soccer ball is above 15 m when t is greater than 1 s and less than 3 s. That is, for 3 s – 1 s = 2 s

The football is above 15 m for 2 s.

We can use a graphing calculator to illustrate the solution of *Example 4*. For the equation $h = -5t^2 + 20t$, we draw the graph of $y = -5x^2 + 20x$. We then use the Trace feature to determine the values of x when $y = 15$.

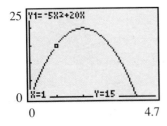   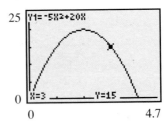

From the graphs, the soccer ball is above 15 m for the 2 s between $x = 1$ and $x = 3$.

1. In the solution of *Example 1a*, when we took the square root of each side, why did we not write $\pm\sqrt{x^2} = \pm\sqrt{25}$, then $\pm x = \pm 5$?

2. In *Example 4*, would it matter if we used x instead of t and y instead of h? Explain.

Practise Your Skills

1. Solve for x.

   a) $x - 3 = 0$ b) $x + 12 = 0$ c) $x - 7 = 0$ d) $x + 8 = 0$

7.8 Exercises

A

1. Solve.

   a) $x^2 = 9$ b) $x^2 = 25$ c) $x^2 = 1$ d) $x^2 = 100$

   e) $3x^2 = 75$ f) $6x^2 = 0$ g) $2x^2 = 32$ h) $4x^2 = 100$

2. Solve.

   a) $3x^2 - 3 = 9$ b) $2p^2 - 5 = 3$ c) $4t^2 + 6 = 10$

   d) $2a^2 = 18$ e) $2n^2 - 64 = n^2$ f) $3b^2 = 98 + b^2$

B

3. Solve and check.

   a) $x^2 + 7x + 10 = 0$ b) $x^2 + 8x + 16 = 0$ c) $x^2 - x - 6 = 0$

   d) $x^2 + 5x - 14 = 0$ e) $x^2 + 7x + 12 = 0$ f) $x^2 - 5x - 24 = 0$

   g) $x^2 - 10x + 25 = 0$ h) $x^2 + 15x + 50 = 0$ i) $x^2 - x - 30 = 0$

4. Solve and check.

   a) $x^2 + 9x + 18 = 0$ b) $x^2 - 6x + 9 = 0$ c) $x^2 + x - 42 = 0$

   d) $x^2 - 10x + 16 = 0$ e) $x^2 - 14x + 49 = 0$ f) $x^2 - 3x - 40 = 0$

   g) $x^2 - 16x + 63 = 0$ h) $x^2 - 2x + 1 = 0$ i) $x^2 + 12x + 36 = 0$

5. Determine the roots of each equation.

   a) $x^2 - 9x + 27 = 7$ b) $x^2 - 16x + 47 = -16$ c) $x^2 - 8x + 20 = 4$

   d) $x^2 + 22x + 25 = 4$ e) $x^2 - 5x - 20 = -6$ f) $x^2 + 8x - 15 = 6x$

   g) $x^2 - x - 18 = 6x$ h) $x^2 - 10x + 16 = 4 - 2x$ i) $x^2 - 5x - 40 = 2 - 4x$

6. a) Solve each equation by first dividing by a common factor.

   i) $2x^2 - 2x - 12 = 0$ ii) $3x^2 - 21x + 18 = 0$ iii) $2x^2 + 2x - 40 = 0$

   b) Choose one equation from part a. Illustrate the solutions. Sketch what you see on the screen.

7. a) Solve the equations $x^2 - 8x + 12 = 0$ and $x^2 - 8x + 16 = 0$.

b) The equations in part a have the form $x^2 - 8x + c = 0$. What are some other values of c for which the equation $x^2 - 8x + c = 0$ can be solved by factoring? For each value of c, what are the roots of the equation?

8. The height, h metres, of an infield fly ball t seconds after being hit is given by the formula $h = 30t - 5t^2$. How long after being hit is the ball at a height of 25 m?

9. In *Example 2b*, the equation had two equal roots. Look through the preceding exercises. Find examples of other quadratic equations with two equal roots. What patterns are there in the equations? Explain.

10. A packaging company makes tin cans with no tops. One type of can has a height of 14.0 cm, and a volume of 1000 cm^3. A formula for the volume of the can is $V = \pi r^2 h$, where r centimetres is the radius of the base of the can, and h centimetres is the height. Use the formula. Substitute $V = 1000$ and $h = 14$. Calculate the radius of the base of the can. Give the answer to 1 decimal place.

C **11.** An object falls d metres in t seconds when dropped from rest. The quantities d and t are related by the formula $d = 4.9t^2$. How long would it take an object to hit the ground when dropped from each height? Give the answers to 1 decimal place.

a) 10 m   **b)** 20 m   **c)** 30 m

12. The formula for the area, A, of an equilateral triangle is $A = \frac{\sqrt{3}}{4}x^2$, where x represents the length of a side. What is the side length of an equilateral triangle with each area? Give the answers to 1 decimal place.

a) 10 cm^2 **b)** 20 cm^2 **c)** 40 cm^2

Communicating
the IDEAS

Suppose you are studying for a quiz on solving quadratic equations. Make a list of the skills you have to review.

Algebra Tools

- Distributive law
  $2x(4x + 5) = 8x^2 + 10x$

- Common factoring
  $6x^2 - 9x = 3x(2x - 3)$

- Multiplying two binomials
  $$(2x - 3)(4x + 5) = (2x)(4x) + (2x)(5) + (-3)(4x) + (-3)(5)$$
  $$= 8x^2 + 10x - 12x - 15$$
  $$= 8x^2 - 2x - 15$$

- Special cases of multiplying two binomials
  $$(2x - 3)^2 = 4x^2 - 12x + 9$$
  $$(2x - 3)(2x + 3) = 4x^2 - 9$$

- Factoring a trinomial
  $x^2 + 3x - 10 = (x + 5)(x - 2)$

- Factoring a difference of squares
  $9x^2 - 4 = (3x + 2)(3x - 2)$

- Zero-Product Property
  If $a \times b = 0$, then either $a = 0$, or $b = 0$, or both

- Solving a quadratic equation by factoring
  See page 386 for examples.

1. Expand.

a) $2x(x - 5)$ b) $(-3x)(4 - x)$ c) $x(x^2 - 3x + 2)$

d) $(-3)(4 - x + 2x^2)$ e) $(-2x)(x^2 - 7)$ f) $4(3 - 4x + 5x^2)$

2. Factor fully.

a) $4x + 14$ b) $12a^2 + 3a$ c) $8y^2 - 6y$

d) $100 - 75x^2$ e) $3x^2 - 12x + 6$ f) $2x^3 - 3x^2 - x$

3. Expand and simplify.

a) $(x + 3)(x - 5)$ b) $(2x - 1)(4x + 5)$ c) $(2x - 3)(x + 9)$

d) $(3x - 4)(2x + 1)$ e) $(5x - 5)(4x - 4)$ f) $(-x - 1)(-x + 1)$

g) $(x - 5)(x + 5)$ h) $(2x - 1)(2x + 1)$ i) $(5c - 3)(5c + 3)$

4. Expand and simplify.

a) $(x + 2)^2$ **b)** $(2x - 1)^2$ **c)** $(4x + 3)^2$

d) $(5 + x)^2$ **e)** $(x - 1)^2$ **f)** $(3x + 2)^2$

5. Simplify.

a) $(3x^2 + 17x) - (12x^2 + 3x)$ **b)** $2(3a - 5) - (4a - 7)$

c) $2x(x + 4) + 3x(2x - 3)$ **d)** $5n(3n - 2) - 4n(n - 1)$

e) $6(2x^2 - 5x) - 14(3x - x^2)$ **f)** $(2x + 3)(x - 1) - (x + 1)^2$

6. Factor.

a) $m^2 + 8m + 16$ **b)** $a^2 - 7a + 12$ **c)** $y^2 - 2y - 8$

d) $n^2 - 4n - 45$ **e)** $m^2 - 2m - 15$ **f)** $x^2 - 5x + 6$

7. Factor.

a) $b^2 - 36$ **b)** $x^2 - 49$ **c)** $4 - 9b^2$ **d)** $81x^2 - 25$

8. Factor if possible.

a) $8d^2 - 12$ **b)** $25x^2 + 16$ **c)** $3x^2 + 6x + 15$

d) $x^2 + x + 1$ **e)** $x^2 - 3$ **f)** $25x^2 - 16$

9. Solve.

a) $x^2 - 4x - 21 = 0$ **b)** $x^2 + x - 56 = 0$ **c)** $x^2 + 8x + 16 = 0$

d) $x^2 - 2x - 8 = 0$ **e)** $x^2 - 4x - 45 = 0$ **f)** $x^2 + x - 20 = 0$

g) $x^2 - 15x + 54 = 0$ **h)** $x^2 - 9x - 90 = 0$ **i)** $x^2 + 20x + 36 = 0$

10. a) Factor each trinomial in the list below. Find as many other polynomials that belong in the list as you can.
$x^2 + 5x - 36$
$x^2 + 9x - 36$
$x^2 + 16x - 36$

b) Write to explain how you determined the polynomials in part a.

c) In part a, all the polynomials begin with a perfect square, x^2, and end with the opposite of a perfect square, -36. Make up another list of examples like these. Each polynomial in your list should begin with a perfect square, end with the opposite of a perfect square, and be factorable.

11. a) Choose any two natural numbers that differ by 2. Multiply the numbers. Calculate the mean of the numbers with which you started. Then square the mean.

b) Repeat part a with other natural numbers. Based on your results, state a probable conclusion.

12. Patterns in Binomial Products: Part II

a) Expand the products in each list.

i) $(x + 5)(x + 5)$ **ii)** $(x - 5)(x - 5)$
$(x + 4)(x + 6)$ $(x - 4)(x - 6)$
$(x + 3)(x + 7)$ $(x - 3)(x - 7)$
$\vdots$ $\vdots$

b) Predict the next three products for each list in part a.

c) Suppose you extended the lists in part a upward. Predict the preceding three products in each list.

13. In the diagram, the radius of the circle is r centimetres.

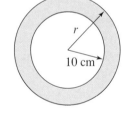

a) What is the area of the circle?

b) What is the side length of the square?

c) What is the area of the square?

d) What is the shaded area?

e) Factor the expression for the shaded area.

f) Suppose the radius of the circle is 2 cm. What is the shaded area?

14. In this diagram, the radius of the large circle is r centimetres. The radius of the small circle is 10 cm.

a) What is the area of the large circle?

b) What is the area of the small circle?

c) What is the shaded area?

d) Factor the expression in part c.

e) Suppose $r = 15$ cm. What is the shaded area?

15. Mathematical Modelling

Suppose you are scheduling a round robin baseball tournament. There are 8 teams. How many games must be scheduled? Explain.

8 Analysing Quadratic Functions

MATHEMATICAL MODELLING

The Basketball Free Throw

In a basketball game, when certain fouls are committed, the player who is fouled takes a "free throw." This is a shot at the basket with no one attempting to stop it. For a team to be successful, its players must score on a high percent of their free throws.

In this chapter, you will study the path of a basketball as it moves through the air.

1. Why do you think free throws are difficult for some players, even professionals?

2. What do you think are some of the physical, emotional, and mental skills that are important for being a good free throw shooter?

3. When a ball is tossed into the air, what is the shape of the flight path of the ball? This flight path is called the trajectory.

4. What factors do you think affect the trajectory of the ball? How do you think these factors can be controlled to ensure that the ball goes into the basket?

You will return to this problem in Section 8.5.

FYI Visit www.awl.com/canada/school/connections

For information about the above problem, click on <u>MATHLINKS</u>, followed by Foundations of Mathematics 10. Then select a topic under The Basketball Free Throw.

Preparing for the Chapter

This section reviews these concepts:

- Simplifying algebraic expressions
- Expanding algebraic expressions
- Evaluating polynomial expressions
- Rearranging a formula
- Graphing an equation using a calculator
- Estimating an equation of best fit for data

INVESTIGATION

A Calendar Square

Choose any month on a calendar. Then choose a 3 × 3 square of 9 dates.

Let x represent the date at the centre of the square.

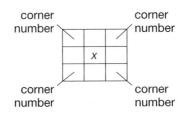

1. Write a polynomial to represent:

   a) each corner number

   b) the square of each corner number

   c) the sum of the squares of the corner numbers

2. Suppose you knew the sum of the squares of the corner numbers. How could you determine the value of x?

3. Check your answer to exercise 2. Determine the value of x when the sum of the squares of the corner numbers is 1356.

4. Simplify.

a) $3x + 5 - 6x - 18$

b) $2x^2 - 8x - 9x - 7x^2$

c) $(5x^2 + 7x + 4) + (5x^2 - 7x + 6)$

d) $(2x^2 - 5x + 3) - (4 - 3x)$

5. Expand, then simplify.

a) $3x(x - 5)$
b) $(x + 4)(x - 3)$
c) $(x + 6)^2$
d) $(x - 5)^2$

e) $2(x + 1)^2$
f) $3(x - 2)^2$
g) $-2(x - 4)^2$
h) $-(2x + 1)^2$

6. Determine the value of each polynomial when $x = 0.5$.

a) $3x + 5$
b) $4x - 8$
c) $13 - 3x$

d) $5x^2 + 6x - 2$
e) $x^2 - 3x - 9$
f) $-x^2 - 4x + 1$

7. Determine the value of each polynomial in exercise 6 when $x = -2$.

8. Solve each equation for y.

a) $5x + y = 3$
b) $2x = 4y - 7$
c) $7x + 4y + 5 = 0$

d) $\frac{1}{2}x + y = 3$
e) $x = \frac{2}{3}y$
f) $\frac{1}{2}x + 2y - 4 = 0$

9. For each equation below:

  i) Graph the function. Copy the graph and state the graphing window you used.

  ii) Determine the y-intercept of the graph.

  iii) Determine any x-intercepts of the graph.

a) $y = 3x + 12$
b) $y = 4x^2 + 10$
c) $y = -5x^2 + 20x$

10. a) Plot these data on a grid. Join the points with a smooth curve.

x	y
−4	−6
−2	−3
0	−2
2	−3
4	−6

b) Estimate the equation of the curve that best fits the data.

c) Check your prediction. Enter the data in lists L1 and L2. Make a scatter plot. On the same screen, graph the equation you estimated in part b. Adjust your equation until the curve goes through the points.

Recall from an earlier grade that two figures are congruent if they have the same shape and size. A tracing of one figure can be made to coincide with the other figure.

Two parabolas are congruent if a tracing of one parabola can be made to coincide with the other parabola. In Chapter 6, you drew graphs of quadratic functions, such as $y = x^2$ and $y = -4x^2 + 1$. In this section, you will draw graphs of quadratic functions and decide if two parabolas are congruent.

INVESTIGATION 1

Transforming Quadratic Graphs

Set-Up

- Press [MODE] and make sure that **Sequential** is selected.

- Press [2nd] [ZOOM] and make sure that **CoordOn** and **ExprOn** are selected.

- Press [WINDOW]. Enter the numbers shown below.

```
WINDOW
 Xmin=-3
 Xmax=3
 Xscl=1
 Ymin=-4
 Ymax=9
 Yscl=1
 Xres=1
```

- Press [Y=]. Use the arrow keys and [CLEAR] to remove any equations in the Y= list. If any of Plot 1, Plot 2, or Plot 3 is highlighted at the top of the screen, move the cursor to the plot then press [ENTER] to turn it off.

Part A Comparing the Graphs of $y = x^2$ and $y = x^2 + q$

Use these functions for exercises 1 to 8.

 i) $y = x^2$ **ii)** $y = x^2 + 2$ **iii)** $y = x^2 + 4$

 iv) $y = x^2 - 1$ **v)** $y = x^2 - 3$

1. Enter the first two functions in the list above as Y1 and Y2.

2. Press [GRAPH] to graph the functions on the same screen. The equation $y = x^2$ is drawn first, then the equation Y2 is drawn.

3. Draw the graphs on the same grid. Label each graph with its equation.

4. What is the relation between the graph of $y = x^2$ and the graph of the equation Y2?

5. Press [Y=]. Enter the next equation in the list above as Y2, then repeat exercises 2, 3, and 4. Draw all the graphs on one grid.

6. Repeat exercise 5 for each function in the list.

7. Copy and complete this table.

Function	Value of q	Direction of opening	Vertex	Axis of symmetry	Congruent to $y = x^2$?
$y = x^2$	0	up	$(0, 0)$	$x = 0$	yes
$y = x^2 + 2$					
$y = x^2 + 4$					
$y = x^2 - 1$					
$y = x^2 - 3$					

8. Predict how you think the graphs of $y = x^2 + 6$ and $y = x^2 - 5$ would appear on your grid. Use a graphing calculator to check your prediction.

9. a) What information about the graph of $y = x^2 + q$ does the value of q provide?

b) What happens to the graph of $y = x^2 + q$ when the value of q is changed? Describe the effect of both positive and negative values of q.

Part B Comparing the Graphs of $y = x^2$ and $y = (x - p)^2$

Press [WINDOW]. Change the settings to those shown below.

```
WINDOW
 Xmin=-7
 Xmax=7
 Xscl=1
 Ymin=-4
 Ymax=9
 Yscl=1
 Xres=1
```

Use these functions for exercises 10 to 13.

i) $y = x^2$ **ii)** $y = (x - 1)^2$ **iii)** $y = (x - 3)^2$

iv) $y = (x + 2)^2$ **v)** $y = (x + 4)^2$

10. Repeat exercises 1 to 6 for the functions in the list above.

11. Copy and complete this table.

Function	Value of p	Direction of opening	Vertex	Axis of symmetry	Congruent to $y = x^2$?
$y = x^2$	0	up	$(0, 0)$	$x = 0$	yes
$y = (x - 1)^2$					
$y = (x - 3)^2$					
$y = (x + 2)^2$					
$y = (x + 4)^2$					

12. Predict how the graphs of $y = (x - 5)^2$ and $y = (x + 6)^2$ would appear on your grid. Use a graphing calculator to check your predictions.

13. a) What information about the graph of $y = (x - p)^2$ does the value of p provide?

b) What happens to the graph of $y = (x - p)^2$ when the value of p is changed? Recall that $-(-2) = +2$, so that $(x - (-2)) = (x + 2)$. Describe the effect of both positive and negative values of p.

In *Investigation 1*, you should have discovered that all the parabolas were congruent to the graph of $y = x^2$.

In Chapter 6, you drew parabolas that were not congruent to $y = x^2$. You will investigate these in *Investigation 2*.

Comparing the Graphs of $y = x^2$ and $y = ax^2$

Set-Up

- Press WINDOW. Enter the numbers shown below.

```
WINDOW
Xmin=-4.7
Xmax=4.7
Xscl=1
Ymin=-2
Ymax=15
Yscl=1
Xres=1
```

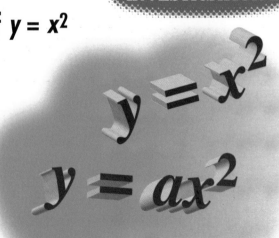

- Press Y=. Use the arrow keys and CLEAR to clear any equations in the Y= list.

Use these functions for exercises 1 to 4.

  i) $y = x^2$ ii) $y = 2x^2$ iii) $y = 0.5x^2$

  iv) $y = -x^2$ v) $y = -2x^2$ vi) $y = -0.5x^2$

1. Enter the first two equations in the list above as Y1 and Y2. Press $\boxed{\text{GRAPH}}$ to graph the functions on the same screen. Sketch the graphs.

2. Press $\boxed{\text{TRACE}}$. The cursor is on the graph whose equation is shown in the upper left corner of the screen. This should be $y = x^2$.

- Press 1 $\boxed{\text{ENTER}}$ to move the cursor to the point on this curve with x-coordinate 1. The coordinates are shown at the bottom of the screen.

- Press $\boxed{\blacktriangle}$ to move the cursor to the other graph, at the point with the same x-coordinate.

a) The point (1, 1) is on the parabola $y = x^2$. Determine the y-coordinate of the point where $x = 1$ on the parabola $y = 2x^2$.

b) The point (2, 4) is on the parabola $y = x^2$. Determine the y-coordinate of the point where $x = 2$ on the parabola $y = 2x^2$.

c) The point (3, 9) is on the parabola $y = x^2$. Determine the y-coordinate of the point where $x = 3$ on the parabola $y = 2x^2$.

d) Describe the relationship between the graph of $y = x^2$ and the graph of $y = 2x^2$.

3. Repeat exercises 1 and 2 for each pair of functions.

a) $y = x^2$; $y = 0.5x^2$ **b)** $y = -x^2$; $y = -2x^2$ **c)** $y = -x^2$; $y = -0.5x^2$

4. Copy and complete this table.

Function	Value of a	Direction of opening	Vertex	Axis of symmetry	Congruent to $y = x^2$?
$y = x^2$	1	up	(0, 0)	$x = 0$	yes
$y = 2x^2$					
$y = 0.5x^2$					
$y = -x^2$					
$y = -2x^2$					
$y = -0.5x^2$					

5. Sketch the graph of $y = x^2$ on a grid. On the same grid, predict then draw the graphs of $y = 3x^2$ and $y = -3x^2$. Use a graphing calculator to check your predictions.

6. a) What information about the graph of $y = ax^2$ does the value of a provide?

   b) What happens to the graph of $y = ax^2$ when the value of a is changed? Consider positive values of a between 0 and 1, and greater than 1, and also the equivalent negative values of a.

The simplest quadratic function is $y = x^2$. Its graph is a parabola, opening up, with vertex (0, 0) and axis of symmetry the y-axis, or $x = 0$. In *Investigation 1* and *2*, you made certain changes to the equation $y = x^2$ and discovered what happened to the graph. The results of the *Investigations* are summarized below.

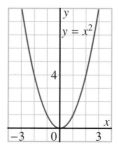

Changing the value of q in y = x² + q

The value of q in the equation $y = x^2 + q$ indicates that the graph of $y = x^2$ moved vertically q units.

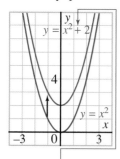

Positive sign, the graph moves up.

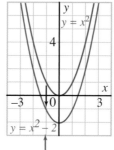

Negative sign, the graph moves down.

Changing the value of p in y = (x – p)²

The value of p in the equation $y = (x - p)^2$ indicates that the graph of $y = x^2$ moved horizontally p units.

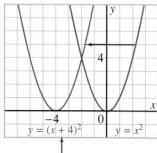

Positive sign, the graph moves left.

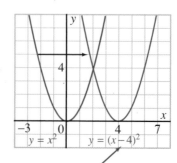

Negative sign, the graph moves right.

Changing the value of a in $y = ax^2$

The value of a in the equation $y = ax^2$ indicates that the graph of $y = x^2$ was stretched or compressed vertically. When a is negative, the graph is also reflected in the x-axis.

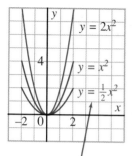

Positive signs, the graphs are stretched or compressed vertically.

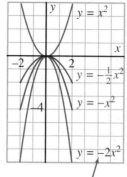

Negative signs, the graphs are stretched or compressed vertically and reflected in the x-axis.

Each of the three types of changes summarized above is called a *transformation*.

Each change corresponds to an equation $y = x^2 + q$, or $y = (x - p)^2$, or $y = ax^2$.

These changes can be combined and represented by a single equation $y = a(x - p)^2 + q$.

We say that $y = x^2 + q$ is the *image* of $y = x^2$ after a *translation* of q units vertically.

Similarly, $y = (x - p)^2$ is the image of $y = x^2$ after a translation of p units horizontally.

And, $y = ax^2$ is the image of $y = x^2$ after a vertical stretch or compression of factor a.

Take Note

Combining Transformations

In the equation $y = a(x - p)^2 + q$, the constants a, p, and q have the following meanings.

$x - p = 0$ is the axis of symmetry.

$$y = a(x - p)^2 + q$$

Congruent to the parabola $y = ax^2$.

Coordinates of the vertex are (p, q).

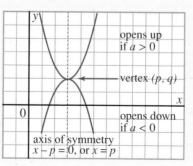

opens up if $a > 0$

vertex (p, q)

opens down if $a < 0$

axis of symmetry $x - p = 0$, or $x = p$

Example

Describe how the graph of $y = x^2$ is transformed to produce the graph of each function. Sketch the graphs.

a) $y = (x - 2)^2$ **b)** $y = -x^2 + 2$

c) $y = (x + 3)^2 - 2$ **d)** $y = 2x^2 - 3$

Solution

a) The equation $y = (x - 2)^2$ has the form
$y = a(x - p)^2 + q$ with $a = 1$, $p = 2$, and $q = 0$.
Since a is positive, the parabola opens up and is congruent to $y = x^2$.
Since $p = 2$, the graph of $y = x^2$ is translated 2 units right.
Since $q = 0$, there is no vertical translation.
The graph of $y = x^2$ is translated 2 units right.

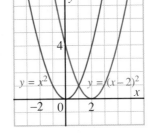

b) The equation $y = -x^2 + 2$ has the form
$y = a(x - p)^2 + q$ with $a = -1$, $p = 0$, and $q = 2$.
Since $a = -1$, the graph of $y = x^2$ is reflected in the x-axis.
Since $p = 0$, there is no horizontal translation.
Since $q = 2$, the graph of $y = -x^2$ is translated 2 units up.
The graph of $y = x^2$ is reflected in the x-axis and translated 2 units up.

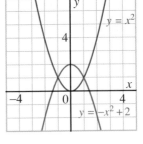

c) The equation $y = (x + 3)^2 - 2$ has the form
$y = a(x - p)^2 + q$ with $a = 1$, $p = -3$, and $q = -2$.
Since a is positive, the parabola opens up and is congruent to $y = x^2$.
Since $p = -3$, the graph of $y = x^2$ is translated 3 units left.
Since $q = -2$, the graph of $y = x^2$ is translated 2 units down.
The graph of $y = x^2$ is translated 3 units left and 2 units down.

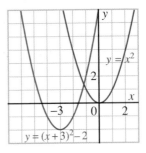

d) The equation $y = 2x^2 - 3$ has the form
$y = a(x - p)^2 + q$ with $a = 2$, $p = 0$, and $q = -3$.
Since $a = 2$, the graph of $y = x^2$ is stretched
vertically by a factor of 2.
Since $p = 0$, there is no horizontal translation.
Since $q = -3$, the graph of $y = 2x^2$ is translated
3 units down.
The graph of $y = x^2$ is stretched vertically by a
factor of 2 and translated 3 units down.

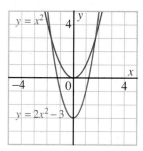

Discussing the IDEAS

1. Suppose you traced the graph of $y = x^2$ in *Investigation 1*. Explain why the tracing could be moved to coincide with each of the other graphs you drew. Describe how you would move the tracing over the grid.

2. How can you determine from the equation of a quadratic function whether its graph opens up or down?

3. Consider the graphs of $y = (x - 2)^2 + 5$ and $y = (x + 2)^2 + 5$. Which features do they have in common? Which features are different? Explain.

4. Explain the difference between translating a parabola vertically and stretching a parabola vertically.

Practise Your Skills

1. Evaluate.

   a) $5 + (-2)$ **b)** $-3 + 9$ **c)** $-6 - (-8)$

   d) $-4 + (-3)$ **e)** $-4 - (-3)$ **f)** $-4 - 3$

2. The expression $x + 3$ can be written as $x - (-3)$. Write each expression below in the same form.

   a) $x + 1$ **b)** $x + 2$ **c)** $x + 4$

   d) $x + 7$ **e)** $x + 9$ **f)** x

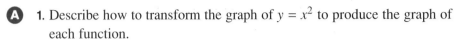
A **1.** Describe how to transform the graph of $y = x^2$ to produce the graph of each function.

a) $y = x^2 - 10$ **b)** $y = x^2 + 10$ **c)** $y = x^2 - 3$ **d)** $y = x^2 + 3$

e) $y = (x - 10)^2$ **f)** $y = (x + 10)^2$ **g)** $y = (x - 3)^2$ **h)** $y = (x + 3)^2$

2. Describe how to transform the graph of $y = x^2$ to produce the graph of each function.

a) $y = 10x^2$ **b)** $y = -10x^2$ **c)** $y = 3x^2$ **d)** $y = -3x^2$

3. Describe how to transform the graph of $y = x^2$ to produce the graph of each function.

a) $y = 2x^2 - 10$ **b)** $y = 2x^2 + 10$ **c)** $y = 4x^2 - 3$

d) $y = 4x^2 + 3$ **e)** $y = -2x^2 - 10$ **f)** $y = -2x^2 + 10$

g) $y = -4x^2 - 3$ **h)** $y = -4x^2 + 3$

4. Describe how to transform the graph of $y = x^2$ to produce the graph of each function.

a) $y = 2(x - 10)^2 + 1$ **b)** $y = 2(x + 10)^2$ **c)** $y = 4(x - 3)^2$

d) $y = 4(x + 3)^2$ **e)** $y = -2(x - 10)^2$ **f)** $y = -2(x + 10)^2$

g) $y = -4(x - 3)^2$ **h)** $y = -4(x + 3)^2$

5. Describe how to transform the graph of $y = x^2$ to produce the graph of each function.

a) $y = 2(x - 10)^2 + 1$ **b)** $y = 2(x - 10)^2 - 1$ **c)** $y = 2(x + 10)^2 + 1$

d) $y = -2(x - 10)^2 + 1$ **e)** $y = -2(x - 10)^2 - 1$ **f)** $y = -2(x + 10)^2 + 1$

B **6.** The parabola $y = x^2$ is transformed as described below. Its image has the form $y = a(x - p)^2 + q$. Determine the values of a, p, and q for each transformation.

a) Translate the parabola 3 units left.

b) Translate the parabola 4 units up.

c) Translate the parabola 3 units right.

d) Translate the parabola 4 units down.

e) Stretch the parabola vertically by a factor of 5.

f) Translate the parabola 2 units right, then reflect it in the x-axis.

g) Stretch the parabola vertically by a factor of $\frac{1}{4}$.

h) Stretch the parabola vertically by a factor of 3, then reflect it in the x-axis.

7. State which graph is represented by each equation. Use a graphing calculator to check your work.

a) $y = (x + 2)^2$ **b)** $y = x^2 + 2$ **c)** $y = 3x^2$ **d)** $y = (x - 3)^2$

i)

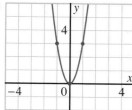

ii)

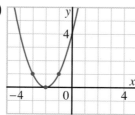

iii)

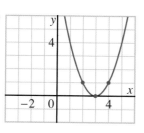

iv)

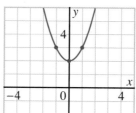

8. State which graph below is represented by each equation.

a) $y = (x + 3)^2 + 1$ **b)** $y = -2(x + 4)^2 + 3$

c) $y = \frac{1}{2}(x - 2)^2 - 5$ **d)** $y = -(x - 3)^2 + 2$

i)

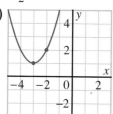

ii)

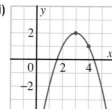

iii)

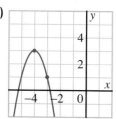

iv)

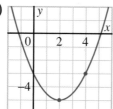

9. For each function below:

a) Describe the transformations needed to change the graph of $y = x^2$ to the graph of the new function.

b) Sketch $y = x^2$ and the new function on the same grid.

c) Check the graphs in part b.

   i) $y = -5x^2$ **ii)** $y = (x - 2)^2 + 3$ **iii)** $y = (x + 1)^2 - 2$

   iv) $y = 3(x - 2)^2$ **v)** $y = -2x^2 + 1$ **vi)** $y = \frac{1}{3}x^2 - 2$

10. The graph of $y = x^2$ has been transformed to the graph of $y = 2(x - 3)^2 + 5$.

a) Describe the transformations. Predict what the graph of $y = 2(x - 3)^2 + 5$ will look like.

b) Check your thinking by plotting the graphs on a graphing calculator.

11. Consider a quadratic function in the form $y = a(x - p)^2 + q$. Explain how its graph will change for each change described.

a) The sign of a is changed.

b) The value of q is decreased by 3.

c) The value of p is increased by 2.

d) The value of a is decreased, but its sign is not changed.

12. For each list of functions, write to explain how their graphs are similar and how they are different.

a) $y = (x - 5)^2 + 4$
$y = (x - 5)^2 + 2$
$y = (x - 5)^2$
$y = (x - 5)^2 - 2$
$y = (x - 5)^2 - 4$

b) $y = (x - 5)^2 + 4$
$y = (x - 3)^2 + 4$
$y = (x - 1)^2 + 4$
$y = (x + 1)^2 + 4$
$y = (x + 3)^2 + 4$

c) $y = 3(x - 5)^2 + 4$
$y = (x - 5)^2 + 4$
$y = 0.5(x - 5)^2 + 4$
$y = -0.5(x - 5)^2 + 4$
$y = -(x - 5)^2 + 4$
$y = -3(x - 5)^2 + 4$

C **13. a)** Visualize what the graph of $y = ax^2$ looks like as a becomes larger.

b) Visualize what the graph of $y = ax^2$ looks like as a becomes smaller.

c) Predict what the graphs of these equations would look like.

$y = 10x^2$
$y = 100x^2$
$y = 1000x^2$
$\vdots$
$y = 1\,000\,000x^2$

$y = 0.1x^2$
$y = 0.01x^2$
$y = 0.001x^2$
$\vdots$
$y = 0.000\,001x^2$

d) Use a graphing calculator to check your predictions.

Communicating the IDEAS

Write to explain why changing the value of p in the quadratic function $y = a(x - p)^2 + q$ moves the parabola right or left, and how you can tell which direction it moves. Include graphs as part of your explanation.

8.2 Analysing the Graph of $y = a(x - p)^2 + q$

In Section 8.1, you learned how changes to the equation of the parabola $y = x^2$ change the position and shape of the parabola. We can use these ideas to identify important features of the graph of a parabola whose equation is written in the form $y = a(x - p)^2 + q$.

The Graph of $y = a(x - p)^2 + q$

Set-Up

- Press WINDOW. Change the settings to Xmin = −4.7, Xmax = 4.7, Xscl = 1, Ymin = −12.4, Ymax = 12.4, Yscl = 2
- Press Y= . Use the arrow keys and CLEAR to remove any equations in the Y= list.
- If any of the statistical plots: Plot 1, Plot 2, or Plot 3 is highlighted, use the arrow keys to move to the plot, then press ENTER to turn it off.

1. Copy this table.

Function	Direction of opening	Maximum/ minimum	Coordinates of vertex
$y = 3x^2$			
$y = (x - 1)^2$			
$y = -x^2 + 2$			
$y = -2(x + 1)^2 - 2$			
$y = 0.5(x - 2)^2 + 1$			

2. Press Y= . Enter the first equation in the table into the graphing calculator as Y1. Graph the function.

3. Determine the coordinates of the vertex.

- Press TRACE and use ◄ or ► to move the cursor to the vertex of the parabola.

Write the coordinates of the vertex in the table.

4. Write the direction of opening of the parabola, in the table.

5. The function, that is, y, has either its greatest value at the vertex or its least value there. Make the appropriate entry in the column headed *Maximum/minimum*; for example, maximum $y = 2$.

6. Repeat exercises 2 to 5 for each function in the table.

7. Look at the completed table. Which number a, p, or q determines:

   a) the direction of opening of the parabola $y = a(x - p)^2 + q$?

   b) the x-coordinate of the vertex of the parabola $y = a(x - p)^2 + q$?

   c) the y-coordinate of the vertex of the parabola $y = a(x - p)^2 + q$?

In the *Investigation,* you reviewed these effects of a, p, and q on the graph of $y = a(x - p)^2 + q$.

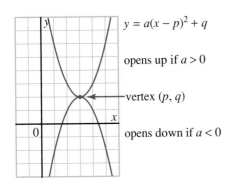

$y = a(x - p)^2 + q$

opens up if $a > 0$

vertex (p, q)

opens down if $a < 0$

Example 1

Without sketching a graph, determine the coordinates of the vertex, the direction of opening of the graph, and the maximum or minimum value of y for each function.

  a) $y = (x - 2)^2 - 1$ **b)** $y = -x^2$ **c)** $y = -2(x + 5)^2 + 3$

Solution

> **Think ...**
>
> Compare each equation with $y = a(x - p)^2 + q$ to identify the values of a, p, and q.
> The sign of a indicates which way the parabola opens.
> The vertex of the parabola is the point (p, q).
> The maximum or minimum value of y is the y-coordinate of the vertex.

  a) The equation $y = (x - 2)^2 - 1$ has $a = 1$, $p = 2$, and $q = -1$.
     Since a is positive, the parabola opens up, and y has a minimum value.
     Since $p = 2$ and $q = -1$, the vertex is $(2, -1)$.
     The minimum value of y is -1.

b) The equation $y = -x^2$ has $a = -1$, $p = 0$, and $q = 0$.

Since a is negative, the parabola opens down, and y has a maximum value.

Since $p = 0$ and $q = 0$, the vertex is $(0, 0)$.

The maximum value of y is 0.

c) The equation $y = -2(x + 5)^2 + 3$ can be written as $y = -2(x - (-5))^2 + 3$, so $a = -2$, $p = -5$, and $q = 3$.

Since a is negative, the parabola opens down, and y has a maximum value.

Since $p = -5$ and $q = 3$, the vertex is $(-5, 3)$.

The maximum value of y is 3.

Example 2

Each graph is a parabola. Its equation has the form $y = a(x - p)^2 + q$. For each parabola:

  i) Determine the values of p and q, and state the sign of a.

  ii) Determine the value of a.

  iii) Determine an equation for the parabola.

a)

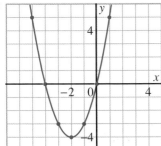

b)

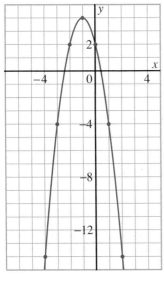

Solution

> **Think ...**
>
> The value of p is the x-coordinate of the vertex of the parabola. The value of q is the y-coordinate. Use the grid to identify these.
> If the parabola opens up, a is positive, and if it opens down, a is negative.
> To calculate a, use the step pattern for consecutive x-values: 1, 3, 5,...

a) The equation of the graph has the form $y = a(x - p)^2 + q$.
From the graph, the vertex is $(-2, -4)$, so $p = -2$ and $q = -4$.
The parabola opens up, so a is positive.
The step pattern for consecutive x-values from the vertex is:
1 right 1 up, 1 right 3 up, 1 right 5 up. This is the standard step pattern so $a = 1$.
Substitute $a = 1$, $p = -2$, and $q = -4$ in $y = a(x - p)^2 + q$.
The equation of the parabola is $y = 1(x - (-2))^2 + (-4)$.
This simplifies to $y = (x + 2)^2 - 4$.

b) The equation of the graph has the form $y = a(x - p)^2 + q$.
From the graph, the vertex is $(-1, 4)$, so $p = -1$ and $q = 4$.
Since the parabola opens down, a is negative.
The step pattern for consecutive x-values from the vertex is:
1 right 2 down, 1 right 6 down, 1 right 10 down.
Each vertical step is two times the standard step; so, $a = -2$.
Substitute $a = -2$, $p = -1$, and $q = 4$ in $y = a(x - p)^2 + q$.
The equation of the parabola is $y = -2(x - (-1))^2 + 4$, or $y = -2(x + 1)^2 + 4$.

Discussing *the* IDEAS

1. Explain how to identify the coordinates of the vertex of a parabola from its equation $y = a(x - p)^2 + q$.

2. From the equation of a parabola written in the form $y = a(x - p)^2 + q$, how can you determine which way it opens?

3. Why is the y-coordinate of the vertex of the parabola $y = a(x - p)^2 + q$ the maximum or minimum value of the function?

4. In *Example 2*, explain how we used the step pattern to calculate a.

A **1.** Identify each component of each quadratic function $y = a(x - p)^2 + q$.
   i) the sign of a
   ii) the coordinates of the vertex.

a)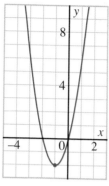

b)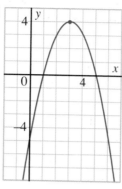

c)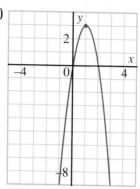

2. Use the word "maximum" or "minimum," and data from each graph.
Copy and complete this sentence for each graph.
"The _____ value of y is _____ when $x =$ _____."

a)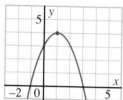

b)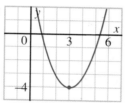

c)

d)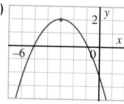

e)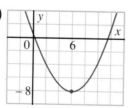

f)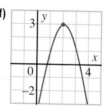

B **3.** For each parabola:
   i) State the coordinates of the vertex.
   ii) State whether the parabola opens up or down.
   iii) State the maximum or minimum value of y, and which it is.

a) $y = (x - 5)^2 + 6$ **b)** $y = (x + 2)^2 + 4$

c) $y = -3(x - 3)^2 + 1$ **d)** $y = -(x + 5)^2 + 2$

e) $y = -\frac{1}{2}x^2 + 7$ **f)** $y = -(x + 4)^2$

4. Determine the equation of each parabola in the form $y = a(x - p)^2 + q$.
   a) with vertex $(0, 2)$ and $a = 1$
   b) with vertex $(0, 5)$ and $a = -2$

c) with vertex $(4, 0)$ and $a = \frac{1}{2}$

d) with vertex $(-3, 0)$ and $a = -1$

5. Does each function have a maximum? If it has a maximum, for what value of x does it occur?

a) $y = -2(x + 5)^2 - 8$ **b)** $y = \frac{1}{4}(x - 2)^2 - 9$ **c)** $y = -0.5(x - 3)^2 + 0.75$

6. For each parabola:
   i) State the maximum or minimum value of y.
   ii) State whether it is a maximum or a minimum.
   iii) State the value of x where it occurs.

a) $y = (x - 3)^2 + 5$ **b)** $y = 2(x + 1)^2 - 3$ **c)** $y = -2(x - 1)^2 + 4$

d) $y = -(x + 2)^2 - 6$ **e)** $y = 0.5x^2 - 9$ **f)** $y = -x^2 + 7$

7. a) Write the equation of a quadratic function that satisfies each set of conditions.
   i) The function has a minimum value of 4 when $x = -2$.
   ii) The function has a minimum value of -7 when $x = 3$.
   iii) The function has a maximum value of -1 when $x = 5$.

   b) Compare your equations with those of your classmates. Did all of you write the same equation for each description? Explain.

8. Each graph below is a parabola with equation of the form $y = a(x - p)^2 + q$.
   For each parabola:
   i) Determine the values of p and q, and state the sign of a.
   ii) Determine the value of a.
   iii) Determine an equation for the parabola.

a)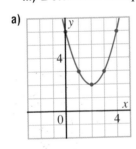

b)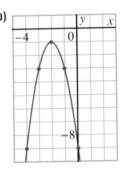

c)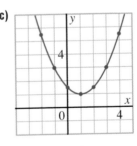

9. For each parabola:
   i) State the coordinates of the vertex.
   ii) State whether the parabola opens up or down.
   iii) Describe the transformation that changed $y = x^2$ to the parabola.

a) $y = 5(x - 3)^2 - 5$ **b)** $y = \frac{1}{3}(x - 6)^2 + 13$

c) $y = (x + 2)^2 + 10$ **d)** $y = -(x + 1)^2 + 8$

10. Create the following images with the given window setting.

a)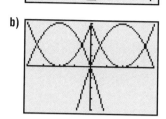

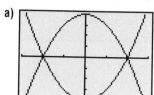
```
WINDOW
 Xmin=-3
 Xmax=3
 Xscl=1
 Ymin=-4
 Ymax=4
 Yscl=1
 Xres=1
```

b)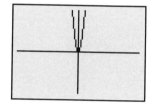

```
WINDOW
 Xmin=-4
 Xmax=4
 Xscl=1
 Ymin=-4
 Ymax=4
 Yscl=1
 Xres=1
```

c) Make similar patterns of your own.

C **11.** The graphs of three quadratic functions are shown below. The scales are not shown, and they are not necessarily the same. Which of these statements are true?

1) In the first graph, the value of a is larger that it is in the third graph.

2) All three graphs could be graphs of the same function.

3) Any of these graphs could be the graph of any function in exercise 13c, page 408.

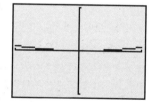

Communicating *the* IDEAS

Your friend is trying to complete her homework without a graphing calculator. She must sketch the graph of $y = -2(x + 1)^2 + 3$. Write a description of the graph that would allow her to complete her homework.

Relating the Graphs of $y = ax^2 + bx + c$ and $y = a(x - p)^2 + q$

In Chapter 2, you learned that the equation of a linear function may be written in different ways. In this section, you will relate the two different forms of the equation of a quadratic function.

INVESTIGATION 1

Equivalent Equations of a Function

Set-Up

• Press WINDOW. Change the settings to Xmin = −4.7, Xmax = 4.7, Xscl = 1, Ymin = −2.3, Ymax = 16.3, Yscl = 2.

• Press MODE and make sure that **Sequential** is selected.

• Press Y= . Use the arrow keys and CLEAR to remove any equations in the Y= list. If any of the statistical plots: Plot 1, Plot 2, or Plot 3 is highlighted, use the arrow keys to move to the plot, then press ENTER to turn the plot off.

1. Enter the equation $y = 2x^2 - 4x$ as Y1 and the equation $y = 2(x - 1)^2 - 2$ as Y2.

2. Compare the graphs of these equations.

• Change the graphing style of Y2 to "animation." Move the cursor to the left of Y2. Press ENTER four times to toggle through different graph styles available. Your screen should look like the one below. Note the symbol to the left of Y2.

```
Plot1 Plot2 Plot3
\Y1∎2X²-4X
•∘Y2∎2(X-1)²-2
\Y3=
\Y4=
\Y5=
\Y6=
\Y7=
```

• Press GRAPH. The graph of Y1 is drawn first, and then the graph of Y2 is traced with a small circle.

What do you notice about the two graphs?

3. Write the equation $y = 2(x - 1)^2 - 2$. Use the techniques you learned in Chapter 7 to expand and simplify the right side of the equation.

4. Compare the equation you obtained in exercise 3 with the equation you entered as Y1. What do you notice?

5. Use the results of exercises 2 and 3 to explain why the two forms of the equation in exercise 1 are called "equivalent."

6. Which equation gives more information about its graph? Explain. When would you find it more convenient to use the other equation?

In Section 8.2, you identified the coordinates of the vertex of a quadratic function when the equation has the form $y = a(x - p)^2 + q$. We will now investigate the graph of a quadratic function when the equation has the form $y = ax^2 + bx + c$.

The Graph of $y = ax^2 + bx + c$

Set-Up

- Press WINDOW. Change the settings to Xmin = −4.7, Xmax = 4.7, Xscl = 1, Ymin = −9, Ymax = 9, and Yscl = 2.

- Press Y=. Use the arrow keys and CLEAR to remove any equations in the Y= list. If any of the statistical plots: Plot 1, Plot 2, or Plot 3 is highlighted, move the cursor to it, then press ENTER to turn the plot off.

1. Copy this table.

	Equation $0 = ax^2 + bx + c$	Roots of the equation	Function $y = ax^2 + bx + c$	x-intercepts of graph	Vertex
a)	$0 = x^2 - 9$		$y = x^2 - 9$		
b)	$0 = x^2 + 4x$		$y = x^2 + 4x$		
c)	$0 = x^2 + 5x + 4$		$y = x^2 + 5x + 4$		
d)	$0 = -x^2 + 4$		$y = -x^2 + 4$		
e)	$0 = -2x^2 + 4x + 6$		$y = -2x^2 + 4x + 6$		

2. Describe the relationship between the equations and the functions in this table.

3. Solve each equation by factoring. Enter the roots of the equations in the table.

4. a) Enter the function in exercise 1a as Y1, then graph it.

   b) Use TRACE to find the x-intercepts of the parabola. Enter these in the table.

   c) What are the y-coordinates of the points of intersection of the parabola and the x-axis?

5. Trace to the vertex of the parabola. Enter the coordinates of the vertex in the table.

6. Repeat exercises 4 and 5 for each function in exercise 1.

7. Use the completed table.

   a) What is the relationship between the roots of an equation and the x-intercepts of a function?

   b) Explain why the x-intercepts are called the "zeros" of a function.

   c) Describe the relationship between the x-intercepts and the x-coordinate of the vertex.

   d) Suppose you knew the x-coordinate of the vertex. How might you determine its y-coordinate?

Here is a review of the results of *Investigation 2*.

Take Note

Quadratic Functions and Quadratic Equations

- The x-intercepts of the graph of $y = ax^2 + bx + c$ are the roots of the equation $ax^2 + bx + c = 0$.

- The x-intercepts of the graph of $y = ax^2 + bx + c$ are called the zeros of the function.

Combine the two previous statements.

- The zeros of the function $y = ax^2 + bx + c$ are the roots of the equation $ax^2 + bx + c = 0$.

- The mean of the x-intercepts is equal to the y-coordinate of the vertex; for example, if the x-intercepts are m and n, the y-coordinate of the vertex is $\frac{m+n}{2}$.

- When the x-coordinate of the vertex is known, the y-coordinate can be determined by substituting the x-coordinate in the equation $y = ax^2 + bx + c$.

Example 1

Determine whether the equations in each pair are equivalent. Check by graphing both equations on the same screen.

a) $y = 2(x + 4)^2 - 20$
$y = 2x^2 + 16x + 12$

b) $y = -(x - 1)^2 + 5$
$y = -x^2 + x + 4$

c) $y = (x + 2)^2$
$y = 2x^2 + 4$

Solution

a) $y = 2(x + 4)^2 - 20$ and $y = 2x^2 + 16x + 12$

Use the equation $y = 2(x + 4)^2 - 20$.

Expand the binomial square.

$y = 2(x + 4)(x + 4) - 20$
$y = 2(x^2 + 4x + 4x + 16) - 20$
$y = 2(x^2 + 8x + 16) - 20$

Simplify.

$y = 2x^2 + 16x + 32 - 20$
$y = 2x^2 + 16x + 12$

The equation $y = 2(x + 4)^2 - 20$ is equivalent to the equation $y = 2x^2 + 16x + 12$.

Graph the two equations. The graphs coincide.

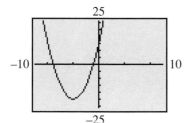
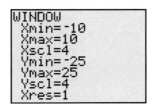

b) $y = -(x - 1)^2 + 5$ and $y = -x^2 + x + 4$

Use the equation $y = -(x - 1)^2 + 5$.

Expand the binomial square.

$y = -(x - 1)(x - 1) + 5$
$y = -(x^2 - x - x + 1) + 5$
$y = -(x^2 - 2x + 1) + 5$

Simplify.

$y = -x^2 + 2x - 1 + 5$
$y = -x^2 + 2x + 4$

The equations are not equivalent.

Graph the two equations. The graphs do not coincide.

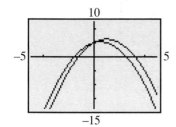

c) $y = (x + 2)^2$ and $y = 2x^2 + 4$

Compare the values of a for the equations:

$a = 1$ for the first equation, $a = 2$ for the second equation.

The values of a must be the same for the equations to be equivalent.

The equations are not equivalent.

Graph the two equations. The graphs do not coincide.

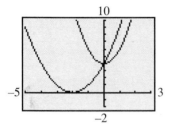

Example 2

Graph the function $y = -x^2 - 2x + 3$.

a) Determine:

   i) the coordinates of the vertex

   ii) the zeros of the function

   iii) the values of x for which y is greater than 3

b) Sketch the graph, then identify the features from part a.

Solution

Set the window to Xmin = −4.7, Xmax = 4.7, Xscl = 1, Ymin = −6.2, Ymax = 6.2, Yscl = 1.

Enter Y1 = $-x^2 - 2x + 3$ in the Y= list. Press [GRAPH] to obtain the first screen on the next page.

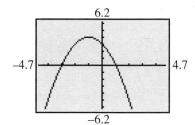

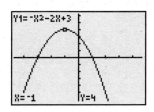

a) **i)** Press [TRACE] and use ◄ and ► to move the cursor along the curve to the vertex (above right).

The coordinates of the vertex are $(-1, 4)$.

ii) Trace to each x-intercept.

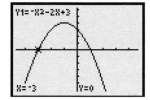

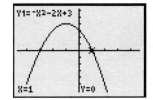

The zeros are -3 and 1.

iii) Trace along the curve to $y = 3$ (below left).

From the screen, when $y = 3$, $x = 0$

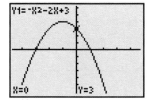

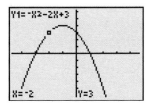

Continue to trace along the curve to the next position where $y = 3$ (above right).

From the screen, when $y = 3$, $x = -2$

So, y is greater than 3 when x is greater than -2 and less than 0.

This can be written $y > 3$ when $x > -2$ and $x < 0$, or $y > 3$ when $-2 < x < 0$.

b)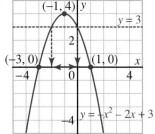

1. Describe how to use algebra to determine whether two equations are equivalent.

2. Suppose you were shown the graphs of two quadratic functions on the same screen. Describe what you would see if the equations are not equivalent.

3. Why can we not use graphs to prove that two equations are equivalent?

4. The graph of $y = 2x^2 + 4$ does not intersect the x-axis. How many zeros has the function $y = 2x^2 + 4$? How many roots has the equation $0 = 2x^2 + 4$? Explain.

Practise Your Skills

1. Expand.

a) $(x + 5)^2$
b) $(x + 2)^2$
c) $(x - 5)^2$
d) $(x + 3)^2$

e) $-(x - 6)^2$
f) $4(x - 1)^2$
g) $-3(x + 5)^2$
h) $2(x - 3)^2$

8.3 Exercises

Use a graphing calculator to complete any exercise that requires graphing.

A 1. Express each equation in the form $y = ax^2 + bx + c$.

a) $y = (x - 3)^2 + 5$
b) $y = -(x + 1)^2 + 4$
c) $y = 2(x - 3)^2$
d) $y = -2(x + 1)^2 + 2$

2. Use a graphing calculator to check your answers to exercise 1.

B 3. a) Graph each function.

b) Determine the zeros of the function and the coordinates of the vertex.

i) $y = x^2 + 4x - 12$
ii) $y = -4x^2 - 31x + 8$
iii) $y = 2x^2 - 11x + 5$
iv) $y = 6x^2 - 6x - 12$

4. Compare the equations in each pair: algebraically and graphically. If the equations are different, describe their differences.

a) $y = -(x + 1)^2 + 3$,
   $y = -x^2 - 2x + 4$

b) $y = 3(x - 2)^2 - 4$,
   $y = 3x^2 - 12x + 8$

c) $y = (x + 2)^2 - 3$,
   $y = -x^2 - 4x - 7$

d) $y = 2(x + 3)^2$,
   $y = 2x^2 + 12x + 18$

5. a) Which functions below have the same graph as the function $y = 2x^2 - 12x + 19$?

    i) $y = 2(x - 3)^2 + 1$

    ii) $y = 2(x - 2)^2 + 11$

    iii) $y = 2(x - 1)^2 + 17$

    Justify your choices algebraically.

b) Determine the coordinates of the vertex of the parabola
$y = 2x^2 - 12x + 19$. Explain how you determined these coordinates.

6. Determine whether the equations in each pair are equivalent. Justify your answer.

a) $y = \frac{1}{2}(x - 6)^2 + 12, \quad y = \frac{1}{2}x^2 - 6x + 30$

b) $y = -(x - 4)^2 + 16, \quad y = -x^2 + 8x$

c) $y = (x + 3)^2 + 9, \quad y = x^2 + 6x + 9$

7. Graph each function. Choose a window setting that displays the vertex and the x-intercepts.

a) $y = 2x^2 + 5x - 3$ **b)** $y = -5x^2 - 10x + 30$

c) $y = x^2 - 7x + 10$ **d)** $y = -3x^2 + 3x + 6$

8. Determine the zeros of each function.

a) $y = x^2 - x - 6$ **b)** $y = x^2 + 2x - 8$

c) $y = -x^2 + 7x$ **d)** $y = -x^2 + 10x - 25$

For exercises 9 to 12, set the window to Xmin = -9.4, Xmax = 9.4, Xscl = 1, Ymin = -25, Ymax = 25, Yscl = 5.

9. a) Graph the function $y = x^2 - 25$.

b) For what values of x is y greater than 0?

c) For what values of y is x greater than 6?

d) For what values of y is $x < 3$ and $x > -3$?

10. a) Graph the function $y = x^2 - 10x + 9$.

b) For what values of x is $y < 0$?

c) For what values of x is $y > 0$?

d) For what values of x is $y < -7$?

11. a) Graph the function $y = -x^2 + 16$.

b) For what values of x is y greater than 12?

c) For what values of y is x less than -5?

d) For what values of y is $x > -3$ and $x < 3$?

12. a) Graph the function $y = 4x^2 - 16x$.

b) For what values of x is y greater than -12?

c) For what values of y is x less than -1?

d) For what values of y is $-1 < x < 1$?

13. Use the Calculate feature of a graphing calculator. For each function:
   i) Determine the zeros.
   ii) Determine the maximum or minimum value, and state which it is.

a) $y = 3x^2 - 12x + 8$ **b)** $y = 2x^2 - 10x + 9$

14. a) Graph these equations on the same screen.

$y = 2x^2 + 4x + 3$ $y = -2x^2 + 4x + 3$

$y = x^2 + 4x + 3$ $y = -x^2 + 4x + 3$

$y = 0.75x^2 + 4x + 3$ $y = -0.75x^2 + 4x + 3$

$y = 0.5x^2 + 4x + 3$ $y = -0.5x^2 + 4x + 3$

$y = 0x^2 + 4x + 3$

b) Write to describe how the graph of $y = ax^2 + 4x + 3$ changes as a changes.

c) What special case occurs when $a = 0$?

⊙ 15. a) Graph these equations on the same screen.

$y = x^2 + 6x + 3$ $y = x^2 - 6x + 3$

$y = x^2 + 4x + 3$ $y = x^2 - 4x + 3$

$y = x^2 + 2x + 3$ $y = x^2 - 2x + 3$

$y = x^2 + 0x + 3$

b) Write to describe how the graph of $y = x^2 + bx + 3$ changes as b changes.

c) What special case occurs when $b = 0$?

d) Determine the x- and y-coordinates of the vertex of each parabola in part a. Enter the coordinates in lists L1 and L2 of a graphing calculator.

e) Draw a scatter plot of the data in the two lists using STAT PLOT. What type of function do you think would represent the points?

Communicating *the* IDEAS

You have worked with two different forms of the equation of a quadratic function: $y = ax^2 + bx + c$ and $y = a(x - p)^2 + q$. Sometimes one form is preferable to the other. When is $y = a(x - p)^2 + q$ preferable? When is $y = ax^2 + bx + c$ preferable? Write to explain your ideas.

One application of quadratic functions is motion under the influence of gravity. For example, when a ball is kicked, its height is a quadratic function of time. The equation of the quadratic function is based on a physics formula. If we ignore air resistance, the path of a ball thrown from one person to another is parabolic. Water squirting from a hose follows a parabolic arc.

If we know the equation of the quadratic function that represents a situation, we can solve problems by interpreting its graph.

Example 1

A soccer ball is kicked from the ground into the air with an initial speed of 20 m/s. The height of the ball, h metres, is represented by the equation $h = -5t^2 + 20t$, where t seconds is the time since the ball was kicked.

a) What is the maximum height of the ball?

b) For how long is the ball in the air?

c) For how long is the ball higher than 15 m?

Solution

Graph the equation $h = -5t^2 + 20t$ as $y = -5x^2 + 20x$.
Set the window to Xmin = 0, Xmax = 4.7, Xscl = 1, Ymin = −5,
Ymax = 25, Yscl = 5.
Enter Y1 = $-5x^2 + 20x$. Press GRAPH to obtain the screen below left.

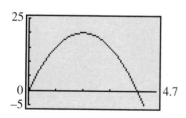
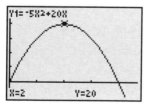

a) Use TRACE to determine the maximum height of the ball. Press ► or ◄ and watch the values of y change. From the screen (above right), the maximum value of y is 20 when $x = 2$.

The maximum height of the ball is 20 m.

b) To determine how long the ball is in the air, use TRACE.
The ball was kicked when $t = 0$; that is, $x = 0$ on the calculator screen
(below left).

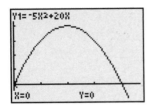
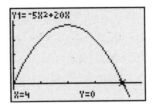

Press ▶ and move the cursor to the point where $y = 0$ to represent $h = 0$,
when the ball landed (above right).
From the screen, when $y = 0$, $x = 4$.
The ball is in the air for 4 s.

c) To determine the time for which the ball is higher than 15 m, use TRACE.
Press ▶ or ◀ to reach the point where $y = 15$ (below left).

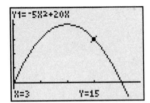
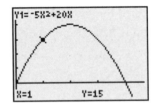

From the screen, when $y = 15$, $x = 3$
Continue to press ◀ until the cursor reaches $y = 15$ again (above right).
From the screen, when $y = 15$, $x = 1$
The ball is higher than 15 m between $t = 1$ and $t = 3$;
that is, for a time of 3 s − 1 s = 2 s.
The ball is higher than 15 m for 2 s.

Example 2

Annie Pelletier won a bronze medal in the
women's springboard competition at the 1996
Summer Olympics in Atlanta.
Pelletier somersaults from a 3-m springboard,
with an initial upward speed of 8.8 m/s.

A formula from physics that represents this situation is $h = -4.9t^2 + 8.8t + 3$, where h metres is the height above the water at time t seconds after Annie leaves the springboard.

a) Graph this equation.

b) Determine Annie's maximum height, and the time taken to reach this height.

c) Determine the time Annie is in the air.

d) Determine the time that Annie was higher than the springboard.

Solution

a) Use these window settings.

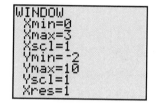
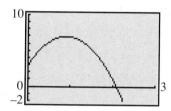

Replace h with y and t with x in the formula. Enter the function $y = -4.9x^2 + 8.8x + 3$ as Y1 in the Y= list. Press [GRAPH]. The graph is shown above right.

b)

> **Think ...**
>
> The graph of the function is a parabola, opening down. The maximum value of the function is the y-coordinate of the vertex.

Press [2nd] [TRACE] for [CALC]. Press **4** to select **maximum**. Move the cursor to the left of the maximum point; press [ENTER]. Move the cursor to the right of the maximum point; press [ENTER].

Press [ENTER] again to display the coordinates as shown on the screen below.

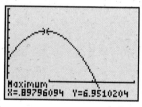

The maximum height above the water is about 6.95 m.
It took about 0.9 s to reach this height.

c) Annie will hit the water when her height is 0 m above the water.
Press ⎡2nd⎤ ⎡TRACE⎤ for ⎡CALC⎤. Press **2** to select **zero**. Move the cursor to the
left of the *x*-intercept; press ⎡ENTER⎤. Move the cursor to the right of the
x-intercept; press ⎡ENTER⎤.

Press ⎡ENTER⎤ again to display the zero as shown on the screen below left.

Annie is in the air for approximately 2.09 s.

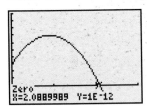 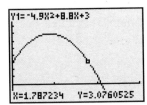

d)

> *Think ...*
>
> The springboard is 3 m above the water. Annie is higher than the board when her
> height above the water is greater than 3 m. Determine the values of *x* for which the
> *y*-coordinates of points on the parabola are greater than 3.

Trace to determine where $y = 3$. The points on the graph with *y*-coordinate
3 are (0, 3) and approximately (1.79, 3.08), above right. The *x*-coordinates
represent time. Annie was higher than the springboard for 1.79 s − 0 s = 1.79 s.
Annie was higher than the springboard for about 1.8 s.

Discussing *the* IDEAS

1. In *Example 1*, why did we set Ymin = −5, when we know we cannot have a
negative height?

2. In *Example 1*, why did we replace *h* with *y* and *t* with *x*?

3. In *Example 2*, why was the graphing window chosen to show only values of *x* greater
than 0?

4. In *Example 2*, how would the equation describing Annie Pelletier's dive change if she
were to use a 5-m springboard with an initial speed of 7.2 m/s?

Use a graphing calculator for these exercises.

B 1. The area, in square metres, of a rectangular lawn with perimeter 14 m is given by the function $A = -x^2 + 7x$, where x metres is the width of the rectangle.

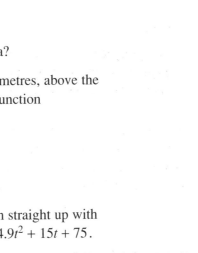

7 − x

a) What is the area of the rectangle when the width is 3 m?

b) What is the length of the rectangle when the width is 3 m?

c) Graph the function. Sketch the graph.

d) What is the maximum area of the rectangle?

e) What is the shape of the rectangle with maximum area?

2. A rock is thrown from a bridge into a river. Its height, h metres, above the river t seconds after it is released, is represented by the function $h = -4.9t^2 + 82$.

a) Graph the function for reasonable values of t.

b) How high is the rock after 2.5 s?

c) When does the rock hit the water?

3. The height, h metres, of a ball t seconds after it is thrown straight up with an initial speed of 15 m/s is given by the function $h = -4.9t^2 + 15t + 75$.

a) Graph the function.

b) What is the maximum height of the ball?

c) Suppose the thrower catches the ball at the same height from which it was thrown. How long is the ball in the air?

4. The path of an Acapulco cliff diver, as he dives into the sea, is given by the function $y = -2.18x^2 + 1.73x + 35$, where y metres is the diver's height above the water, and x metres is the horizontal distance travelled by the diver.

a) Graph the function.

b) Determine the maximum height of the diver above sea level.

c) Assume the cliff is vertical. How far from the base of the cliff will the diver enter the water?

5. The power, P watts, supplied to a circuit by a 9-V battery is given by the formula $P = -0.5I^2 + 9I$, where I is the current in amperes.

a) Graph the function.

b) For what value of the current will the power be a maximum?

c) What is the maximum power?

6. A juggler tosses a ball from her hand at A, along the arc ACB to her hand at B. She then returns the ball from her hand at B to her hand at A along the arc BDA.

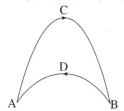

The height of the ball on arc ACB is given by the function $h = -4.9t^2 + 3.5t$. The height of the ball on arc BDA is given by the function $h = -4.9t^2 + 2t$. The variable t is the time in seconds since the ball left the juggler's hand.

a) For how long is the ball in the air as it travels along arc ACB?

b) For how long is the ball in the air as it travels along arc BDA?

c) How long does it take the ball to complete one cycle: from A to C to B to D and back to A?

7. The height of a flare above the ground is a function of the time it is in the air. An equation representing the height of a flare, h metres, above the release position, after t seconds, is $h = -5t^2 + 100t$.

a) What is the height of the flare after 3 s?

b) What is the maximum height reached by the flare?

c) What is the height of the flare after 25 s?

d) Does your answer in part c make sense? Explain.

e) Determine the approximate time for which the flare is higher than 80 m.

8. When a flare is fired vertically upward, its height, h metres, after t seconds is modelled by the function $h = -5t^2 + 153.2t$.

a) Graph the function. Adjust the window setting if necessary to show the vertex and zeros of the function.

b) What are the coordinates of the vertex?

c) Determine the approximate time for which the flare is higher than 1 km.

The Rising Fastball

Some baseball fans believe that pitchers can throw a baseball that rises as it approaches the batter. This seems to contradict common sense, that the ball should fall because of gravity. However, a ball pitched with backspin is subject to a lifting force that acts against gravity. This lifting force is about 20% of the force of gravity. This means that a ball with backspin falls about 80% of the distance that a ball without backspin falls.

A rising fastball is an illusion. A ball thrown with backspin may appear to rise simply because it does not fall as much as the batter expects it to. You can confirm this by completing these exercises.

9. It is about 18.4 m from the pitcher's mound to home plate. A fast pitch travels about 150 km/h. Convert 150 km/h to metres per second. Use the formula $time = \frac{distance}{speed}$ to calculate how long it takes the ball to reach the batter at this speed, to the nearest hundredth of a second.

10. Assume the pitcher throws a ball horizontally without backspin. After the ball leaves the pitcher's hand, the ball begins to fall. The distance, d metres, that it falls is given by the equation $d = 4.9t^2$, where t is the elapsed time in seconds. Use your answer from exercise 9. Determine how far the ball has fallen by the time it reaches the batter. Give your answer to the nearest centimetre.

11. Suppose the ball is thrown with backspin. How far would it have fallen by the time it reaches the batter?

Communicating *the* IDEAS

In the examples and exercises of this section, it is the shape of the parabola that makes the quadratic function suitable as a mathematical model. List some features of the parabola and explain why they make the quadratic function a good choice to represent real situations.

On page 394, you considered the trajectory of a basketball during a "free throw." In this section, we will create a mathematical model to represent the path of a successful throw.

The free throw line is 4.5 m from the point on the floor directly below the centre of the basket. The rim of the basket is 3 m above the floor. We will assume that the player releases the ball at a point 2 m above the floor.

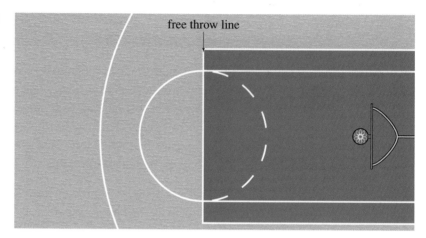

free throw line

An Algebraic Model

1. Draw coordinate axes on grid paper. Horizontal distances are measured along the *x*-axis, and the heights of the ball are measured along the *y*-axis. Mark the point from which the ball will be thrown at A(0, 2). Mark the centre of the basket at the point B(4.5, 3).

2. Sketch several possible free throw trajectories on your grid. Recall that the trajectory must be parabolic with its axis of symmetry parallel to the *y*-axis.

3. Your trajectories have an equation of the form $y = ax^2 + bx + c$. The point A(0, 2) lies on all the trajectories. Substitute the coordinates of this point into the equation $y = ax^2 + bx + c$. Use the result to determine the value of *c*.

4. The point B(4.5, 3) also lies on all the trajectories. Substitute the coordinates of this point into the equation $y = ax^2 + bx + 2$. Simplify the result to obtain an equation containing *a* and *b* only.

5. Look at your graphs of free throw trajectories. Estimate some possible values for *a*.

6. Substitute each value of *a* into the equation you found in exercise 4. Use the equation to determine the value of *b* for this trajectory. Copy and complete the table below. List the values of *a*, *b*, and the corresponding equation of the trajectory.

a	b	$y = ax^2 + bx + 2$

7. Graph each equation $y = ax^2 + bx + 2$ from the table. Use ⟨TRACE⟩ to check to see if each trajectory represents a successful free throw. Adjust the list of possible values of *a* as necessary.

8. Copy a graph that represents a successful free throw. Label it with its equation.

9. Write to describe any assumptions you made to create this model.

A Geometric Model

If you have access to a computer with *The Geometer's Sketchpad*, you can create a model that takes account of two factors under the control of the thrower: the angle at which the ball is thrown and the initial speed (velocity).

Ask your teacher for the sketch **freethro.gsp.**

10. Start *The Geometer's Sketchpad*. Open the sketch **freethro.gsp** by clicking on **File**, then **Open**. Choose the appropriate drive.

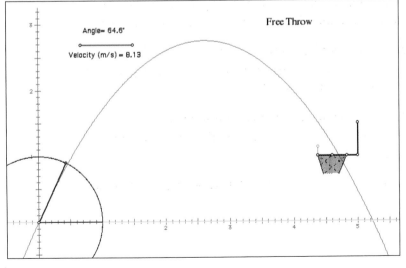

Free Throw
Angle= 64.6°
Velocity (m/s) = 8.13

11. When you open the sketch, you will see: a circle, centre the origin, with a blue radius; a red parabola; a horizontal green line segment; and a basketball net. Move the endpoint of the blue radius slowly around the circle. You will see the measurement of the angle and the red parabola change.

   The origin is at the point A from which the ball is thrown.

   Adjust the length of the horizontal green segment above the word "velocity." You will see the velocity and the red parabola change.

   The angle is that at which the basketball player throws a free throw; that is, the angle between the radius and the horizontal axis. The velocity is the speed with which the ball is thrown.

12. Adjust the angle and the velocity to make the red parabola pass through the basket. Try this a few times.

13. a) In this model, the ball is represented by a point "running along" the parabola. Is this realistic? When the parabola passes through the net, would the player always score a point? Explain.

   b) Describe some restrictions that would ensure that the shot scores.

14. In the sketch, there is a small vertical turquoise segment in front of the net. The length of this segment is one-half the diameter of a basketball. Describe how this segment can help determine if a shot will score.

15. When you have decided on your rules for scoring a point, use the sketch to create several "scoring trajectories." Record the angle and the velocity of each scoring trajectory.

16. Collect the data from at least 8 different scoring trajectories. Input the angles and velocities in lists L1 and L2, respectively. Create a scatter plot of the data. Do the points appear to lie on a curve? Estimate the equation of this curve. Graph the curve. Copy the graph and write its equation.

Communicating the IDEAS

Prepare a report for a coach to advise her or him of the ideal conditions for free throw shooting. Make reference in the report to your calculations.

Quadratic Function Tools

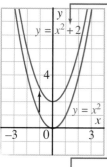

Positive sign, the graph moves up.

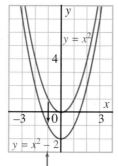

Negative sign, the graph moves down.

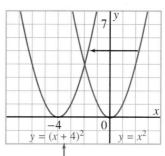

Positve sign, the graph moves left.

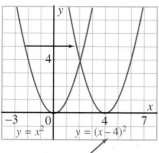

Negative sign, the graph moves right.

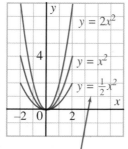

Positive signs, the graphs are stretched or compressed vertically.

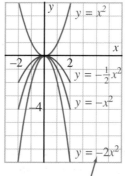

Negative signs, the graphs are stretched or compressed vertically and reflected in the x-axis.

$x - p = 0$ is the axis of symmetry.

$$y = a(x - p)^2 + q$$

Congruent to the parabola $y = ax^2$ Coordinates of the vertex are (p, q).

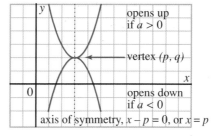

opens up if $a > 0$

vertex (p, q)

opens down if $a < 0$

axis of symmetry, $x - p = 0$, or $x = p$

- The maximum or minimum value of a quadratic function is the y-coordinate of the vertex of its graph.

- The x-intercepts of the graph are the zeros of the function.

1. Describe how to transform the graph of $y = x^2$ to produce the graph of each function.

   a) $y = 5x^2$ **b)** $y = (x - 3)^2 + 2$ **c)** $y = -x^2 - 4$

   d) $y = -2(x + 5)^2$ **e)** $y = -(x + 2)^2 - 2$ **f)** $y = 0.5(x - 4)^2 - 6$

2. Write the equation of the image graph after $y = x^2$ has undergone each transformation.

   a) Translate 2 units left and 3 units up.

   b) Stretch by a factor of 2, then translate 1 unit right.

   c) Reflect in the x-axis, then translate 4 units down.

3. For each parabola:

   i) State the maximum or minimum value of y.

   ii) State whether it is a maximum or a minimum.

   iii) State the value of x where it occurs.

   a) $y = -2(x + 5)^2 - 8$ **b)** $y = \frac{1}{4}(x - 2)^2 - 9$ **c)** $y = -0.5(x - 3)^2 + 7.5$

   d) $y = 2x^2 + 7$ **e)** $y = 5(x - 8)^2 + 12$ **f)** $y = -(x - 7)^2$

4. For each quadratic function, identify:

   i) the coordinates of the vertex   **ii)** the equation of the axis of symmetry

   iii) the x- and y-intercepts   **iv)** the maximum or minimum value

   a)

   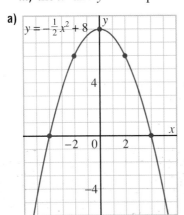

   b)

   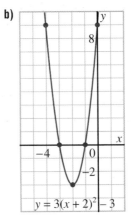

5. Write the equation of each parabola.

   a) vertex $(0, 0)$ with $a = -1$

   b) vertex $(4, 0)$ with $a = \frac{1}{2}$

   c) vertex $(4, -1)$ with $a = 2$

   d) vertex $(-4, 1)$ with $a = -2$

6. For each parabola, identify:

    i) the coordinates of the vertex   **ii)** the x- and y-intercepts

    iii) the direction of opening   **iv)** the maximum or minimum value

  a) $y = 3(x + 7)^2$ **b)** $y = -2(x - 3)^2 + 4$ **c)** $y = -\left(x - \frac{1}{2}\right)^2 - \frac{3}{4}$

7. Determine whether the equations in each pair are equivalent.

  a) $y = 2(x - 3)^2 - 7$ **b)** $y = (x - 5)^2 - 25$ **c)** $y = -2(x - 20)^2 + 800$

    $y = 2x^2 - 12x + 11$ $y = x^2 - 10x$ $y = -2x^2 + 80x$

8. Graph each function. Identify:

    i) the coordinates of the vertex

    ii) the x-intercepts

  a) $y = x^2 - 6x + 5$ **b)** $y = 2x^2 - 8x - 5$ **c)** $y = -3x^2 + 18x - 20$

9. a) For the function in exercise 8a, for what values of x is y less than -3?

  b) For the function in exercise 8b, for what values of y is x greater than 2?

  c) For the function in exercise 8c, for what values of y is $1 < x < 4$?

10. The height, h metres, of an infield fly ball t seconds after being hit is given by the equation $h = 30t - 5t^2$.

  a) Graph the equation.

  b) Determine how high the ball is after 2.3 s.

  c) Determine when the ball is at a height of 25 m.

  d) Determine the maximum height of the ball.

11. Mathematical Modelling The trajectory of a certain basketball shot can be represented by the function $h = -0.088d^2 + 0.92d + 2.5$, where h is the height in metres and d is the horizontal distance, in metres, of the ball from the player.

  a) Graph the function.

  b) What is the maximum height of the ball?

  c) How far had the ball travelled horizontally when it reached its maximum height?

  d) What was the initial height of the ball?

  e) Assume the ball hits the ground before it touches any object. To the nearest centimetre, how far from the thrower does it land?

1. An oil tanker requires 25 min to connect to the piping system at an oil refinery. The time, T minutes, spent docked also depends on x, the number of barrels of oil pumped to the refinery. The equation is $T = 25 + \frac{x}{1600}$.

   a) Graph the equation for up to 300 000 barrels of oil.

   b) How long must the tanker be docked to pump 160 000 barrels?

   c) How many barrels can be pumped when the tanker is docked for 75 min?

2. A student makes hats to sell at a craft fair. The table shows the numbers of hats that can be made from different areas of material.

Material (m²)	2	4	6	8
Number of hats	8	17	26	35

   a) Does the relationship represent a direct or a partial variation? Explain.

   b) Write an equation that relates n, the number of hats as a function of a, the area of material.

3. Determine the equation of each line.

   a) slope 7, passes through A(7, −2)   b) passes through B(−2, −5) and C(2, 7)

4. Solve each equation.

   a) $\frac{x}{2} + 4 = 1$

   b) $\frac{x}{2} - 3 = 2x - 4$

   c) $\frac{1}{4}(2x + 5) = 2x + 1$

   d) $\frac{2}{3}(x - 1) = 3x + 4$

5. A courier company charges its customers depending on the distance a parcel is transported. These are the charges.
   Up to 10 km: $2/km
   Any distance greater than 10 km, and less than 20 km: $1.50/km
   Any distance greater than 20 km: $1/km
   Sketch a graph to display these data.

6. Write each equation in the form $y = mx + b$.

   a) $2x - y + 6 = 0$ b) $5x - 2y - 10 = 0$ c) $12x + 5y - 60 = 0$

7. Solve each linear system.

   a) $y = -3x + 15$
      $x + y = 1$

   b) $3y = 11x$
      $2x - y = -5$

   c) $y = x$
      $x + y = 4$

   d) $y = 10x$
      $12x - y = 4$

   e) $x + y = 1$
      $2x + 3y = 7$

   f) $3x + y = -1$
      $x - 2y = 16$

8. A taxi charges a flat rate, plus an additional charge per kilometre. A 9-km trip costs $11.80. A 20-km trip costs $23.90.

a) What is the flat rate?

b) What is the charge per kilometre?

9. An electrical transformer has 2 coils, wrapped around an iron core. The voltages across the input and output coils are proportional to the number of turns in each coil. A transformer has 100 turns in the input coil and 350 turns in the output coil. The voltage across the input coil is 3000 V. What is the voltage across the output coil?

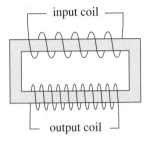

10. Rachel downloaded a 979.2-kb file in 102 s. Suppose Rachel downloads a 1132.8-kb file at the same speed. How much time will be required?

11. The scale diagram shows a plan of a student apartment. The bedroom and study doorways are 80 cm wide.

a) What is the scale of the drawing?

b) What are the dimensions of the apartment?

c) What is the area of the study and guest room, in square metres?

d) What is the area of the bedroom?

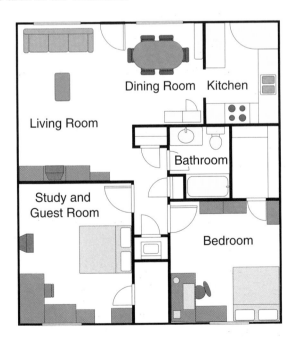

12. Calculate each value of x and y.

a)

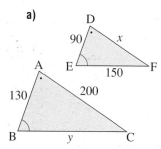

b)

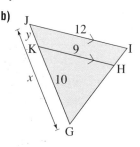

c)

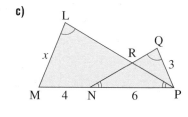

13. Don's driveway has an area of 48 m². He had it sealed at a cost of $60. Leanne's driveway measures 3.2 m by 11.0 m. She employs the company that Don used. The company's costs are proportional to the area of the driveway. What will the cost be to seal Leanne's driveway?

14. From an altitude of 600 m, the angle of depression to the start of a runway is 12°. What is the horizontal distance to the runway?

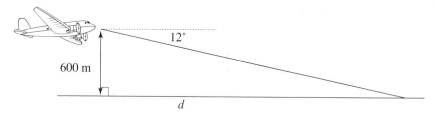

15. A ship's radar indicates that the ship is 200 m from a breakwater. The breakwater is 76 m long. The ship's current course is toward the centre of the breakwater. What is the minimum angle through which the ship must turn to avoid hitting the breakwater?

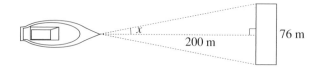

16. From a point 27 m from the base of a building, the angle of elevation of the top of the building is 44°. The angle of elevation of the top of a flagpole on the roof of the building is 55°.

a) Calculate the height of the building.

b) Calculate the height of the flagpole.

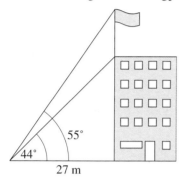

17. Copy each set of points on grid paper. The points in each set lie on a parabola. Use the step pattern to add 4 more points to each parabola.

a)

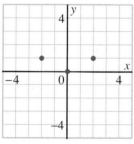

b)

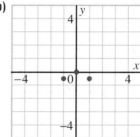

18. The data in each table represent a function whose graph is a parabola. Copy each table. Use the step pattern to complete each table.

a)

x	y
−3	
−2	12
−1	3
0	
1	3
2	
3	

b)

x	y
−6	
−4	
−2	−2
0	0
2	
4	−8
6	

19. For each function, state whether its graph opens up or down, and the coordinates of the vertex.

a) $y = 5x^2$

b) $y = 0.2x^2$

c) $y = -8x^2$

d) $y = 2x^2 + 3$

e) $y = 0.1x^2 - 5$

f) $y = -0.5x^2 + 4$

20. Graph each function. Determine the coordinates of the vertex and the zeros of each function.

a) $y = x^2 - 2x - 8$

b) $y = 0.5x^2 - 4x + 6$

c) $y = 5x^2 - 20x + 15$

21. Expand.

a) $2(x + 3)$

b) $5(x + 7)$

c) $x(x + 4)$

d) $(-2)(2x + 6)$

e) $3x(5x + 2)$

f) $(-5x)(2x + 3)$

22. Factor.

a) $3x^2 + 24$

b) $-5x - 10$

c) $7x^2 + 21$

d) $x^3 + 2x$

e) $2x^2 + 16x$

f) $-10x^2 + 15x$

23. Expand and simplify.

a) $(x + 3)(x - 2)$

b) $(x + 5)(x - 8)$

c) $(2x + 3)(5x - 1)$

d) $(2x + 3)^2$

e) $(3x - 1)(8x - 7)$

f) $(15 - x)(5 + x)$

g) $2(x + 3) + x(x + 5)$

h) $(x + 2)(x + 5) - (x + 3)^2$

i) $(3x + 3)(x + 5) - (x + 1)(2x - 3)$

24. Factor. Check the answers by expanding.

a) $x^2 - 9x - 10$

b) $x^2 + 2x - 15$

c) $x^2 - 12x + 32$

d) $x^2 + 14x + 13$

e) $x^2 - x - 42$

f) $x^2 - 5x - 24$

g) $x^2 - 36$

h) $25x^2 - 49$

i) $100 - 9x^2$

25. Solve each equation.

a) $x^2 + 5x - 24 = 0$

b) $x^2 - 8x - 9 = 0$

c) $25x^2 - 49 = 0$

26. For each parabola:

i) State the coordinates of the vertex.

ii) State whether the parabola opens up or down.

iii) Describe the transformations needed to change the graph of $y = x^2$ to the given parabola.

a) $y = -(x - 3)^2 + 5$

b) $y = 3(x - 6)^2 - 20$

c) $y = \frac{1}{2}(x + 1)^2 + 2$

27. Match each equation with the appropriate graph (below left).

a) $y = (x - 3)^2$ **b)** $y = x^2 - 3$ **c)** $y = x^2 + 3$ **d)** $y = (x + 3)^2$

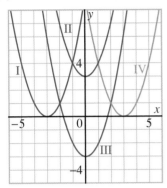 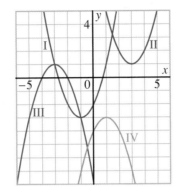

28. Match each equation with the appropriate graph (above right).

a) $y = (x - 3)^2 + 1$ **b)** $y = -(x + 3)^2 + 1$

c) $y = -(x - 1)^2 - 3$ **d)** $y = (x + 1)^2 - 3$

29. Eighty metres of fencing were used to construct a pen with 3 equal parts, as shown. The width is x metres. The area, A square metres, of the entire pen is given by the equation $A = -2x^2 + 40x$.

a) Graph the function.

b) Determine the maximum area of the pen.

c) What length and width will produce the maximum area?

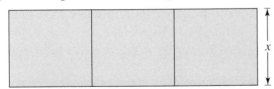

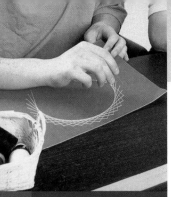

Project

Curve Stitching Designs

Suggested Group Size: 3 (or 4, if you complete the optional extension)

Materials: cotton string, embroidery thread, or wool, stiff cardboard needle, scissors, tape, ruler, protractor, *The Geometer's Sketchpad* (optional)

This project highlights an application of the mathematics you learned in earlier chapters. You will work as part of a team to create a presentation. The ability to work as part of a team is highly prized in industry. Team members listen to different ideas on how to approach a problem.

THE TASK

In this project, you will work with other members of your group to design and make decorations using curve stitching. You will make a presentation to your class, explaining the mathematics behind the art.

ASSESSMENT

There are many valid responses to this project. Your teacher may choose to use the rubric on page 451 to assess your work.

BACKGROUND

In Chapter 6, you used a ruler to construct a parabola. The idea of creating curves with straight lines has interested artists, as well as mathematicians, for many years.

You see curves created with straight lengths of wire, wood, string, yarn, and even brick. Some curves are parabolic, other curves are circular, or oval. Look for examples of this technique.

You may find these in:

- art books
- furniture catalogues
- decorating magazines
- CD covers
- Internet sites
- your home: on dishes, clothing, toys, or ornaments

1. Collect, draw, or photograph 4 examples of straight lines used to produce curves. These examples should use different materials or techniques.

2. For each of your curves, write to explain how the lines were used to make the curve, and what type of curve was created.

GETTING STARTED

You will make a parabola and a circle with thread, string, or wool on cardboard. When you have learned the basics, you can improve your design by trying some of the ideas described under *Variations*.

To make a parabola

1. Draw an angle with two equal arms on a piece of heavy cardboard.

2. Mark each arm at 1-cm intervals.

3. Number the marks. Start with 1 at the vertex on one arm, then reverse the order of the numbers on the other arm.

4. Carefully push a pin through each mark to make a hole.

5. Tape one end of the string to the back of the cardboard. Thread the other end through the needle. Choose one arm as the first arm. Push the needle up through hole #1 on the first arm, then down through hole #1 on the second arm. Push the needle up through hole #2 on the second arm, then down through hole #2 on the first arm. Continue in this manner until you have used all the holes. When you have finished, you should see a parabola.

Project

Variations

- Cut out an equilateral triangle, a square, a regular hexagon, or other figure with equal sides. Make a parabola at each vertex.
  If you cut out the figure, you can make tiny cuts evenly spaced along its edges. The thread can be pulled into the cut and you will not require a needle.

- Vary the colours of the thread you use. Cover the cardboard with coloured paper, paint, sparkles, or cloth.

- Write the numbers on the back of the figure so the numbers do not show in your finished design.

- Try putting the marks closer together.

- Use different angles or make the arms longer.

- What happens if you use two arms of different lengths? (Remember to put the same number of marks on each arm, so on one arm the marks will be closer together.)

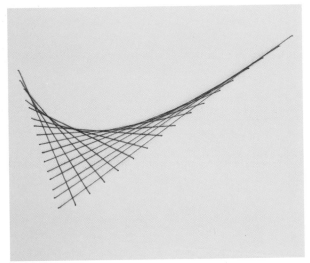

To make a circle

1. On a piece of heavy cardboard, draw a circle with compasses (or draw around a plate).

2. Use a protractor to mark every 10° along the circumference of the circle. Make a hole at each mark. Number the holes.

3. Tape one end of the string to the back of the circle. Thread the other end through a needle and come up through hole #1. Go down through hole #8. Push the needle up through hole #9, then down through hole #2, up through hole #10, then down through hole #3. Continue until you have used all the holes.

Variations

- Cut out the circle and make tiny cuts equally spaced along the circumference.

- Use coloured paper or paint to disguise the cardboard.

- Trying bringing the thread up through hole #1, then down through hole #13.

- Try doing two patterns on the same circle with two different colours.

  - First pattern: with red, begin up through hole #1, then down through hole #7.

  - Second pattern: with yellow, begin up through hole #1, then down through hole #12.

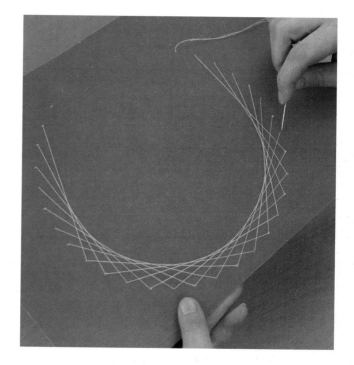

Optional Extension

If you have a fourth person in your group and access to a computer with *The Geometer's Sketchpad*, one of you can use this to investigate the topic:

To make a parabola

1. Open a new sketch in *The Geometer's Sketchpad*.

2. Use the Segment tool to construct $\angle ABC$.

   - Shift-click A, B, and C, then from the Transform menu, choose Mark Angle.

3. Use the Segment tool to construct segment DE elsewhere on the screen.

   - Double-click on point D to make it a centre of rotation.

   - Shift-click on point D, segment DE, and point E. From the Transform menu, choose Rotate. When the dialog box appears it will show "By Marked Angle?" Click OK.

4. Segment DE will rotate to segment DE′. Change the label E′ to F.

5. Construct point G on DE and point H on DF.

- Shift-click G and H, then from the Construct menu, choose Segment.
- While segment GH is selected, from the Display menu, choose Trace Segment.

6. Drag point G until it is very close to D. Drag point H until it is very close to F.

7. Shift-click in this order: G, DE, H, DF. From the Edit menu, choose Action Button, Animation. When the dialog box appears, click Animate.

- You should see a small rectangular button on the screen labelled Animate.
- Double-click the Animate button. Point G should move along DE, point H should move along DF, and segment DH is traced on the screen. The trace should form a parabola as the thread in the curve stitching does.

8. Click anywhere to stop the animation. Before trying the animation again, change the size of ∠ABC at the top of your screen. Why does ∠FDE also change?

- Double-click the Animate button to see the curve drawn in a new angle.

9. Print two of your sketches. They should have equal angles but with the points at different starting places on the arms. Cut out the parabolas and place them on a grid. How does the starting place affect the curve? How does a different angle affect the curve?

To make a circle

1. Use the Circle tool to make a circle on the screen.

2. Construct two points on the circle. Change their labels to M and N.

- Shift-click M and N, then from the Construct menu, choose Segment.
- While MN is selected, from the Display menu, choose Trace Segment.

3. Shift-click in this order: M, N, the circle. From the Edit menu, choose Action Button, Animation. When the dialog box appears, click Animate.

- You should see a small rectangular button on the screen labelled Animate.

4. Double-click the Animate button. Points M and N should move around the circle and MN is traced on the screen.

- What do you notice about the shape in the middle of the circle?

5. Click anywhere to stop the animation.

- Drag N closer to M and animate again. What effect did this have? What do you think the pattern will look like if you move N farther from M?

- Drag N farther from M and animate again. Was your prediction correct?

Create your own design.

PROJECT PRESENTATION

Your finished project should include:

1. A poster display of the objects you found in the background study; this will include a paragraph explaining the type of curve and how it was made. Include a description of the materials used.

2. Three examples of curved stitching — two parabolas and one circle

3. An explanation of the effect of using a larger or smaller angle on the shape of the parabola

4. An explanation of how the choice of holes on the circle affects the result; include a description of how you would make a circle with radius 5 cm from a circle with radius 7 cm.

5. A decoration or useful object of your own design; you must create curves with straight lines, but you may use other materials, such as wire or toothpicks.

6. If your group completed *The Geometer's Sketchpad* exercises, present the animations to the class.

   - Explain how the animations were constructed.

   - Explain and demonstrate how a change in angle affects the shape of the parabola.

   - Demonstrate the effect of choosing a shorter and longer length for MN in the construction of the circle.

Career Opportunities

Artists do not simply paint pictures to display in houses and art galleries. Every magazine and book publisher uses artists for layout and design. Labels, road signs, and containers for products of all descriptions have lettering and symbols done by graphic artists. Artists sketch patterns for material, wallpaper, carpet, and flooring before these are manufactured.

An eye for design is an asset in many careers. Interior decorating, set design, landscaping, and fashion are familiar examples. Tile layers, stone masons, metal workers, and carpenters produce interesting patterns.

Rubric for the Curve Stitching Designs project

Level 1	Level 2	Level 3	Level 4
Knowledge/Understanding			
The student: • draws partially correct starting figures. • creates partially correct designs.	• draws almost entirely correct starting figures. • creates designs with most parts correct.	• draws the starting figures correctly. • creates the designs correctly.	• draws the starting figures perfectly. • creates the designs perfectly.
Thinking/Inquiry/Problem Solving			
The student: • states that varying starting figures affects the designs. • if *The Geometer's Sketchpad* is used, makes a few correct statements about the construction.	• gives partially correct descriptions of some effects of variations, such as changes in angles or lengths. • if *The Geometer's Sketchpad* is used, gives a partially correct explanation of the construction.	• correctly describes some effects of variations, such as changes in angles or lengths. • if *The Geometer's Sketchpad* is used, correctly explains most of the construction.	• correctly explains reasons for effects of variations, such as changes in angles or lengths. • if *The Geometer's Sketchpad* is used, correctly explains details of the construction.
Communication			
The student: • sometimes uses terms related to quadratic functions correctly. • communicates explanations so they could be related to the project.	• generally uses terms related to quadratic functions correctly. • communicates explanations so they could be followed by someone familiar with the project.	• uses terms related to quadratic functions correctly. • communicates explanations so they could be followed by someone familiar with curve stitching.	• correctly uses terms related to quadratic functions whenever appropriate. • communicates explanations so they could be followed easily.
Application			
The student: • researches an example of a curve that could be related to straight lines. • relates lines to the curve in the example. • follows some directions for creating curves and completing the project correctly.	• researches an example of a curve made with straight lines. • gives some explanation of how lines form the curve in the example. • follows most directions for creating curves and completing the project correctly.	• researches examples of curves made with straight lines. • gives a reasonable explanation of how lines form the curves in the examples; names a curve correctly. • follows the directions for creating curves and completing the project correctly.	• researches excellent examples of curves made with straight lines. • expertly explains how lines form the curves in the examples; names the curves correctly. • follows the directions for creating curves and completing the project perfectly.

Answers

Chapter 1 Introduction to Functions

Practise Your Skills, page 9

1. a) 15   b) 100   c) 7.2   d) 86   e) 170   f) 148.52

2. a) 27%   b) 82%
   c) 75%   d) Approximately 33%
   e) Approximately 29%   f) Approximately 58%

1.1 Exercises, page 10

1. a) The axis between 0 and 4.50 has been compressed.
   b) i) 5.00%   ii) 5.00%   iii) 5.00%
   c) 5.00% for the first part of August, then 6.00% for the rest of August
   e) No

2. Estimates may vary.
   a) 1 square represents 0.2%.
   b) 43.6, 43.9, 44.1, 44.3, 44.4, 44.6, 44.8, 44.7, 44.9
   c) 56.4, 56.1, 55.9, 55.7, 55.6, 55.4, 55.2, 55.3, 55.1

3. a) 1 square represents 2%.
   b) 2 squares represent 3 months.

4. b) $6.50, $8.75

5. a) $7, $10, $13   b) $8.50, $10.75

6. a) The time it takes the driver to brake once he or she decides to brake
   b) Distances may vary.

Speed (km/h)	Reaction-time distance (m)	Braking distance (m)	Total stopping distance (m)
20	4	3	7
40	8	10	18
60	12	22	34
80	16	40	56
100	21	68	89
120	25	109	134

   e) Estimates may vary. 70 m, 112 m

7. a) 3.5 million   b) 0.5 million
   c) $\frac{0.5}{3.5}$ d) Approximately 14%
   e) Approximately 14%

8. a) 6.8 million   b) 3.5 million
   c) Approximately 51%

Practise Your Skills, page 21

1. a) −1   b) −6   c) 7   d) −4   e) 10   f) −12
2. a) 9   b) 2   c) 18   d) 3   e) 0   f) −8
3. a) 4   b) −8   c) 3   d) 0.5   e) 5   f) 7
4. a) 13   b) 4   c) −5   d) −14   e) −17
5. a) 10   b) 16   c) 4   d) 0   e) −10

1.2 Exercises, page 21

1. b) i) Linear   ii) Linear   iii) Linear
2. a) i) $y = x + 3$ ii) $y = 3x$ iii) $x + y = 5$
4. a) $y = 3x − 3$ b) $y = 2(x + 1)$ c) $y = 12 − x$
8. $35.08

9. a)

Tree (6 ft)	Retail price ($)	Amount you pay ($)
Crab apple	50.00	35.08
Ash	60.00	44.85
Red oak	70.00	54.63
Purple leaf birch	80.00	64.40
Balsam fir	90.00	74.18

   c) $52.18
10. b) $600   c) 37
11. c) i) Approximately 2.0 s   ii) Approximately 2.9 s
12. All the graphs intersect the y-axis at $y = 5$.
13. a) Statement 3 is always true.
14. c) $2938.33   d) $13 889   f) $3333.33, $41 666.78
15. a) 4°F   b) 2°F   c) −2°F

Testing Your Knowledge, page 46

1. a) The gate price is the cost to enter the exhibition.
   b), c),

Year	Gate Price ($)	Attendance (millions)	Revenue From Ticket Sales (millions of $)
1987	5.00	2.12	10.6
1988	5.00	2.01	10.05
1989	5.00	2.05	10.25
1990	7.50	1.70	12.75
1991	7.50	1.98	14.85
1992	7.50	1.85	13.875
1993	8.50	1.71	14.535
1994	8.50	1.70	14.45
1995	9.00	1.56	14.04
1996	9.00	1.71	15.39
1997	16.00	1.75	28.0
1998	16.00	1.80	28.8

3. b) i) −30.5°C   ii) About 2.3 km
4. b) 125 m   c) 113 km/h
5. b) 206 beats/min, 146 beats/min, 188 beats/min
6. b) Approximately 70°C
7. a) Measurements may vary.

Base (cm)	Height (cm)
3.2	9.8
0.9	7.1
2.0	8.2
6.5	3.6
2.2	3.6
8.4	4.0
8.4	3.3
9.0	7.8
0.3	7.6

Chapter 2 Linear Functions

Preparing for the Chapter, page 52

6. a) $12x + 20$ **b)** $10x - 15$ **c)** $-12x - 24$ **d)** $-14x + 6$

7. a) -4 **b)** 8 **c)** -4 **d)** 3

8. a) 3 **b)** 9 **c)** 11 **d)** 1 **e)** 3 **f)** 1

2.1 Exercises, page 58

1. a)

Input x	Output y	Differences
0	-4	
1	-1	3
2	2	3
3	5	3

b)

Input x	Output y	Differences
0	2	
1	3	1
2	4	1
3	5	1

c)

Input x	Output y	Differences
0	3	
1	-2	-5
2	-7	-5
3	-12	-5

2. a) $y = x + 2$, $y = -5x$, $y = -2x - 1$,
$y = 0.1x - 1.3$, $y = \frac{2}{5}x - \frac{1}{5}$

3. $y = 4x - 5$

4. a) Multiplies the input by -2, then adds 5.
b) Multiplies the input by 3, then adds 3.
c) Subtracts the product of -5 and the input from 2.

5. a) Tables may vary.

Hours Worked	Earnings ($)	Differences
0	0	
1	6.80	6.80
2	13.60	6.80
3	20.40	6.80
4	27.20	6.80
5	34.00	6.80
6	40.80	6.80
7	47.60	6.80
8	54.40	6.80
9	61.20	6.80
10	68.00	6.80

Hours Worked	Earnings ($)	Differences
11	74.80	
12	81.60	6.80
13	88.40	6.80
14	95.20	6.80
15	102.00	6.80
16	108.80	6.80
17	115.60	6.80
18	122.40	6.80
19	129.20	6.80
20	136.00	6.80

c) Approximately $105

6. a)

Input x	Output y	Differences
0	0	
1	3	3
2	6	3
3	9	3
4	12	3
5	15	3

b) Multiply the input by 3. $y = 3x$

7. a)

Input x	Output y	Differences
0	3	
1	7	4
2	11	4
3	15	4

b) Multiply the input by 4, then add 3. $y = 4x + 3$

8. a) $52.50 **b)** $C = 35 + 5h$ **c)** $5/h

9. a)

Weekly Sales ($)	10% of weekly sales ($)	Earnings ($)
200	$0.1 \times 200 = 20$	$325 + 20 = 345$
400	$0.1 \times 400 = 40$	$325 + 40 = 365$
600	$0.1 \times 600 = 60$	$325 + 60 = 385$
800	$0.1 \times 800 = 80$	$325 + 80 = 405$
1000	$0.1 \times 1000 = 100$	$325 + 100 = 425$
1200	$0.1 \times 1200 = 120$	$325 + 120 = 445$

b) $E = 325 + 0.1s$

10. a) 13 **b)** -11 **c)** 4 **d)** 1

11. a)

Input x	Output y	Differences
0	4	
1	6	2
2	8	2
3	10	2
4	12	2

b) $y = 2x + 4$

12. b) i) $y = -4x + 16$ **ii)** $y = 4$

Practise Your Skills, page 65

1. a) 5 **b)** $\frac{1}{-5}$ **c)** -1 **d)** 5
e) $\frac{5}{3}$ **f)** $\frac{1}{2}$ **g)** $\frac{1}{3}$ **h)** $\frac{5}{2}$

2. a) 5 **b)** 1 **c)** $\frac{9}{8}$ **d)** -2 **e)** 1 **f)** 1

2.2 Exercises, page 65

1. a) 3, $(4, 1)$ **b)** 4, $(4, -3)$ **c)** 2, $(-1, 3)$
d) -2, $(3, 2)$ **e)** -3, $(-2, 6)$ **f)** -0.5, $(-3, -6)$

2. a) $y = 4(x - 3) + 2$ **b)** $y = 2(x + 1) + 4$
c) $y = 3(x - 3) - 5$ **d)** $y = \frac{1}{2}(x - 5) - 6$
e) $y = \frac{2}{3}(x - 2) + 3$ **f)** $y = -\frac{2}{5}(x - 3) + 5$

3. a) $y = 2.5(x - 4) - 2$ **b)** $y = -(x - 4)$
c) $y = 3(x + 6) + 3$ **d)** $y = -2(x + 2) - 3$

4. The form of an equation may vary.
a) $y = \frac{2}{5}(x - 3) + 4$ **b)** $y = (x - 3) - 2$
c) $y = \frac{1}{2}(x + 4) + 2$ **d)** $y = -(x + 4)$
e) $y = x + 2$ **f)** $y = \frac{2}{3}(x - 5)$

5. a) 12 **b)** -3 **c)** 0 **d)** -11 **e)** 1
f) 9 **g)** 3 **h)** -2.5 **i)** -0.8

6. L_1: $y = \frac{2}{3}x + 2$; L_2: $y = -\frac{1}{2}x + 1$

7. The form of an equation may vary.
AB: $y = \frac{3}{5}(x + 4) + 2$, or $y = \frac{3}{5}x + \frac{22}{5}$
BC: $y = -\frac{5}{3}(x - 1) + 5$, or $y = -\frac{5}{3}x + \frac{20}{3}$
CD: $y = \frac{3}{5}(x + 1) - 3$, or $y = \frac{3}{5}x - \frac{12}{5}$
DA: $y = -\frac{5}{3}(x + 1) - 3$, or $y = -\frac{5}{3}x - \frac{14}{3}$

8. b) i) $y = 2x$ **ii)** $y = 5x - 17$
iii) $y = -4x$ **iv)** $y = 2x + 3$

9. The form of an equation may vary.
AB: $y = \frac{1}{4}x + \frac{5}{4}$; BC: $y = -\frac{1}{3}x + \frac{16}{3}$; CA: $y = 2x - 4$

10. $y = -4x + 16$

11. a) $y = -3x + 9$ **b)** $y = -2x + 4$ **c)** $y = -x + 1$

$= 4x + 16;\ y = \pm 3.5x + 12.25;\ y = 3x + 9;$
$y = \pm 2.5x + 6.25;\ y = 2x + 4;\ y = \pm 1.5x + 2.25;$
$y = x + 1;\ y = \pm 0.5x + 0.25$

Practise Your Skills, page 72

1. a) 2   **b)** −2   **c)** −2   **d)** 2
   e) 8   **f)** −8   **g)** −8   **h)** 8

2.3 Exercises, page 72

1. a) 8   **b)** $-\frac{2}{5}$ **c)** 5   **d)** $-\frac{15}{2}$ **e)** $-\frac{11}{2}$ **f)** −1

2. a) \$2350   **b)** 300

3. a) \$140   **b)** 150   **c)** $n = \frac{P + 60}{2}$; 205

4. a) D is the diameter in centimetres; t is the time in years after the first measurement.
   b) 94 cm   **c)** 6 years

5. a) 13   **b)** 2   **c)** −1   **d)** −3   **e)** $-\frac{1}{6}$ **f)** −4

6. a) 63 m/s   **b)** 11 s   **c)** $t = \frac{s - 14}{9.8}$; 7 s

7. a) C is the circumference in any units; d is the diameter in the same units as the circumference.
   b) Approximately 9.4 m   **c)** Approximately 62.8 cm
   d) 2.4 cm

8. a) 18   **b)** −18   **c)** 40   **d)** 10   **e)** 40   **f)** −16

9. b) 32 L   **c)** 687.5 km   **d)** 3
   e) $d = \frac{V - 60}{-0.08}$; 750 km

10. a) −4   **b)** 4   **c)** 2   **d)** 0   **e)** 4   **f)** 2

11. a) −3   **b)** 3   **c)** −8.5   **d)** −2   **e)** 18   **f)** $\frac{5}{2}$

12. a) Approximately 31°C   **b)** $F = \frac{9}{5}C + 32$; 190.4°F
   c) 88°C

13. a) i) 49 km/h   **ii)** 48 km/h   **iii)** 47 km/h
   iv) 46 km/h   **v)** 45 km/h   **vi)** 44 km/h
   b) $\left(50 - \frac{d}{2}\right)$ km/h   **c)** $s = 50 - \frac{d}{2}$
   d) $d = -2(s - 50)$; 100 cm

14. b) The acceleration due to gravity that slows down the ball when it is moving up
   c) 2.2 m/s
   d) −7.6 m/s; the speed is negative, the ball has changed direction and is falling
   e) $t = \frac{s - 12}{-9.8}$; approximately 1.2 s

Practise Your Skills, page 80

1. a) \$4.80   **b)** \$32   **c)** \$24

2. a) \$32.40   **b)** \$200   **c)** \$252

2.4 Exercises, page 80

1. a) 45 min after they started
   b) 1 h 45 min after they started
   c) After lunch, up to the time they turned round to return (from 1 h 15 min to 1 h 45 min after they started)

2. a) The car was being driven.
   b) The car was stopped, and its gas tank was filled to 40 L of gasoline.
   c) The car was parked, the engine turned off.
   d) The car was being driven.

3. a) Aug.–Oct.   **b)** Oct.–Dec.

5. a) iv   **b)** i   **c)** iii

6. It would only make a difference in Pool C.

7. a) i) \$2.00   **ii)** \$2.00   **iii)** \$2.40   **iv)** \$3.60
   v) \$3.00   **vi)** \$3.30   **vii)** \$4.50   **viii)** \$6.00

8. a) i) \$45   **ii)** \$80   **iii)** \$120
   iv) \$140   **v)** \$175   **vi)** \$210
   c) i) Approximately \$75   **ii)** Approximately \$111
   iii) Approximately \$186

9. a) i) \$15.00   **ii)** \$18.00   **iii)** \$19.00   **iv)** \$19.95
   v) \$18.00   **vi)** \$18.90   **vii)** \$22.50   **viii)** \$27.00

b)

Total price ($)	Cost of order ($)
15.00	15.00
18.00	18.00
19.00	19.00
19.95	19.95
20.00	18.00
21.00	18.90
25.00	22.50
30.00	27.00

d) Approximately \$24.00

11. a) $\frac{1}{2}$ dozen donuts + 3 donuts

c)

Number of donuts	Price ($)	Number of donuts	Price ($)
1	0.70	13	5.20
2	1.40	14	5.90
3	2.10	15	6.60
4	2.80	16	7.30
5	3.50	17	8.00
6	3.15	18	7.65
7	3.85	19	8.35
8	4.55	20	9.05
9	5.25	21	9.75
10	5.95	22	10.45
11	6.65	23	11.15
12	4.50	24	9.00

12. a) Buy 1 dozen donuts.

b)

Number of donuts	Most economical price ($)	Number of donuts	Most economical price ($)
1	0.70	13	5.20
2	1.40	14	5.90
3	2.10	15	6.60
4	2.80	16	7.30
5	3.15	17	7.65
6	3.15	18	7.65
7	3.85	19	8.35
8	4.50	20	9.00
9	4.50	21	9.00
10	4.50	22	9.00
11	4.50	23	9.00
12	4.50	24	9.00

Practise Your Skills, page 90

1. a) $y = 2(x - 5) + 8$ **b)** $y = 0.7(x - 9) + 6$
   c) $y = 1.75(x - 28.3) + 650$ **d)** $y = 2.5(x - 57.2) + 476$

2.5 Exercises, page 90

2. b) $537.88, $806.82

3. b) $487.03, $645.15

5. a) 49¢, 78¢
  a) 0 – 30 g : 49¢; 30 – 50 g : 78¢

6. b) $125.85

9. a) 0　　**b)** $1700　　**c)** $3400
  d) $5030.30　**e)** $7736.90　**f)** $10 336.90
  g) $12 723.70　**h)** $15 861.50　**i)** $18 761.50

10. b) i) $3170.50　ii) $8592.30　iii) $10 811.40

11. a) i) Between $9 and $13.99　ii) Between $13 and $17.99
  iii) Between $17 and $21.99　iv) Between $21 and $25.99
  v) Between $25 and $54.99　vi) $50 and over

2.6 Mathematical Modelling: When Will the Competitors Appear?, page 93

3. a), b) Times may vary. The elapsed times are in hours.

Location	Distance from start (km)	Elapsed time (hours after 8 A.M.)	Arrival time (clock time)
Penticton	0	0	8:00 A.M.
Okanagan Falls	24	0.75	8:45 A.M.
Oliver	44	1.20	9:12 A.M.
Osoyoos	64	1.80	9:48 A.M.
Richter Mountain	79	2.20	10:12 A.M.
Keremeos	132	3.50	11:30 A.M.
Olalla	138	3.80	11:48 A.M.
Yellow Lake	149	4.10	12:06 P.M.
Kaleden	163	4.50	12:30 P.M.
Penticton	180	5	1:00 P.M.

4. a), b) Times may vary.

Location	Distance from start of bike race(km)	Elapsed time (hours after 8 A.M.)	Arrival time (clock time)
Penticton	180	5	1:00 P.M.
Lakeshore Highlands	194	6	2:00 P.M.
Half-way point	201	6.5	2:30 P.M.
Lakeshore Highlands	208	7	3:00 P.M.
Penticton	222	8	4:00 P.M.

5. Bike race: $y = 36x$
  Run: $y = 14x + 100$

7. a), b) Times may vary. The elapsed times are in hours and minutes.

Location	Distance from start (km)	Elapsed time (hours after 8 A.M.)	Arrival time (clock time)
Penticton	0	0	8:00 A.M.
Okanagan Falls	24	0:40	8:40 A.M.
Oliver	44	1:13	9:13 A.M.
Osoyoos	64	1:47	9:47 A.M.
Richter Mountain	79	2:12	10:12 A.M.
Keremeos	132	3:40	11:40 A.M.
Olalla	138	3:50	11:50 A.M.
Yellow Lake	149	4:08	12:08 P.M.
Kaleden	163	4:32	12:32 P.M.
Penticton	180	5:00	1:00 P.M.

8. a), b) Times may vary. The elapsed times are in hours and minutes.

Location	Distance from start of bike race(km)	Elapsed time (hours after 8 A.M.)	Arrival time (clock time)
Penticton	180	5:00	1:00 P.M.
Lakeshore Highlands	194	6:00	2:00 P.M.
Half-way point	201	6:30	2:30 P.M.
Lakeshore Highlands	208	7:00	3:00 P.M.
Penticton	222	8:00	4:00 P.M.

12. OR: $y = 25t$
  RT: $y = 52.5t - 82.5$

Practise Your Skills, page 105

1. a) $-3x - 10$　　**b)** $-3x + 8$　　**c)** $-2x - 15$

2. a) $2x - 3$　　**b)** $\frac{2}{3}x + 3$　　**c)** $-\frac{3}{4}x + \frac{1}{2}$

2.7 Exercises, page 105

1. a) $2x + y - 3 = 0$　**b)** $x - y + 1 = 0$　**c)** $3x - y - 2 = 0$

2. a) $y = -x + 3$　**b)** $y = -2x - 4$　**c)** $y = 2x - 8$

3. a) $x - 2y - 8 = 0$　**b)** $2x + 3y + 6 = 0$　**c)** $3x - 5y - 40 = 0$
  d) $8x - 2y - 5 = 0$　**e)** $3x + 2y - 8 = 0$　**f)** $2x + 3y - 5 = 0$

4. a) $y = -\frac{1}{2}x - 4$　**b)** $y = \frac{1}{3}x + 4$　**c)** $y = -\frac{2}{3}x + 6$
  d) $y = \frac{3}{2}x + 3$　**e)** $y = -\frac{4}{3}x - 8$　**f)** $y = -\frac{5}{2}x - 10$

5. a) $-1, 4$　**b)** $2, 8$　**c)** $\frac{1}{2}, 3$
  d) $-\frac{3}{2}, 3$　**e)** $-\frac{3}{2}, -6$　**f)** $-\frac{2}{3}, -6$

7. L_1: $2x - 5y + 15 = 0$; L_4: $4x - 10y - 10 = 0$

8. L_1: $y = -\frac{1}{3}x + 4$ and L_5: $x + 3y - 12 = 0$
  L_2: $y = -\frac{2}{5}x + 2$ and L_6: $2x + 5y - 10 = 0$
  L_3: $y = \frac{2}{7}x + 1$ and L_4: $2x - 7y + 7 = 0$

9. a) $x - y + 4 = 0$; $x + y + 4 = 0$; $x - y - 4 = 0$
  b) $2x + 3y + 12 = 0$; $2x - 3y + 12 = 0$; $2x - 3y - 12 = 0$

Testing Your Knowledge, page 108

2. a) 35¢, 78¢　　**b)** 1982–1983

3. b)

Number of copies required	Number of copies for cheapest price	Cost ($)
1	1	0.05
80	80	4.00
81	101	4.04
100	101	4.04
101	101	4.04
188	188	7.52
189	251	7.53
250	251	7.53
251	251	7.53
300	300	9.00
334	334	10.02
335	501	10.02
500	501	10.02
550	550	11.00
600	600	12.00

6. a) $y = 3(x - 3) + 4$　　**b)** $y = -(x - 2) - 3$
  c) $y = \frac{1}{2}(x + 3) + 2$

4. a) 90¢　　**b)** $1.05; $1.25
  c) Small: $P = 8.25 + 0.90n$
  Medium: $P = 10.50 + 1.05n$
  Large: $P = 13.55 + 1.25n$

6. a) Assumptions and answers may vary.

For men – assume swim begins at 7:00 A.M.
– assume bike race begins at 8:00 A.M.
– assume run begins at 12:30 P.M.
– assume run ends at 3:10 P.M.
For women – assume swim begins at 7:00 A.M.
– assume bike race begins at 8:10 A.M.
– assume run begins at 1:20 P.M.
– assume run ends at 4:06 P.M.

Location	Distance from start (km)	Arrival time (clock time) Men	Women
Penticton	0	8:00 A.M.	8:10 A.M.
Okanagan Falls	24	8:40 A.M.	8:50 A.M.
Oliver	44	9:10 A.M.	9:30 A.M.
Osoyoos	64	9:40 A.M.	10:00 A.M.
Richter Mountain	79	10:00 A.M.	10:30 A.M.
Keremeos	132	11:15 A.M.	11:55 A.M.
Olalla	138	11:25 A.M.	12:10 P.M.
Yellow Lake	149	11:45 A.M.	12:25 P.M.
Kaleden	163	12:00 P.M.	12:45 P.M.
Penticton	180	12:30 P.M.	1:20 P.M.
Penticton	180	12:30 P.M.	1:20 P.M.
Lakeshore Highlands	194	1:20 P.M.	2:10 P.M.
Half-way point	201	1:45 P.M.	2:40 P.M.
Lakeshore Highlands	208	2:15 P.M.	3:10 P.M.
Penticton	222	3:10 P.M.	4:06 P.M.

c) i)

Location	Distance from start (km)	Arrival time (clock time) Men	Women
Penticton	0	8:00 A.M.	8:10 A.M.
Okanagan Falls	24	8:36 A.M.	8:50 A.M.
Oliver	44	9:06 A.M.	9:24 A.M.
Osoyoos	64	9:36 A.M.	9:59 A.M.
Richter Mountain	79	9:58 A.M.	10:25 A.M.
Keremeos	132	11:18 A.M.	11:56 A.M.
Olalla	138	11:27 A.M.	12:06 P.M.
Yellow Lake	149	11:44 A.M.	12:25 P.M.
Kaleden	163	12:05 P.M.	12:49 P.M.
Penticton	180	12:30 P.M.	1:19 P.M.
Penticton	180	12:30 P.M.	1:20 P.M.
Lakeshore Highlands	194	1:23 P.M.	2:15 P.M.
Half-way point	201	1:50 P.M.	2:43 P.M.
Lakeshore Highlands	208	2:17 P.M.	3:11 P.M.
Penticton	222	3:10 P.M.	4:06 P.M.

7. a) 48.4 m/s **b)** 5.5 s **c)** $t = \frac{s - 19}{9.8}$; 11.4 s

8. a) $y = 3x - 5$ **b)** $y = -x - 1$ **c)** $y = \frac{1}{2}x - \frac{1}{2}$

9. $y = \frac{3}{4}x + \frac{11}{4}$, or $3x - 4y + 11 = 0$

10. a) $y = -2x + 3$ **b)** $y = -\frac{1}{2}x - 2$ **c)** $y = -\frac{3}{4}x - \frac{1}{4}$

11. a) $3x - y - 2 = 0$ **b)** $2x + 3y + 15 = 0$ **c)** $3x - 4y + 8 = 0$

12. a) $-3, \frac{3}{2}$ **b)** $-3, 2$ **c)** $\frac{1}{2}, -\frac{1}{2}$

Chapter 3 Linear Systems
Preparing for the Chapter, page 114

1. a) 3 **b)** −5 **c)** 10 **d)** −4 **e)** −24 **f)** −12

2. a) 2 **b)** $-\frac{2}{3}$ **c)** 1 **d)** 1 **e)** −15 **f)** 2

g) −1 **h)** 8 **i)** 3

3. a) $3 - y$ **b)** $y + 7$ **c)** $-8 - 2y$
d) $y - 10$ **e)** $-3y - 5$ **f)** $13 - 4y$

4. a) $4 - x$ **b)** $x + 6$ **c)** $5 - 3x$
d) $7 - x$ **e)** $2x + 1$ **f)** $-4x - 3$

5. a) 7 **b)** −10 **c)** 5 **d)** −10 **e)** 1 **f)** 6

6. a) 3 **b)** 2 **c)** −5 **d)** −4 **e)** −1 **f)** 8.5

7. a) −3 **b)** 2 **c)** −3 **d)** 2.5 **e)** −5 **f)** $\frac{8}{3}$

3.1 Exercises, page 120

1. a) i) The number of guests and the cost, for which the halls cost the same
ii) Hall C is cheaper for fewer than 150 guests.
Hall D is cheaper for more than 150 guests.
b) i) The number of guests and the cost, for which the halls cost the same
ii) Hall E is cheaper for fewer than 240 guests.
Hall F is cheaper for more than 240 guests.

2. a) The number of pizzas sold for which the daily cost is equal to the daily income
b) 31

3. The time and distance from St. Catharines when the two cars meet

4. a) (12 000, 96 000)
b) The number of discs sold for which the income is equal to the production costs
c) 12 001

5. a) (32, 160)
b) The number of hours the student works for her or his costs to match the annual fee from the snow-removal company
c) When the total time for snow removal is less than 5 h

6. a) Approximately (2.3, 186.7)
b) The time and distance from Gander when the two cars meet

7. Estimates may vary.
a) (1, 2.4) **b)** (4, −2)
c) (2, 5) **d)** (2, −1)

8. a) (1, 2) **b)** (3, 1) **c)** (4, −3)

9. a) (−1, −3.5) **b)** (−2.67, 0.2) **c)** (3.08, 2.38)

10. (6, 37)

11. a) (10, 6) **b)** (−5, 50) **c)** (−10, −5)

12. a) (95, 63.75)
b) For both systems, for 95 min of calls, the cost is $63.75.
c) Plan B **d)** Plan A
e) When a person uses fewer than 95 min per month

3.2 Exercises, page 129

1. a) (1, 2) **b)** (2, 1) **c)** (2, 3)

2. a) (3, 2) **b)** (4, 0) **c)** (1, 3)

3. a) (−1, 4) **b)** (−2, −5) **c)** (3, −4)

4. a) (0, 5) **b)** (4, −1) **c)** (−3, −3)

5. a) (3, −1) **b)** (2, 4) **c)** (−5, 2)

6. a) i) (2, 2) **ii)** (−3, 2) **iii)** (2, −4)
iv) (1, 4) **v)** (−3, −4) **vi)** (5, −3)

7. b) $(3, 60)$
   c) For each stables, 3 h riding costs $60.
   d) Pyramid

8. b) $(4, 7)$
   c) Four hours were spent sailing and 7 h were spent fishing.
   d) 7 h

9. a) $(0, -4)$ **b)** $(-14, -9)$ **c)** $(1.5, 4.5)$
   d) $(-3, 0)$ **e)** $(3, 4)$ **f)** $(10, -8)$

10. a) $(-1, 5)$ **b)** $(0, 2)$ **c)** $(-2, 3)$

11. a) $(-2, -4)$ **b)** $(5, -2)$ **c)** $(4, 6)$
   d) $(2.5, -2.5)$ **e)** $(0, 4.5)$ **f)** $(4.5, 3.5)$

12. a) $(2, 5)$ **b)** $(1, 3)$ **c)** $(2, 1)$
   d) $(4, -2)$ **e)** $(1.5, 0)$ **f)** $(-1.5, 0.5)$

Practise Your Skills, page 135

1. a) 9   **b)** 3   **c)** -3 **d)** -9
   e) 14   **f)** -6 **g)** 6   **h)** -14

2. a) $7x$ **b)** $3x$ **c)** $-3y$ **d)** $-7y$
   e) $11x$ **f)** $-5x$ **g)** $5y$ **h)** $-11y$

3.3 Exercises, page 135

1. a) $(4, -1)$ **b)** $(1, -8)$ **c)** $(0, -6)$
   d) $(3, 7)$ **e)** $(0, -1)$ **f)** $(0, 0)$

2. a) $(4, -3)$ **b)** $(2, -1)$ **c)** $(3, -3)$
   d) $(-2, 3)$ **e)** $(4, 2)$ **f)** $(-14, -9)$

3. a) $(0, 4)$ **b)** $(2, -2)$ **c)** $(-1, 1)$
   d) $(3, 3)$ **e)** $(-0.5, -4)$ **f)** $(3, -7)$

4. a) $(1, 3)$ **b)** $(-2, 0)$ **c)** $(-4, 5)$
   d) $(2, 8)$ **e)** $(-2, -6)$ **f)** $(7, -1)$

6. a) $(4, 2)$ **b)** $(-5, 2)$ **c)** $(3, -4)$
   d) $(-6, -2)$ **e)** $(6, -1)$ **f)** $(-5, 5)$

7. b) $(31, 52)$
   c) Sundin had 31 goals and 52 assists.

8. a) $(2, 0)$ **b)** $(-3, -3)$ **c)** $(-1, -1)$
   d) $(-2.5, -2.5)$ **e)** $(2, -1)$ **f)** $(1, -3)$

9. a) $(-1.5, -2.5)$ **b)** $(-3, -4)$ **c)** $(1, -1)$
   d) $(3, -1)$ **e)** $(3, -4)$ **f)** $(-3, -1)$

10. P lies on all the lines.

11. Answers may vary. For example:
   a) $x + 4y = 7$, $3x + 6y = 9$ **b)** $(-1, 2)$

Practise Your Skills, page 142

1. a) $36   **b)** $3.60   **c)** $675   **d)** $67.50
   e) $320   **f)** $32   **g)** $80.50   **h)** $8.05

3.4 Exercises, page 143

1. a) $(9, -2)$ **b)** $(6, 2)$ **c)** $(-7, -3)$
   d) $(1, 1)$ **e)** $(0, -2)$ **f)** $(-1, 3)$

2. a) $(5, 3)$ **b)** $(-1, 5)$ **c)** $(3, -5)$
   d) $(3, 4)$ **e)** $(-2, 5)$ **f)** $(-7, -2)$

3. a) $s = 650$, $w = 90$
   b) The speed of the plane with no wind is 650 km/h. The wind speed is 90 km/h.
   c) 90 km/h

4. b) $t = 11$, $c = 15$ **c)** $15   **d)** $11

5. b) $(20\ 588.24, 2058.82)$

6. b) $(500, 3.55)$
   c) When each bulb is burned for 500 h, the cost is $3.55.

7. $35.00, $50.00

8. a) $150.00   **b)** $6.00

9. If weekly sales exceed $4445, then plan A is better.

10. Variables may differ.
   a) $w + c = 300$ **b)** $w + 2c = 500$
   d) 100   **e)** 200

11. Variables may differ.
   a) $v + t = 100$ **b)** $8v + 10t = 980$
   d) 10   **e)** 90

12. Variables may differ.
   a) $t = 0.35d + 2.50$ **b)** $t = 35$
   c) Approximately 93 km for $35
   d) It is cheaper to use a taxi if you travel less than 93 km.

13. a) 1200
   b) Either company could be used – the costs are the same for 1200 books.

14. a) Approximately 467 km   **b)** Approximately 133 km

3.5 Mathematical Modelling: When Will the Skydivers Meet?, page 146

1. a) 3200 m, 2800 m   **b)** 2400 m; 80 m/s
   c) 2000 m; 50 m/s   **d)** Beth; Alison
   e) Approximately 13 s
   f) They opened their parachutes.

2. a) $h = -80t + 3200$ **b)** $h = -50t + 2800$
   c) Approximately 13.3 s

3. a) Change her position to spread stable position.
   b) Approximately 13.3 s   **c)** Approximately 26.7 s

5. a) $h = -55t + 2950$; $h = -85t + 3350$
   b) Approximately 13.3 s   **c)** Approximately 25.8 s

Testing Your Knowledge, page 148

1. a) i) The cost in dollars to rent the hall
   ii) The number of guests
   b) $(100, 17\ 500)$
   c) The number of guests and the cost, for which the halls cost the same
   d) Hall X is cheaper when fewer than 100 guests attend. Hall Y is cheaper when more than 100 guests attend.

2. b) $(5.5, 467.5)$
   c) The time in hours and the distance in kilometres from Thunder Bay when the cars meet

3. a) $(15, 120)$ **b)** $(113, -57)$ **c)** $(-36, -17)$

4. a) $(3, -2)$ **b)** $(1, 4)$ **c)** $(-2, 3)$
   d) $(2, 5)$ **e)** $(2, 1)$ **f)** $(-48, 96)$

5. b) $(4, 8)$
c) There are four 30-s breaks and eight 60-s breaks.
d) 4

6. a) $(2, 0.5)$ **b)** $(1, 1)$ **c)** $(4, -2)$
d) $(0, 2)$ **e)** $(3, -2)$ **f)** $(-4, 3)$

7. b) $(150, 100)$
c) 150 deluxe shirts and 100 standard shirts are produced.
d) 100

8. a) $(-2, -3)$ **b)** $(4, -3)$ **c)** $(-1, 4.5)$
d) $(0, 1)$ **e)** $(-2, 7)$ **f)** $(-3, 1)$

9. Variables may differ.
a) $120r + 100f = 49\,000$ **b)** $180r + 120f = 64\,200$
c), d) 150 racing, 310 freestyle

10. For 7 or fewer visits, plan B is cheaper. For 8 or more visits, plan A is cheaper.

11. For monthly sales greater than $10\,000, plan A is better. For monthly sales less than $10\,000, plan B is better.

12. a) $S = x + 1000y$, $S = (x - 10\,000) + 1500y$
b) 20 goals

13. a) d is the height in metres of the skydiver above the ground. t is the time in seconds since leaving the plane.
b) A: 2700 m, B: 3000 m
c) A   **d)** B
e) Approximately 18 s after they jump
f) Yes, one skydiver passes the other at a height of approximately 1765 m.

Cumulative Review, page 152

1. a) One graph shows the sales of PC software each month. The other graph shows the sales of Mac software each month.
b) The sales were greatest in March.
c) January: $30\,000; February: $70\,000
d) $40\,000.   **e)** $30\,000.

3. Each graph represent a function because for each value of x, there is no more than one value for y.

4. b) 55 s   **c)** 28 L

5. a)

Time (min)	0	1	2	3	4	5	6	7	8
Length (cm)	8.5	8	7.5	7	6.5	6	5.5	5	4.5
Time (min)	9	10	11	12	13	14	15	16	17
Length (cm)	4	3.5	3	2.5	2	1.5	1	0.5	0

b) After 6 min   **c)** 3.0 cm
d) 17 min   **e)** $L = 8.5 - 0.5t$

6. a) $y = 0.2x + 3$, $y = 2 - x$, $y = \frac{1}{3}x + \frac{2}{3}$, $y = -\frac{2}{3}x$

7. a) $\frac{1}{4}$, -5 **b)** 3, -5 **c)** -2, -16

8. a) $y = 3(x + 2) + 3$ **b)** $y = \frac{2}{5}(x - 1) + 4$

9. Forms of an equation may vary.
a) $y = 5(x - 2) + 5$ **b)** $y = -4(x + 6) + 5$

10. a) $150   **b)** $1850   **c)** $n = \frac{P + 2400}{8.5}$; 400 books

11. a) $\frac{13}{9}$ **b)** $\frac{12}{5}$ **c)** -3
d) 2   **e)** 1   **f)** $-\frac{5}{8}$

12. a) $6

14. a) $42\,500   **b)** $113\,000

15. a) $2x - y - 8 = 0$ **b)** $2x - 3y + 15 = 0$ **c)** $3x - 6y - 4 = 0$

16. a) $y = 5x + 4$ **b)** $y = -3x + 10$ **c)** $y = \frac{7}{2}x + 6$

18. a) The number of days of rental and the income for which the cost to own the cars is equal to the rental income
b) 27 days   **c)** $1000

19. a) $(1.11, 1.44)$ **b)** $(-2.94, -1.53)$

20. a) $(2, 1)$ **b)** $(-1, 4)$ **c)** $(8, 2)$
d) $(2, -3)$ **e)** $(-1, 2)$ **f)** $(3, 10)$

21. a) $(2, 15)$ **b)** $(10, 100)$ **c)** $(-6, 20)$

22. a) $(4, 2)$ **b)** $(1, 7)$ **c)** $(-10, 3)$
d) $(-1, -2)$ **e)** $(5, 6)$ **f)** $(2, -5)$

23. b) $x = 188$, $y = 112$; the number of youth tickets sold and the number of adult tickets sold
c) 188 youth tickets   **d)** 112 adult tickets

24. a) $2.50   **b)** $1.25

Chapter 4 Proportional Reasoning
Preparing for the Chapter, page 162

1. b) i) $8   **ii)** $24   **iii)** $40   **iv)** $68
c) i) $15   **ii)** $25   **iii)** $50   **iv)** $75

2. a), b)

Amount pledged per kilometre (¢)	Amount of money raised ($)
40	4.00
75	7.50
50	5.00
60	6.00
25	2.50

3. Estimates may vary. Approximately $50\,000

4. a) i) $6 \times 7 = 42$ **ii)** $42 \div 6 = 7$; $42 \div 7 = 6$

5. a) 16   **b)** 46.4, 32

6. a) $\frac{1}{3}$ **b)** $\frac{5}{3}$ **c)** $\frac{3}{7}$ **d)** $\frac{3}{2}$ **e)** $\frac{9}{16}$

7. a) $\frac{10}{100}$, or $\frac{1}{10}$ **b)** $\frac{20}{100}$, or $\frac{1}{5}$ **c)** $\frac{40}{100}$, or $\frac{2}{5}$ **d)** $\frac{70}{100}$, or $\frac{7}{10}$
e) $\frac{90}{100}$, or $\frac{9}{10}$ **f)** $\frac{110}{100}$, or $\frac{11}{10}$ **g)** $\frac{120}{100}$, or $\frac{6}{5}$ **h)** $\frac{140}{100}$, or $\frac{7}{5}$
i) $\frac{170}{100}$, or $\frac{17}{10}$ **j)** $\frac{190}{100}$, or $\frac{19}{10}$

8. a) 0.1   **b)** 0.2   **c)** 0.4   **d)** 0.7   **e)** 0.9
f) 1.1   **g)** 1.2   **h)** 1.4   **i)** 1.7   **j)** 1.9

9. a) 50%   **b)** 25%
c) 75%   **d)** Approximately 33%
e) Approximately 67%   **f)** 70%
g) 80%   **h)** Approximately 67%
i) 37.5%   **j)** Approximately 42%

10. a) 0.5   **b)** 0.25   **c)** 0.75   **d)** $0.\overline{3}$ **e)** $0.\overline{6}$
f) 0.7   **g)** 0.8   **h)** $0.\overline{6}$ **i)** 0.375   **j)** $0.41\overline{6}$

11. a) $\frac{1}{5}$ **b)** $\frac{5}{1}$ **c)** $\frac{3}{1}$ **d)** $\frac{1}{3}$
e) $\frac{5}{7}$ **f)** $\frac{7}{5}$ **g)** $\frac{5}{12}$ **h)** $\frac{12}{5}$

12. a) 5   **b)** 4   **c)** 3.5   **d)** 2.4
e) 2.5   **f)** $\frac{10}{3}$ **g)** 14   **h)** 1.4

13. a) 12 **b)** 20 **c)** 20 **d)** 12
e) 2 **f)** 4 **g)** $\frac{20}{3}$ **h)** 2.4

Practise Your Skills, page 170

1. a) 0.8 **b)** 0.85 **c)** 30 **d)** 6 **e)** $13.\overline{3}$ **f)** 0.4725
2. a) 50 **b)** 2.4 **c)** 3.4 **d)** 360 **e)** 120 **f)** 2.4675
3. a) $10 **b)** $20 **c)** $30 **d)** $160 **e)** $200 **f)** $220
4. a) $8 **b)** $24 **c)** $24 **d)** $48 **e)** $40 **f)** $72

4.1 Exercises, page 171

1. a) 50¢ **b)** 75¢ **c)** $1.00 **d)** $1.50 **e)** $2.25 **f)** $3.00
2. a) 80¢ **b)** $2.40 **c)** $3.20 **d)** $6.40 **e)** $12.80 **f)** $16
3. a) i) 85¢ **ii)** $1.70 **iii)** $2.55 **iv)** $3.40
v) $5.10 **vi)** $5.95 **vii)** $6.80 **viii)** $7.65
4. a) $19.00 **b)** $380.00 **c)** $665.00 **d)** $1710.00
5. a) $12.10 **b)** $96.80 **c)** $726.00 **d)** $12 100.00
6. a) 360 words **b)** 20 min
7. a) $120 **b)** 25 h
8. Estimates may vary. Approximately 3.4 million
9. a) 24 pages **b)** 5 min
c) Multiplication and division **d)** Yes
10. a) $6.00 **b)** 10 km
c) Addition, multiplication, subtraction, and division
d) No
11. a) Approximately 67 km **b)** 7.5 h
12. a) Approximately 1.54 h, or 1 h 32 min
b) Approximately 26 144 km
13. a) Problem 1: solution is incorrect.
Problem 2: solution is correct.
b) Problem 1: correct answer is 14 laps.
14. a) Problem 1: 60 hits; Problem 2: 66 hits; Problem 3:
approximately 103 hits
15. a) $20 **b)** $60 **c)** $5 **d)** $75
16. a) For $500
i), ii)

Interest rate (%)	Interest earned after one year ($)	Total amount after one year ($)
0	0	500
2	10	510
4	20	520
6	30	530
8	40	540

For $2000

Interest rate (%)	Interest earned after one year ($)	Total amount after one year ($)
0	0	2000
2	40	2040
4	80	2080
6	120	2120
8	160	2160

c) In each case, the graph of *Interest earned* against *Interest rate* represents a proportional situation. The graphs of *Total amount* against *Interest rate* do not represent proportional situations because the graphs do not pass through the origin.

17. a) For 4%
i), ii)

Principal invested ($)	Interest earned after one year ($)	Total amount after one year ($)
200	8	208
400	16	416
600	24	624
800	32	832
1000	40	1040

For 6%
i), ii)

Principal invested ($)	Interest earned after one year ($)	Total amount after one year ($)
200	12	212
400	24	424
600	36	636
800	48	848
1000	60	1060

18. 525-g box
19. 950 mL
22. a) 22.5 s **b)** 90 km; 90 km/h

Practise Your Skills, page 182

1. a) 30 **b)** 12.5 **c)** 2.1
d) $\frac{400}{7}$ **e)** 200 **f)** 2.868 75
2. a) 25% **b)** 50%
c) 30% **d)** 80%
e) 75% **f)** 20%
g) Approximately 44% **h)** 37.5%
i) Approximately 133% **j)** 62.5%
k) 125% **l)** 90%

4.2 Exercises, page 182

1. a) Approximately 0.75 **b)** Approximately 75%
2. a) 0.81 **b)** 81% **c)** 0.62 **d)** 62%
3. a) 3.2 cm **b)** Approximately 94 m
4. Approximately 24.8 million
5. a) 96¢ **b)** Approximately 182 g
6. a) $6.25 **b)** Approximately 360 g
7. a) $73.80 **b)** Approximately 43 h
8. 250
9. a) 4.8 km **b)** Approximately 167 min
10. Estimates may vary between 1.20 m and 1.80 m.
11. a) Approximately 52
12. a) Solutions are incorrect for both problems.
b) Problem 1: 5 h 19 min; Problem 2: $285.71
13. Answers may vary.
a) Approximately 888 000 hot dogs
b) Approximately 8 million hot dogs
14. a) 10.125 L **b)** Approximately 667 km
15. a) Measurements may vary.

		Length on photograph (cm)	Actual length (m)
i)	Height of one person	2.4	1.80
ii)	Height of one tire	4.6	3.45
iii)	Height of the truck	9.7	7.28
iv)	Width of the truck	11.2	8.40
v)	Length of a staircase	4.6	3.45
vi)	Width of a staircase	0.6	0.45

c) 1.35 m **d)** 3.90 m

16. Estimates may vary.
   a) 3211 m **b)** 963 300 floppy disks

17. 19.27 m

20. Measurements may vary.
   a) 0.9 cm by 17.7 cm **b)** 177 cm
   d) Approximately 180 times as high

21. 22.5 g

22. **a)** 22.5 cm **b)** 3.75 cm

24. Approximately 5:36 P.M. on December 31

4.3 Mathematical Modelling: Food for a Healthy Heart, page 189

1. **a)** 49.5 Cal
   b) i) $\frac{49.5}{139}$ ii) Approximately 36%

2. No

3. **a)** Approximately 20%; yes
   b) 30%; no **c)** Approximately 47%; no

5. **a)** 528; no **b)** 78; yes **c)** 15; yes
   d) 48; no **e)** 21; yes **f)** 3; yes
   g) 240; no **h)** 276; no **i)** 204; no
   j) 81; yes **k)** 480; no **l)** 810; no

4.4 Exercises, page 195

1.

	Length in drawing (cm)	Actual length (cm)	Scale
a)	5	50	1 : 10
b)	5	10	1 : 2
c)	10	5	1 : 0.5
d)	10	50	1 : 5
e)	4.5	225	1 : 50
f)	3.7	92.5	1 : 25
g)	38	3.8	1 : 0.1
h)	420	8.4	1 : 0.02
i)	5.2	52	1 : 10
j)	12.7	317.5	1 : 25
k)	37	370	1 : 10
l)	30	7500	1 : 250

2.

	Actual distance (km)	Distance on map (cm)	Scale
a)	120	60	1 : 200 000
b)	120	40	1 : 300 000
c)	120	10	1 : 1 200 000
d)	120	6	1 : 2 000 000
e)	90	3.6	1 : 2 500 000
f)	0.215	4.3	1 : 5000
g)	0.33	13.2	1 : 2500
h)	75	15	1 : 500 000
i)	150	300	1 : 50 000
j)	99	3.3	1 : 3 000 000
k)	1.4	3.5	1 : 40 000
l)	3.6	18	1 : 20 000
m)	56	28	1 : 200 000
n)	3.78	10.8	1 : 35 000
o)	0.616	15.4	1 : 4000

   c) Between 1 h 15 min and 1 h 45 min

3. **a)** 1 cm to 24 m **b)** 1 cm to 85 cm **c)** 1 cm to 3 cm

4. **a)** Approximately 50 cm **b)** 1.76 m
   c) 0.9 cm

5. **a)** Answers may vary. These are internal dimensions.
   Living room: 7.60 m by 5.40 m;
   Bedroom B: 5.80 m by 5.00 m;
   Bathroom: 4.00 m by 2.40 m;
   Bedroom A: 7.00 m by 4.60 m;
   Dining room: 6.60 m by 4.60 m;
   Kitchen: 7.40 m by 6.00 m
   b) External dimensions: 22.00 m by 12.80 m

6. **a)** 2.40 m **b)** 6.3 mm

7. Answers may vary.
   a) 117 cm **b)** 62 cm **c)** 415 cm

8. **a)** 4 mm **b)** 2 mm **c)** 4 mm

9. About 3950 m

10. **a)** 13 m **b)** 41 m

11. **a)** 37 m **b)** 16 m **c)** 24 m

13. Estimates may vary.
   a) 3400 km **b)** 195 km **c)** 0.3 mm
   d) 0.1 mm **e)** 0.7 mm **f)** 0.02 mm

14. Estimates may vary.
   a) 4800 km **b)** 60 km **c)** 32 days

15. Estimates may vary.
   a) Estimate: 405 km; actual: 441 km
   b) Estimate: 510 km; actual: 535 km

16. **a)** 69 m; approximately 10 m
   b) 99 m; approximately 12 m

18. **a)** 750 days **b)** 37.5 months
   c) About 70 cm **d)** Approximately 130

19. **a)** 1.7 cm **b)** 0.1 cm

4.5 Exercises, page 207

1. **a)** 25 cm **b)** 50 cm **c)** 75 cm
   d) 100 cm **e)** 125 cm **f)** 150 cm

2. **a)** 3000 m **b)** 7000 m **c)** 10 000 m
   d) 14 000 m **e)** 25 000 m **f)** 2500 m

3. **a)** 1 cm **b)** 0.5 cm **c)** 2 cm
   d) 2.5 cm **e)** 3 cm **f)** 5 cm

4. **a)** 1 cm **b)** 1.9 cm **c)** 2.9 cm
   d) 29 cm **e)** 35 cm **f)** 3.5 cm

12. Original: 3.9 cm by 1.3 cm; stretched horizontally: 5.9 cm by 1.3 cm; stretched vertically: 3.9 cm by 2.0 cm; stretched horizontally and vertically: 5.9 cm by 2.0 cm

Testing Your Knowledge, page 211

1. **a)** 9.3 million **b)** 18.6 million

2. **a)** i) 900 pages ii) 21 600 pages
   b) i) Approximately 67 min ii) Approximately 11 h 7 min

3. **a)** $382 500 **b)** Approximately 8 h

4. About 330 seats

5. Approximately 138 kg

6. Estimates may vary.
   a) Red: 1500, White: 900

7. 9 cm by 5.25 cm

8. Estimates may vary. 3400 km

9. 16%

10. a) \$4 355 000   **b)** \$2 010 000   **c)** \$335 000

11. a) \$480 000   **b)** 12 000 m, or 12 km

12. Approximately 9

13. a) i) 5840   **ii)** 315 360
  b) i) 60 min   **ii)** 1440 min
   iii) 525 600 min   **iv)** 3 416 400 min
  c) Yes, approximately correct

14. Estimates may vary.
  a) 103 km   **b)** 1 cm to 32 km

15. Estimates may vary. 7.7 km

16. a) 9.2 m   **b)** 15.4 m   **c)** 11.8 m   **d)** 7.8 m

17. 231 250 km

18. a) Yes   **b)** No   **c)** Yes

19. a) i) 0.89 mm   **ii)** 0.14 mm
  b) 70 mm to 1 mm

20. Estimates may vary.
  a) 3 m   **b)** 8.5 m

Chapter 5 Similar Triangles and Trigonometry

Preparing for the Chapter, page 220

1. a) 2.5 cm   **b)** 2.9 cm

2. 9.8 m

3. a) 135°   **b)** 80°   **c)** 77°

4. a) $x = 130°, y = 50°, z = 130°$
  b) $x = 122°, y = 128°, z = 128°$
  c) 107.5°

5. a) 87°   **b)** 64°   **c)** 48°

8. a) 55°   **b)** 50°   **c)** 75°   **d)** 120°
  e) 120°   **f)** 60°   **g)** 45°   **h)** 135°

Practise Your Skills, page 235

1. a) 41°   **b)** 77°   **c)** 108°

2. a) 2   **b)** 2   **c)** 25   **d)** $\frac{5}{2}$ **e)** 10   **f)** 3

5.3 Exercises, page 235

1. R

2. a) $\angle A = \angle E$, $\angle B = \angle D$, $\angle C = \angle F$; AB corresponds to
  ED; BC corresponds to DF; CA corresponds to FE.
  b) $\angle P = \angle Z$, $\angle Q = \angle X$, $\angle Y = \angle R$; PQ corresponds to
  ZX; QR corresponds to XY; RP corresponds to YZ.

3. a) $\angle A = 34°$, $\angle P = 56°$, $\angle Q = 90°$
  b) $\angle D = 70°$, $\angle F = 40°$, $\angle J = 70°$, $\angle G = 70°$
  c) $\angle K = 60°$, $\angle M = 60°$, $\angle S = 60°$, $\angle T = 60°$, $\angle U = 60°$

4. Yes

5. a) $\frac{FD}{DE}$ **b)** $\frac{DE}{EF}$ **c)** $\frac{ED}{DF}$

6. a) $x = 3.0, y = 5.0$ **b)** $x = 2, y = \frac{8}{3}$

7. a) $\frac{AB}{PQ} = \frac{AC}{PR} = \frac{CB}{RQ}$; 4
  b) $\frac{ST}{HR} = \frac{TU}{RB} = \frac{BH}{US}$; 28
  c) $\frac{EF}{EN} = \frac{FG}{NS} = \frac{GE}{SE}$; 2.5

8. a) $\triangle GJH \sim \triangle MJR$; $x = 8, y = 12$
  b) $\triangle TCS \sim \triangle TDM$; $x = 2.5, y = 7$

10. a) $x = 21, y = 36$ **b)** $x = 12, y = 20$

11. a) $x = 7$ **b)** $x = 4.2$ **c)** $x = 7.5$ **d)** $x = \frac{8}{3}$

15. $\triangle DGE \sim \triangle FDE \sim \triangle FGD$

16. a) $\triangle DAC \sim \triangle BAE$; $x = 7.2$, $y = 7$
  b) $\triangle FGI \sim \triangle HGF \sim \triangle HFI$; $x \doteq 7.0$, $y \doteq 3.8$

5.4 Exercises, page 243

1. a) $\triangle WXY \sim \triangle VTU$; $x = 9, y = 7.5$
  b) $\triangle CDE \sim \triangle BZA$; $x \doteq 26.7, y = 24$
  c) $\triangle FGH \sim \triangle KIJ$; $x = 14.4, y = 10$

2. 22.4 m

3. Approximately 18.7 m

4. 8 m

5. Approximately 34.9 m

6. 32 m

7. 14 m

8. 225 m

9. 30 m

10. 38.4 m

11. 22.8 m

12. 9.85 m

5.6 Exercises, page 254

1. a) i) PQ   **ii)** MN   **b) i)** PA   **ii)** MA

2. i) $\frac{PQ}{PA}$ **ii)** $\frac{MN}{MA}$

3. a) i) 0.700   **ii)** 1.000   **iii)** 2.145
  c) Greater than 1

4. a) $\angle B = 63°$

5. a) i) $\frac{9}{15}$; $\frac{15}{9}$ **ii)** $\frac{9}{6}$; $\frac{6}{9}$ **iii)** $\frac{19}{22}$; $\frac{22}{19}$
  b) i) 17.5 cm   **ii)** 10.8 m   **iii)** 29.1 mm

7. a) i) 3.9 cm   **ii)** 25.7 cm   **iii)** 4.4 cm   **iv)** 12.0 cm
  b) i) 12.6 cm   **ii)** 28.4 cm   **iii)** 12.8 cm   **iv)** 17.0 cm

8. a) i) 32.1 cm   **ii)** 268.7 cm   **iii)** 60.4 cm   **iv)** 178.2 cm
  b) i) 78.8 cm   **ii)** 278.2 cm   **iii)** 94.0 cm   **iv)** 192.2 cm

9. a) 4.1 m   **b)** 4.4 m

10. 0.73 m

11. a) The angle between the horizontal and the line of sight to
  an object above the horizontal
  b) 6435 m   **c)** 3202 m

12. a) The angle between the horizontal and the line of sight to
  an object below the horizontal
  b) Approximately 185 m   **c)** Approximately 340 m

13. a) 6.5 cm   **b)** 15.5 cm   **c)** 37.3 m   **d)** 4.1 m
e) 3.5 cm   **f)** 1.7 m   **g)** 7.7 m   **h)** 2.0 cm
i) 1.5 m

5.7 Exercises, page 259

1. a) 14°   **b)** 35°   **c)** 54°   **d)** 73°
e) 29°   **f)** 43°   **g)** 64°   **h)** 78°

2. a) 27°   **b)** 22°   **c)** 53°   **d)** 51°
e) 37°   **f)** 31°   **g)** 68°   **h)** 55°

3. a) i) $\frac{2.5}{1.5}$, 59°   **ii)** $\frac{12}{45}$, 15°
    b) i) $\frac{1.5}{2.5}$, 31°   **ii)** $\frac{45}{12}$, 75°

4. a) 2, 63°   **b)** 1.5, 56°   **c)** 0.75, 37°

5. a) 1.6, 58°   **b)** 2.4, 67°   **c)** 2.8, 70°   **d)** 4.2, 77°

6. a) Approximately 67°   **b)** 9.3 m

7. a) $\angle A \doteq 28°$, $\angle B \doteq 62°$ **b)** $\angle A \doteq 53°$, $\angle B \doteq 37°$
   c) $\angle A \doteq 44°$, $\angle B \doteq 46°$

9. Approximately 42°

10. Approximately 71°

11. d) $\angle A = 45°$, $\angle B \doteq 18°$, $\angle C \doteq 76°$, $\angle D \doteq 27°$

5.9 Exercises, page 268

1. a) i) RG   **ii)** LQ
   b) i) RA   **ii)** LA
   c) i) GA   **ii)** AQ

2. a) i) $\frac{RG}{GA}$ **ii)** $\frac{LQ}{AQ}$ **b) ii)** $\frac{RA}{GA}$ **ii)** $\frac{LA}{AQ}$

3. a) 0.970, 0.242   **b)** 0.682, 0.731   **c)** 0.326, 0.946
d) 0.988, 0.156   **e)** 0.883, 0.469   **f)** 0.242, 0.970
g) 0.731, 0.682   **h)** 0.946, 0.326   **i)** 0.156, 0.988
j) 0.469, 0.883

4. a) 5.3 cm, 8.5 cm   **b)** 14.1 cm, 5.1 cm
   c) 3.5 cm, 11.5 cm   **d)** 13.4 m, 14.9 m

6. $AB \doteq 15.0$ m, $BC \doteq 20.0$ m

7. $PQ \doteq 31.7$ m, $QR \doteq 27.6$ m

8. a) i) 0.848   **ii)** 0.766   **iii)** 0.500   **iv)** 0.309

9. a) 9.6 m   **b)** 2.9 m

10. Approximately 118 m

11. a) $PQ \doteq 7.9$ cm, $RQ \doteq 12.7$ cm
   b) $PQ \doteq 10.0$ cm, $RQ \doteq 11.1$ cm

12. a) 29.7 m   **b)** 6.9 m

13. 2.8 m

14. 13.3 m

15. a) 23.3 km   **b)** 12.7 km

5.10 Exercises, page 273

1. a) 37°   **b)** 29°   **c)** 46°   **d)** 72°   **e)** 78°   **f)** 77°

2. a) 65°   **b)** 22°   **c)** 31°   **d)** 38°   **e)** 66°   **f)** 52°

3. a) 42°   **b)** 76°   **c)** 51°   **d)** 25°   **e)** 37°   **f)** 13°

4. a) $\cos A = \frac{5}{12}$, 65°   **b)** $\sin A = \frac{6}{9}$, 42°
   c) $\sin A = \frac{3}{20}$, 9°   **d)** $\cos A = \frac{12}{16}$, 41°

5. a) $\frac{8}{14}$, 34.8°   **b)** $\frac{6}{11}$, 33.1°

6. a) 55.2°   **b)** 56.9°

7. a) 24°   **b)** 34°   **c)** 27°   **d)** 51°

8. a) 49°, 41°   **b)** 2.6 cm

9. a) 58°, 32°   **b)** 59°, 31°

10. Approximately 53°

12. a) Approximately 50°   **b)** 11.5 m

13. a) Approximately 73°   **b)** 5.7 m

14. a) The angle between the horizontal and a line above the horizontal
   b) Approximately 12°

15. a) Approximately 31°   **b)** Approximately 118°

Practise Your Skills, page 281

1. a) 15   **b)** 42   **c)** 1.95
   d) 8.2095   **e)** 0.9093   **f)** 0.8908

2. a) 0.6   **b)** 1.167   **c)** 18.519
   d) 5.146   **e)** 4.850   **f)** 3.244

5.12 Exercises, page 281

1. a) 21.3 cm   **b)** 14.2 cm

2. a) 53°   **b)** 88°   **c)** 21°

3. 51.4 m

4. a) 15.7 m   **b)** 8.3 m

5. 2215 m

6. a) 1893 m   **b)** 4407 m

7. Approximately 3.7 m

10. 53 m

Testing Your Knowledge, page 285

1. a) 3.255   **b)** Approximately 2.29

2. $x = 11.25$, $y \doteq 9.2$, $w = 20$, $z \doteq 29.8$

3. $\triangle PNR$ is similar to $\triangle KLM$.

4. a) 23°   **b)** 57°   **c)** 59°   **d)** 35°

5. a) 0.2250, 0.9744   **b)** 0.6947, 0.7193   **c)** 0.3907, 0.9205
   d) 0.9613, 0.2756   **e)** 0.5000, 0.8660

6. a) 52°   **b)** 43°   **c)** 53°   **d)** 39°   **e)** 26°   **f)** 7°

7. a) 6.3 cm   **b)** 11.0 cm

8. a) $\frac{7.1}{14.2}$; 27°   **b)** 15.9 cm

9. 63°, 27°

10. a) $\frac{7.5}{9.0}$, 56.4°   **b)** $\frac{8.2}{12.4}$, 41.4°

11. a) 3.0 m   **b)** 26.0 cm   **c)** 54.5°   **d)** 33.5°

12. Approximately 141 m

13. Approximately 16 m

14. Approximately 142 m

14. a) 75.5°   **b)** 5.8 m   **c)** 70.5°, 5.7 m

15. Approximately 53.7 m

Cumulative Review, page 288

1. All graphs pass through (0, 2).

2. b) During week 5.

c)

Week	1	2	3	4	5	6	7	8	9	10
Cumulative total	375	765	1155	1460	1550	2105	2485	2880	3270	3660

3. a) The difference between successive pairs of entries
is constant.
c) Student A: slope = 6.5. The student's hourly rate
of pay in dollars
Student B: slope = 5. The student's hourly rate
of pay in dollars
d) A: $E = 6.5n$; B: $E = 5n + 40$
e) Student A: direct variation; Student B: partial variation

4. a) 7   **b)** 11   **c)** 1

5. a) $y = 4(x - 2) + 7$ or $y = 4(x - 3) + 11$
b) $y = 2(x + 2) + 3$ or $y = 2(x - 2) + 11$
c) $y = \frac{5}{2}(x - 2) + 3$ or $y = \frac{5}{2}(x - 4) + 8$

6. b) $112   **c)** 8 people

9. a) (3, 1)   **b)** (−4, 2)

10. a) (5, 6)   **b)** (3, 14)   **c)** (−3, −16)
d) (1, −2)   **e)** (−5, 6)   **f)** (−10, 2)

11. a) (2, 7)   **b)** (−2, 8)   **c)** (8, 2)
d) (−1, 4)   **e)** (3, −2)   **f)** (−7, −2)

12. 7 videos and 4 DVDs

13. a) 159 minutes   **b)** 10 games

14. $54.53

15. Approximately 79 cm

16. a) Answers may vary. Approximately 1:300.
b) Answers may vary. The fighter is about 16.5 m long.

17. Answers may vary.
a) 1 cm to 21.4 cm **b)** 29 cm   **c)** 21 cm

18. 874.5 V

19. b) 14.4

20. a) $x \doteq 10.67$, $y = 22$ **b)** $x = 4$, $y = 8.5$

21. a) $x \doteq 8.0$, $y \doteq 10.2$ **b)** $x \doteq 11.5$, $y \doteq 14.6$
c) $x \doteq 55.7°$, $y \doteq 13.0$

22. Approximately 125 m

Chapter 6 Introduction to Quadratic Functions

Preparing for the Chapter, page 300

7.a), 8. a)

x	y	Difference
0	2	
1	3	1
2	4	1
3	5	1
4	6	1

7.b), 8. b)

x	y	Difference
0	−1	
1	1	2
2	3	2
3	5	2
4	7	2

7.c), 8. b)

x	y	Difference
0	4	
1	3	−1
2	2	−1
3	1	−1
4	0	−1

8. c) For each function, the differences are constant.
d) In each case, since x increases by 1, the difference is the
slope of the line.

9. a) A, D, E

6.1 Exercises, page 309

1. a) No   **b)** Yes   **c)** Yes   **d)** No

4. a)

x	y
−3	9
−2	4
−1	1
0	0
1	1
2	4
3	9

b)

x	y
−3	10
−2	5
−1	2
0	1
1	2
2	5
3	10

c)

x	y
−3	12
−2	7
−1	4
0	3
1	4
2	7
3	12

d)

x	y
−3	8
−2	3
−1	0
0	−1
1	0
2	3
3	8

e)

x	y
−3	13
−2	8
−1	5
0	4
1	5
2	8
3	13

5. a) The parabolas have the same size and shape.
b) The vertices are different.

7. a)

x	y
−3	18
−2	8
−1	2
0	0
1	2
2	8
3	18

b)

x	y
−3	19
−2	9
−1	3
0	1
1	3
2	9
3	19

c)

x	y
−3	21
−2	11
−1	5
0	3
1	5
2	11
3	21

d)

x	y
−3	17
−2	7
−1	1
0	−1
1	1
2	7
3	17

8. a) (0, 0); (0, 1); (0, 3); (0, −1)
b) The parabolas have the same size and shape.
c) The step pattern for the graphs in part a is 2, 6, 10, ...
The step pattern for the graphs in exercise 4 is 1, 3, 5, ...

6.2 Exercises, page 317

1. $y = 3x^2$; $h = 2t^2 + 1$; $A = s^2$

2. a) Up, (0, 0)   **b)** Down, (0, 0)   **c)** Up, (0, 0)
d) Up, (0, 1)   **e)** Down, (0, 2)   **f)** Down, (0, −2)

3. e) $y = 0.25x^2$

4. a) $y = 2x^2$ **b)** $y = -x^2$ **c)** $y = \frac{x^2}{9}$

5. a) $y = x^2$ **b)** $y = x^2 + 1$

   c) $y = -2x^2$ **d)** $y = -\frac{1}{2}x^2 - 1$

6. a)

Edge length (cm)	Surface area (cm²)
0	0
1	6
2	24
3	54
4	96
5	150

   c) Yes   **e)** $y = 6x^2$ **f)** 600 cm²

7. c) $y \doteq 0.0047x^2$ **d)** Approximately 92 m

6.3 Exercises, page 326

1. a) i) $(0, 0)$ **ii)** Positive   **b) i)** $(0, 3)$ **ii)** Positive

   c) i) $(0, 0)$ **ii)** Negative   **d) i)** $(0, 5)$ **ii)** Negative

   e) i) $(0, -2)$ **ii)** Positive   **f) i)** $(0, -1)$ **ii)** Negative

2. a)

x	y
−4	16
−3	9
−2	4
−1	1
0	0
1	1
2	4
3	9
4	16

b)

x	y
−8	17
−6	10
−4	5
−2	2
0	1
2	2
4	5
6	10
8	17

c)

x	y
−8	−16
−6	−9
−4	−4
−2	−1
0	0
2	−1
4	−4
6	−9
8	−16

d)

x	y
−4	−13
−3	−6
−2	−1
−1	2
0	3
1	2
2	−1
3	−6
4	−13

3. a) $(0, 0)$ **b)** $(0, 1)$ **c)** $(0, 0)$ **d)** $(0, 3)$

4. e) $(0, 0)$; $y = ax^2$ **g)** $y = \frac{1}{16}x^2$

   h) i) 625 cm²   **ii)** 1.5625 cm²

   iii) Approximately 17.9 cm   **iv)** 225 m²

   v) No

5. a) In each table, the data represent a function.

b) i)

x	y	Difference
0	0	
1	1	1
2	4	3
3	9	5
4	16	7
5	25	9

ii)

x	y	Difference
0	0	
2	1	1
4	4	3
6	9	5
8	16	7
10	25	9

iii)

x	y	Difference
0	0	
1	2	2
2	8	6
3	18	10
4	32	14
5	50	18

iv)

x	y	Difference
0	0	
2	2	2
4	8	6
6	18	10
8	32	14
10	50	18

v)

x	y	Difference
0	0	
1	3	3
2	12	9
3	27	15
4	48	21

vi)

x	y	Difference
0	0	
2	3	3
4	12	9
6	27	15
8	48	21

6. i) $y = x^2$ **ii)** $y = 0.25x^2$ **iii)** $y = 2x^2$

   iv) $y = 0.5x^2$ **v)** $y = 3x^2$ **vi)** $y = 0.75x^2$

7. a) $y = 40x$ **b)** $y = -2x^2 + 40$ **c)** $y = -x + 40$

   d) $y = 40x^2$ **e)** $y = 40 + x$

6.4 Mathematical Modelling: Projectile Motion, page 330

1., 2., 3., 4., 5., 8. Answers may vary.

Image number	Height of ball on picture (cm)	Time (s)	Actual height of ball (m)	Differences
1	0	0	0	0.76
2	3.80	0.1	0.76	0.66
3	7.10	0.2	1.42	0.58
4	10.00	0.3	2.00	0.46
5	12.30	0.4	2.46	0.38
6	14.20	0.5	2.84	0.26
7	15.50	0.6	3.10	0.17
8	16.35	0.7	3.27	0.08
9	16.75	0.8	3.35	−0.02
10	16.65	0.9	3.33	−0.13
11	16.00	1.0	3.20	−0.21
12	14.95	1.1	2.99	−0.33
13	13.30	1.2	2.66	−0.41
14	11.25	1.3	2.25	−0.51
15	8.70	1.4	1.74	−0.61
16	5.65	1.5	1.13	−0.71
17	2.10	1.6	0.42	

7. The quadratic model

9. Approximately $(0.8, 3.35)$; the vertex is not on the vertical axis.

6.5 Exercises, page 336

1. a) i) $(6, -4)$ **ii)** 32   **iii)** 4, 8   **iv)** $(12, 32)$

   b) i) $(3, 4)$ **ii)** −5   **iii)** 1, 5   **iv)** $(6, -5)$

   c) i) $(-4, -9)$ **ii)** 7   **iii)** −1, −7   **iv)** $(-8, 7)$

2. The vertex of a parabola lies on its axis of symmetry.

3. a) i) $(0, -8)$ **ii)** −8   **iii)** −2, 2

   b) i) $(3, -1)$ **ii)** 8   **iii)** 2, 4

   c) i) $(-3, 9)$ **ii)** 0   **iii)** −6, 0

   d) i) $(-1.5, -4.5)$ **ii)** 0   **iii)** 0, −3

   e) i) $(2, 5)$ **ii)** 7   **iii)** None

   f) i) $(1.5, 0.5)$ **ii)** 5   **iii)** None

4. a) 1, 2   **b)** 1, 7   **c)** 0, 4

   d) −1, 4   **e)** −1, 4   **f)** −2, 0

5.

Length (cm)	Width (cm)	Area (cm²)
14	0	0
13	1	13
12	2	24
11	3	33
10	4	40
9	5	45
8	6	48
7	7	49
6	8	48
5	9	45
4	10	40
3	11	33
2	12	24
1	13	13
0	14	0

   b) Yes   **c)** $(7, 49)$ **d)** $x = 7$

   e) $(0, 0)$, $(14, 0)$ **f)** 49   **g)** 7 cm, 7 cm

   h) $(4, 40)$, $(10, 40)$ **i)** 5

6. a)

Ticket price ($)	Number of People	Revenue ($)
10	1600	16 000
15	1300	19 500
20	1015	20 300
25	760	19 000
30	517	15 510

d) Approximately \$20   **f)** \$12.50–\$27

7. b) i) Approximately 431.8 m   **ii)** Approximately 9.4 s
iii) Approximately 18.8 s   **iv)** Approximately 0.6 s, 18.2 s
c) Approximately 17.6 s

8. b) (1.0, 5.0)   **c)** 0.3 s
d) 0, 2.0; the difference between the horizontal intercepts is the time the ball was in the air.
e) The first coordinate of the vertex is one-half the time the ball was in the air.

10. a) (−1.75, −18.125); −4.760, 1.260
b) (−3.5, 22.25); −8.217, 1.217
c) (3.5, −8.25); 0.628, 6.372

Testing Your Knowledge, page 341

6. (0, 3)

8. a) (0, 0), minimum   **b)** (0, 0), maximum
c) (0, 2), minimum   **d)** (0, −4), maximum
e) (0, 0), minimum

9. a) i) (0, −5)   **ii)** −5   **iii)** −1.29, 1.29
b) i) (0, 2)   **ii)** 2   **iii)** −1.41, 1.41
c) i) (0, 1)   **ii)** 1   **iii)** No x-intercepts

10. a) i) (−3, −1)   **ii)** 8   **iii)** −4, −2   **iv)** (−6, 8)
b) i) (3.5, 4.5)   **ii)** −20   **iii)** 2, 5   **iv)** (7, −20)

11. a) −1, −4   **b)** 0, 3   **c)** −2, 5

12. a) Coordinates may vary. (7, 1), (11, 2), (14, 3), (−7, 1), (−11, 2) or (189, 27), (297, 54), (378, 81), (−189, 27), (−297, 54)
f) $y \doteq 0.0005x^2 + 9$

13. a)

Length (cm)	Width (cm)	Area (cm²)
0	0	0
9	1	9
8	2	16
7	3	21
6	4	24
5	5	25
4	6	24
3	7	21
2	8	16
1	9	9

c) Parabola   **e)** (5, 25)   **f)** $x = 5$
g) 0, 10; the widths in centimetres of the rectangle when its area is zero
h) The area is a maximum (25 cm²) and the rectangle is a square, side length 5 cm.
i) (7, 3), (3, 7)   **j)** 3

14. b) 20.4 m   **c)** 4.1 s   **d)** 3.8 s

15. a), b), c) Answers may vary.

Number of dot	Time (s)	Distance on photo (cm)	Actual distance (cm)
1	0	0	0
2	0.1	0.5	5
3	0.2	2.0	20
4	0.3	4.4	44
5	0.4	7.8	78

g) $y \doteq 480x^2$

16. a) 45.9 m   **b)** 3.06 s   **c)** 6.12 s

Chapter 7 From Algebra to Quadratic Equations

Preparing for the Chapter, page 348

5. a) 980 cm²; 980 cm²
b) i) $A = (15 \times 15 + 15 \times 5)$ cm²; $A = (15 \times 20)$ cm²
ii) $A = (19 \times 30 + 19 \times 10)$ cm²; $A = (19 \times 40)$ cm²

6. a) $3x + 6$ **b)** $3x - 6$ **c)** $4x + 20$ **d)** $4x - 20$
e) $-3x - 6$ **f)** $-3x + 6$ **g)** $-4x - 20$ **h)** $-4x + 20$

7. a) $3(x + 2)$ **b)** $3(x - 2)$
c) $-3(x + 2)$ **d)** $-3(x - 2)$

8. a) $2a - 9$ **b)** $a - 6b + 4c$ **c)** $6x + 2y + 5$
d) $6x^2 + 5x - 2$ **e)** $5m^2 - 11m + 3$ **f)** $2t - 5w$

9. a) $10x + 6$ **b)** $10x - 6$ **c)** $-10x + 6$ **d)** $-10x - 6$
e) $3x^2 + x + 5$ **f)** $4x^2 - x + 1$ **g)** $-7x - 1$ **h)** $-x^2 + 2x$

10. a) $4x + 2$ **b)** $4x - 2$ **c)** $-4x + 2$
d) $-4x - 2$ **e)** $x^2 + 11x + 3$ **f)** $-2x^2 - 3x + 3$
g) $-2x^2 - 3x - 3$ **h)** $3x^2 + 4x + 2$

11. a) $16a^2$ **b)** $-27x^2$ **c)** $-28a^3$ **d)** $15x^2$

12. a) 6   **b)** 4   **c)** −11   **d)** −3

Practise Your Skills, page 351

1. a) 28   **b)** −6   **c)** −15   **d)** 48
e) −24   **f)** −35   **g)** 42   **h)** 0.5

2. a) $28x$ **b)** $-6x$ **c)** $-15x$ **d)** $48x$
e) $-24x$ **f)** $-35x$ **g)** $42x$ **h)** $0.5x$

3. a) $28x^2$ **b)** $-6x^2$ **c)** $-15x^2$ **d)** $48x^2$
e) $-24x^2$ **f)** $-35x^2$ **g)** $42x^2$ **h)** $0.5x^2$

7.1 Exercises, page 352

1. a) $(x)(x)$ **b)** $(x + 1)(x)$ **c)** $(x + 1)(2x)$
d) $(2x)(2x)$ **e)** $(3x)(x)$ **f)** $(2x)(2x + 2)$

2. a) $2x + 6$ **b)** $6x + 3$ **c)** $2x^2 + 2x$ **d)** $3x^2 + 6x$

3. a) C   **b)** E   **c)** B   **d)** A   **e)** F   **f)** D

4. a) $12x + 8$ **b)** $15b - 18$ **c)** $-5x - 10$ **d)** $-3x + 6$
e) $4a + 10$ **f)** $-2n - 14$ **g)** $-y + 1$ **h)** $-6x + 10$

5. a) $x^2 + 7x$ **b)** $x^2 + 3x$ **c)** $3a^2 - 4a$ **d)** $-4x^2 - x$
e) $-4a^2 - 2a$ **f)** $y - y^2$ **g)** $3b^2 + 5b$ **h)** $-2n^2 + n$

6. a) $3x^2 + 12x$ **b)** $6x^2 - 4x$ **c)** $-10n^2 + 15n$ **d)** $-21a^2 - 6a$
e) $-7y^2 - 21y$ **f)** $-15x^2 - 20x$ **g)** $-8x^2 + 6x$ **h)** $12x - 24x^2$

7. a) $2x^2 + 6x - 8$ **b)** $-2x^2 - 6x + 8$ **c)** $x^3 + 3x^2 - 4x$
d) $-x^3 - 3x^2 + 4x$ **e)** $-3a^3 + 9a^2 + 6a$ **f)** $4m^3 + 4m^2 - 4m$
g) $-2x^3 + 3x^2 - 4x$ **h)** $2x^3 + 4x^2 + 2x$ **i)** $4x^3 + 4x^2 - 12x$

8. a) $5(30 + 4) = 170$ **b)** $15x + 20$; 170

9. a) 4   **b)** −2   **c)** 2   **d)** −4

10. a) 2   **b)** −4   **c)** 4   **d)** −2

11. a) 14   **b)** −2   **c)** 2   **d)** −14   **e)** 6   **f)** 6

12. a) 2   **b)** −14   **c)** 14   **d)** −2   **e)** −6   **f)** −6

Practise Your Skills, page 357

1. a) $2 \times 2 \times 2 \times 2 \times 3$ **b)** $3 \times 3 \times 3 \times 3$
c) $3 \times 3 \times 5$ **d)** $2 \times 2 \times 3 \times 3 \times 3$
e) $2 \times 2 \times 2 \times 2 \times 2 \times 5 \times 5$ **f)** $2 \times 2 \times 2 \times 3 \times 5$

2. a) 2   **b)** 3   **c)** 5
   d) 16   **e)** 4   **f)** 7
   g) 10   **h)** 13   **i)** 20

3. a) 4   **b)** 3   **c)** 4   **d)** 4   **e)** 5

4. a) 2   **b)** 3   **c)** 10   **d)** 4

7.2 Exercises, page 358

1. a) 3   **b)** 4   **c)** 25   **d)** 12   **e)** 4

2. a) $2(x)$ **b)** $2(-2x)$ **c)** $2(x^2)$ **d)** $2(-2x^2)$ **e)** $2(4x^3)$

3. a) $3x(-2)$ **b)** $3x(x)$ **c)** $3x(2x)$ **d)** $3x(-4x^2)$ **e)** $3x(x^2)$

4. a) $2a^2(1)$ **b)** $2a^2(-1)$ **c)** $2a^2(2)$ **d)** $2a^2(-4)$ **e)** $2a^2(-4a)$

5. a) 2   **b)** $2x$ **c)** $2x^2$ **d)** 2
   e) 3   **f)** $3x$ **g)** $6x$ **h)** 3

6. a) $3(x + 1)$ **b)** $4(a + 2)$ **c)** $2(3 + 2x)$
   d) $3(5 + 7x)$ **e)** $-2(x + 5)$ **f)** $-1(4 + 3x)$

7. a) $3(x - 1)$ **b)** $4(a - 2)$ **c)** $2(3 - 2x)$
   d) $3(5 - 7x)$ **e)** $2(x - 5)$ **f)** $2(x^2 - 3)$

8. a) x **b)** x **c)** a **d)** 4   **e)** $2x$ **f)** $3x$

9. a) $3(x + 1)$ **b)** $2(y + 2)$ **c)** $3(2a + 1)$
   d) $5(x + 3)$ **e)** $4(a + 3)$ **f)** $2(2n + 5)$

10. a) $3(x - 1)$ **b)** $-2(y - 2)$ **c)** $-3(2a + 1)$
   d) $5(x - 3)$ **e)** $-4(a - 3)$ **f)** $2(2n - 5)$

11. a) $x(1 + 3x)$ **b)** $y(4 - y)$ **c)** $2(a - 3)$
   d) $3b(3b - 2)$ **e)** $-5b(1 + 3b)$ **f)** $x(-4 + 3x)$
   g) $2b(5 + 6b)$ **h)** $x(-4 + 9x)$ **i)** $-2x^2(3 + 4x)$

12. a) $5(y + 2)$ **b)** $8(m^2 + 3)$ **c)** $x(6 + 7x)$
   d) $5a(7 + 2a)$ **e)** $9d(5d^2 + 4)$ **f)** $7b(7b + 1)$
   g) $3x(x + 2)$ **h)** $4y(2y + 1)$ **i)** $5p^2(p + 3)$
   j) $8m(3m + 2)$ **k)** $6a(2 + 3a)$ **l)** $-7y(4y + 5)$

14. a) $6(6x^2 + 2x - 1)$ **b)** $x(8x + 5x^2 + 1)$ **c)** $a(a^2 + 12a - 4)$
   d) $3x(3x + 2x^2 - 4)$ **e)** $8y(y - 4 + 3y^2)$ **f)** $16x(x - 2 + x^2)$

15. a) $3(3b^2 - b + 4)$ **b)** $y(13y^2 + 6y + 3)$
   c) $4x^4(x^2 + 8x + 12)$ **d)** $12y^2(y^2 - y + 2)$
   e) $a(10a^2 + 17a + 18)$ **f)** $5z(2z^3 - 3z + 6)$

16. a) $A = 2\pi r(r + h)$
   c) i) 534.1 cm^2 **ii)** 31.42 cm by 17.00 cm

17. a) 100π cm^2 **b)** πr^2 cm^2 **c)** $(100\pi - \pi r^2)$ cm^2
   d) $\pi(100 - r^2)$ **e)** 235.6 cm^2

7.3 Mathematical Modelling: Round Robin Scheduling, page 360

1. a) 10   **b)** 15   **c)** 21

2. $g = \frac{1}{2}n(n - 1)$

3. a) 10   **b)** 15   **c)** 21

4. $g = \frac{1}{2}(n^2 - n)$

---

6. a) They are identical.
b)

Number of Teams	Number of Games
8	28
9	36
10	45
11	55
12	66

7.

Number of Teams	Number of Games
8	28
9	36
10	45
11	55
12	66

9. a) i) 9   **ii)** 36
   b) i) 12   **ii)** 66

Practise Your Skills, page 365

1. a) -8 **b)** -8 **c)** -17 **d)** 9   **e)** -2 **f)** -2

2. a) 20   **b)** -28 **c)** 32   **d)** -15 **e)** 45   **f)** -30

3. a) $-4x$ **b)** $-4x$ **c)** $-3a$ **d)** $-3a$ **e)** a **f)** $-a$

7.4 Exercises, page 366

1. a) i) $(x + 4)(x + 3)$ **ii)** $x^2 + 7x + 12$
   b) i) $(x + 3)(x + 3)$ **ii)** $x^2 + 6x + 9$
   c) i) $(x + 5)(x + 1)$ **ii)** $x^2 + 6x + 5$

2. a) i) $(x + 5)(x + 2)$ **ii)** $x^2 + 7x + 10$
   b) i) $(x + 6)(x + 3)$ **ii)** $x^2 + 9x + 18$

3. a) i) $(x + 1)(x + 2)$ **ii)** $(x + 2)(x + 6)$ **iii)** $(2x + 3)(2x + 5)$
   iv) $(2x)(x + 3)$ **v)** $3x(x + 1)$ **vi)** $(3x + 2)(2x + 1)$
   b) i) $x^2 + 3x + 2$ **ii)** $x^2 + 8x + 12$ **iii)** $4x^2 + 16x + 15$
   iv) $2x^2 + 6x$ **v)** $3x^2 + 3x$ **vi)** $6x^2 + 7x + 2$

4. a) i) $(x + 7)(x + 3)$ **ii)** $(x + 2)(2x)$ **iii)** $(a + 2)(a + 4)$
   iv) $(x + 2)(x + 8)$ **v)** $(x + 3)(3x)$ **vi)** $(2m + 5)(m + 8)$
   b) i) $x^2 + 10x + 21$ **ii)** $2x^2 + 4x$ **iii)** $a^2 + 6a + 8$
   iv) $x^2 + 10x + 16$ **v)** $3x^2 + 9x$ **vi)** $2m^2 + 21m + 40$

5. a) $x^2 + 5x + 4$ **b)** $n^2 + 7n + 10$
   c) $a^2 + 6a + 9$ **d)** $y^2 + 6y + 5$

6. a) $a^2 - a - 2$ **b)** $b^2 - 2b - 15$ **c)** $n^2 - 5n + 6$
   d) $y^2 + y - 20$ **e)** $b^2 - 3b - 18$ **f)** $a^2 - 16a + 60$
   g) $z^2 - 11z + 30$ **h)** $b^2 + 15b + 50$

7. a) 1   **b)** 12   **c)** -10 **d)** -6

8. a) $2a^2 - 3a - 2$ **b)** $9b^2 + 6b + 1$ **c)** $2x^2 - 7x + 6$
   d) $4x^2 + 9x + 5$ **e)** $6a^2 - 5a - 6$ **f)** $15x^2 + 14x - 8$

9. a) $x^2 + 10x + 21$ **b)** $x^2 - 10x + 21$
   c) $x^2 + 4x - 21$ **d)** $x^2 - 4x - 21$

10. a) $6a^2 + 7a + 2$ **b)** $6a^2 - 7a + 2$
   c) $6a^2 + a - 2$ **d)** $6a^2 - a - 2$

11. a) $a^2 + 4a + 4$ **b)** $9n^2 + 6n + 1$ **c)** $x^2 - 12x + 36$
   d) $a^2 - 6a + 9$ **e)** $4y^2 - 20y + 25$ **f)** $9b^2 + 30b + 25$

12. a) i) $x^2 + 2x + 1$ **ii)** $x^2 - x - 2$ **iii)** $x^2 + 3x + 2$
   $x^2 + 3x + 2$ $x^2 - 4$ $x^2 + 4x + 4$
   $x^2 + 4x + 3$ $x^2 + x - 6$ $x^2 + 5x + 6$

c) i) $(x + 1)(x + 4)$ **ii)** $(x + 4)(x - 2)$ **iii)** $(x + 2)(x + 4)$
$(x + 1)(x + 5)$ $(x + 5)(x - 2)$ $(x + 2)(x + 5)$
$(x + 1)(x + 6)$ $(x + 6)(x - 2)$ $(x + 2)(x + 6)$

d) i) $(x + 1)(x - 2)$ **ii)** $(x - 2)(x - 2)$ **iii)** $(x + 2)(x - 2)$
$(x + 1)(x - 1)$ $(x - 1)(x - 2)$ $(x + 2)(x - 1)$
$(x + 1)(x)$ $(x)(x - 2)$ $(x + 2)(x)$

13. a) $x^2 - 9$ **b)** $4a^2 - 1$ **c)** $64n^2 - 9$
d) $16a^2 - 9$ **e)** $9x^2 - 4$ **f)** $25x^2 - 1$

15. a) $12x^2 - 36x + 15$ **b)** $9b^2 - 4$ **c)** $20a^2 - 31a - 7$
d) $8a^2 + 65a + 8$ **e)** $4a^2 - 12a + 9$ **f)** $6a^2 - a - 12$

16. a) $6 - 5x + x^2$ **b)** $15 + 8a + a^2$ **c)** $12 + m - m^2$
d) $18 - 3t - t^2$ **e)** $49 - 14x + x^2$ **f)** $49 - x^2$

17. a) $n + 5$ **b)** $x - 4$ **c)** $x - 2$ **d)** $a + 2$
e) $x + 3$ **f)** $t + 5$ **g)** $x + 5, x + 4$ **h)** $a - 7, a - 2$

7.5 Exercises, page 371

1. a) $x^2 + 8x + 10$ **b)** $x^2 + 4x + 10$
c) $x^2 + 4x + 10$ **d)** $x^2 - 8x + 10$

2. a) $2x^2 + 6x$ **b)** $-2x^2 - 6x$ **c)** $2x^2 - 6x$ **d)** $-2x^2 + 6x$

3. a) $5x^2 + 5x + 9$ **b)** $5x^2 + x - 9$
c) $5x^2 - 5x + 1$ **d)** $5x^2 - x - 1$

4. a) $x^2 + 7xy + 10y^2$ **b)** $a^2 - ab - 6b^2$
c) $6m^2 - 5mn + n^2$ **d)** $20x^2 + 7xy - 3y^2$
e) $6r^2 - 17rs - 3s^2$ **f)** $56a^2 + 113ab + 56b^2$

5. a) $3x^2 + 6x - 24$ **b)** $2y^2 - 14y + 20$
c) $-5m^2 + 35m - 60$ **d)** $-t^2 + 4t + 5$

6. a) $2x^2 + 14x + 20$ **b)** $3m^2 + 9m - 12$ **c)** $5x^2 + 20x - 60$
d) $7x^2 - 70x + 175$ **e)** $-3x^2 - 6x + 24$ **f)** $-2x^2 + 8x + 120$

7. a) $2x^2 - 4x + 2$ **b)** $3y^2 + 24y + 48$
c) $-9k^2 - 6k - 1$ **d)** $2t^2 - 12t + 18$

8. a) $x^2 + 6x - 4$ **b)** $x^2 - 2x$ **c)** $2x^2 - 5x - 17$
d) $3x^2 - 10x - 27$ **e)** $-18x + 12$ **f)** $-10x - 13$

9. a) $2x^2 - 10x + 13$ **b)** $2y^2 - 10y - 3$ **c)** $2y^2 - 4y + 10$
d) $2x^2 - 14x + 25$ **e)** $-21y^2 - 22y + 8$ **f)** $-12x^2 + 28x - 8$

10. a) $x^2 + 3x - 1$ **b)** $3x^2 + 13x + 13$

11. a) $x^2 + 8x + 12$ **b)** $10x^2 + 23x + 12$ **c)** $13x^2 + 37x + 20$

7.6 Exercises, page 376

1. a) 2, 4   **b)** $-2, -4$ **c)** 2, 6   **d)** $-2, -6$ **e)** $-2, 6$ **f)** $-6, 2$

2. a) i) $x^2 + 6x + 8$ **iii)** $(x + 2)(x + 4)$
b) i) $x^2 + 8x + 7$ **iii)** $(x + 1)(x + 7)$
c) i) $x^2 + 8x + 12$ **iii)** $(x + 2)(x + 6)$

3. a) i) $(x + 1)^2$ **ii)** $(x + 1)(x + 2)$
iii) $(x + 1)(x + 3)$ **iv)** $(x + 1)(x + 4)$
d) $x^2 + 6x + 5, x^2 + 7x + 6, x^2 + 8x + 7$

4. a) $(x + 2)(x + 5)$ **b)** $(x + 1)(x + 10)$
c) $(x + 6)(x + 1)$ **d)** $(t + 1)(t + 8)$

5. a) i) $(x + 3)(x + 4)$ **ii)** $(x - 3)(x - 4)$
iii) $(x + 5)^2$ **iv)** $(x - 5)^2$

6. a) i) $(x - 4)(x + 5)$ **ii)** $(x - 5)(x + 4)$
iii) $(x - 4)(x + 6)$ **iv)** $(x - 6)(x + 4)$

7. a) $(x - 2)^2$ **b)** $(x - 8)(x - 3)$
c) $(x - 6)(x - 3)$ **d)** $(x - 9)(x - 2)$
e) $(x - 18)(x - 1)$ **f)** $(x - 9)(x - 5)$
g) $(x - 4)^2$ **h)** $(x - 3)(x - 1)$

8. a) $(x - 1)(x + 6)$ **b)** $(x - 6)(x + 1)$
c) $(q - 1)(q + 7)$ **d)** $(q - 7)(q + 1)$
e) $(x - 2)(x + 10)$ **f)** $(n - 10)(n + 2)$
g) $(a - 2)(a + 12)$ **h)** $(a - 12)(a + 2)$

9. a) $(x - 4)(x - 2)$ **b)** $(x + 3)(x + 6)$
c) $(a - 9)(a - 2)$ **d)** $(m + 4)^2$
e) $(n - 5)^2$ **f)** $(n - 10)(a - 3)$
g) $(p + 8)^2$ **h)** $(y - 7)(y - 6)$
i) $(x + 7)(x + 8)$ **j)** $(x - 14)(x + 4)$
k) $(x + 1)(x + 5)$ **l)** $(a + 2)(a + 6)$
m) $(m + 3)^2$ **n)** $(x - 2)(x + 1)$
o) $(x - 3)(x - 2)$

10. a) $(r - 7)(r - 2)$ **b)** Not possible
c) Not possible   **d)** $(m - 5)(m - 4)$
e) $(k - 5)(k - 3)$ **f)** Not possible
g) $(a - 5)(a + 3)$ **h)** $(m + 4)(m + 5)$
i) Not possible

11. a) $(x - 1)(x + 8)$ **b)** $(a - 2)(a + 7)$
c) $(r - 9)(r + 4)$ **d)** $(a - 9)(a + 5)$
e) $(n - 9)(n + 6)$ **f)** $(m - 8)(m + 6)$
g) $(k - 9)(k + 7)$ **h)** $(x - 10)(x + 3)$
i) $(a - 9)^2$ **j)** $(m + 11)^2$
k) $(t - 3)(t + 1)$ **l)** $(n + 6)(n + 7)$
m) $(x - 15)(x - 2)$ **n)** $(c - 6)(c - 5)$
o) $(m + 11)(m - 5)$

12. a) 2750   **b)** $(x - 9)(x - 4)$ **c)** 2750

13. a) i) $(x + 1)^2, (x + 2)^2, (x + 3)^2$
ii) $(x - 1)^2, (x - 2)^2, (x - 3)^2$
c) i) $x^2 + 8x + 16, x^2 + 10x + 25$
ii) $x^2 - 8x + 16, x^2 - 10x + 25$

14. a) $(x + 9)(x + 7)$ **b)** Not possible   **c)** Not possible
d) $(t + 8)(t + 3)$ **e)** $(n - 7)(n - 5)$ **f)** Not possible

15. e) $y = 0.25x^2$

16. a) 7   **b)** -12 **c)** 7   **d)** 6   **e)** 3   **f)** 1

17. $x^2 + 11x + 30, x^2 + 13x + 30, x^2 + 17x + 30, x^2 + 31x + 30,$
$x^2 - 11x + 30, x^2 - 13x + 30, x^2 - 17x + 30, x^2 - 31x + 30$

Practise Your Skills, page 382

1. a) 1^2 **b)** 2^2 **c)** 3^2 **d)** 4^2 **e)** 5^2 **f)** 6^2
g) 7^2 **h)** 8^2 **i)** 9^2 **j)** 10^2 **k)** 11^2 **l)** 12^2
m) 13^2 **n)** 14^2 **o)** 15^2 **p)** 20^2 **q)** 25^2 **r)** 30^2

2. a) $(3x)^2$ **b)** $(8m)^2$ **c)** $(11y)^2$ **d)** $(10x)^2$ **e)** $(9a)^2$

7.7 Exercises, page 383

1. a) $(x - 7)(x + 7)$ **b)** $(b - 11)(b + 11)$
c) $(m - 8)(m + 8)$ **d)** $(9 - x)(9 + x)$
e) $(y - 10)(y + 10)$ **f)** $(x - 6)(x + 6)$
g) $(2 - y)(2 + y)$ **h)** $(4 - t)(4 + t)$

2. a) $(10m - 7)(10m + 7)$ **b)** $(8b - 1)(8b + 1)$
c) $(11a - 10)(11a + 10)$ **d)** $(6b - 5)(6b + 5)$
e) $(5p - 9)(5p + 9)$ **f)** $(12m - 1)(12m + 1)$
g) $(3x - 11)(3x + 11)$ **h)** $(5q - 1)(5q + 1)$

3. a) $(11m - 2)(11m + 2)$ **b)** $(2x - 13)(2x + 13)$
c) $(3x - 20)(3x + 20)$ **d)** $(8 - 3b)(8 + 3b)$
e) $(1 - 10m)(1 + 10m)$ **f)** $(7 - 2p)(7 + 2p)$
g) $(8s - 3)(8s + 3)$ **h)** $(10 - 7y)(10 + 7y)$

4. a) i) $(4s - 3)(4s + 3)$ **ii)** $(2x - 7)(2x + 7)$
iii) $(9a - 5)(9a + 5)$ **iv)** $(10 - 11d)(10 + 11d)$
v) $(5 - 6q)(5 + 6q)$ **vi)** $(25 - 9z)(25 + 9z)$
vii) $(7 - 20n)(7 + 20n)$ **viii)** $(2 - 15f)(2 + 15f)$

5. a) $\pi(320)^2$ **b)** $\pi(150)^2$
c) $\pi(320)^2 - \pi(150)^2$ **d)** $251\,013$ cm^2

6. a) 53 **b)** 197 **c)** 128 **d)** 381

7. a) 1352 cm^2 **b)** 2220 cm^2 **c)** $6x + 9$

8. a) $(x)(10 - x)$; $(10 - x)(10)$; $100 - x^2$
b) $x(10 - x)$; $x(10 - x)$; $(10 - x)(10 - x)$; $100 - x^2$
c) $10 - x$; $10 + x$; $100 - x^2$

9. a) $2(x - 3)(x + 3)$ **b)** $3(2x - 1)(2x + 1)$
c) $5(1 - 3x)(1 + 3x)$ **d)** $4(5 - x)(5 + x)$
e) $6(2 - 5x)(2 + 5x)$ **f)** $2(5 - 3x)(5 + 3x)$
g) $3(4x - 1)(4x + 1)$ **h)** $3(21x^2 - 16)$

Practise Your Skills, page 389

1. a) 3 **b)** -12 **c)** 7 **d)** -8

7.8 Exercises, page 389

1. a) -3 or 3 **b)** $-5, 5$ **c)** $-1, 1$ **d)** $-10, 10$
e) -5 or 5 **f)** 0 **g)** $-4, 4$ **h)** $-5, 5$

2. a) $-2, 2$ **b)** $-2, 2$ **c)** $-1, 1$ **d)** $-3, 3$ **e)** $-8, 8$ **f)** $-7, 7$

3. a) $-5, -2$ **b)** -4 **c)** $-2, 3$
d) $-7, 2$ **e)** $-4, -3$ **f)** $-3, 8$
g) 5 **h)** $-10, -5$ **i)** $-5, 6$

4. a) $-6, -3$ **b)** 3 **c)** $-7, 6$
d) 2, 8 **e)** 7 **f)** $-5, 8$
g) 7, 9 **h)** 1 **i)** -6

5. a) 4, 5 **b)** 7, 9 **c)** 4
d) $-21, -1$ **e)** $-2, 7$ **f)** $-5, 3$
g) $-2, 9$ **h)** 2, 6 **i)** $-6, 7$

6. a) i) $-2, 3$ **ii)** 1, 6 **iii)** $-5, 4$

7. a) 2, 6; 4
b) Answers may vary. Possible values for c include 7, -9, 15, -20. Corresponding roots are 1, 7; 9, -1; 3, 5; -2, 10.

8. 1 s and 5 s

10. 4.8 cm

11. a) 1.4 s **b)** 2.0 s **c)** 2.5 s

12. a) 4.8 cm **b)** 6.8 cm **c)** 9.6 cm

Testing Your Knowledge, page 391

1. a) $2x^2 - 10x$ **b)** $-12x + 3x^2$ **c)** $x^3 - 3x^2 + 2x$
d) $-12 + 3x - 6x^2$ **e)** $-2x^3 + 14x$ **f)** $12 - 16x + 20x^2$

2. a) $2(x + 7)$ **b)** $3a(4a + 1)$ **c)** $2y(4y - 3)$
d) $25(4 - 3x^2)$ **e)** $3(x^2 - 4x + 2)$ **f)** $x(2x^2 - 3x - 1)$

3. a) $x^2 - 2x - 15$ **b)** $8x^2 + 6x - 5$ **c)** $2x^2 + 15x - 27$
d) $6x^2 - 5x - 4$ **e)** $20x^2 - 40x + 20$ **f)** $x^2 - 1$
g) $x^2 - 25$ **h)** $4x^2 - 1$ **i)** $25c^2 - 9$

4. a) $x^2 + 4x + 4$ **b)** $4x^2 - 4x + 1$ **c)** $16x^2 + 24x + 9$
d) $25 + 10x + x^2$ **e)** $x^2 - 2x + 1$ **f)** $9x^2 + 12x + 4$

5. a) $-9x^2 + 14x$ **b)** $2a - 3$ **c)** $8x^2 - x$
d) $11n^2 - 6n$ **e)** $26x^2 - 72x$ **f)** $x^2 - x - 4$

6. a) $(m + 4)^2$ **b)** $(a - 3)(a - 4)$ **c)** $(y - 4)(y + 2)$
d) $(n - 9)(n + 5)$ **e)** $(m - 5)(m + 3)$ **f)** $(x - 3)(x - 2)$

7. a) $(b + 6)(b - 6)$ **b)** $(x + 7)(x - 7)$
c) $(2 + 3b)(2 - 3b)$ **d)** $(9x + 5)(9x - 5)$

8. a) $4(d^2 - 3)$ **b)** Not possible **c)** $3(x^2 + 2x + 5)$
d) Not possible **e)** Not possible **f)** $(5x + 4)(5x - 4)$

9. a) $-3, 7$ **b)** $-8, 7$ **c)** -4 **d)** $-2, 4$ **e)** $-5, 9$
f) $-5, 4$ **g)** 6, 9 **h)** $-6, 15$ **i)** $-2, -18$

10. a) $(x - 4)(x + 9)$; $(x - 3)(x + 12)$; $(x - 2)(x + 18)$
$x^2 + 0x - 36$, $x^2 - 5x - 36$, $x^2 - 9x - 36$, $x^2 - 16x - 36$,
$x^2 + 35x - 36$, $x^2 - 35x - 36$

12. a) i) $x^2 + 10x + 25$ **ii)** $x^2 - 10x + 25$
$x^2 + 10x + 24$ $x^2 - 10x + 24$
$x^2 + 10x + 21$ $x^2 - 10x + 21$
b) i) $(x + 2)(x + 8) = x^2 + 10x + 16$
$(x + 1)(x + 9) = x^2 + 10x + 9$
$(x)(x + 10) = x^2 + 10x$
ii) $(x - 2)(x - 8) = x^2 - 10x + 16$
$(x - 1)(x - 9) = x^2 - 10x + 9$
$(x)(x - 10) = x^2 - 10x$
c) i) $(x + 6)(x + 4) = x^2 + 10x + 24$
$(x + 7)(x + 3) = x^2 + 10x + 21$
$(x + 8)(x + 2) = x^2 + 10x + 16$
ii) $(x - 6)(x - 4) = x^2 - 10x + 24$
$(x - 7)(x - 3) = x^2 - 10x + 21$
$(x - 8)(x - 2) = x^2 - 10x + 16$

13. a) πr^2 cm^2 **b)** $2r$ cm
c) $4r^2$ cm^2 **d)** $(4r^2 - \pi r^2)$ cm^2
e) $r^2(4 - \pi)$ **f)** Approximately 3.4 cm^2

14. a) πr^2 cm^2 **b)** 100π cm^2
c) $(\pi r^2 - 100\pi)$ cm^2 **d)** $\pi(r + 10)(r - 10)$
e) Approximately 393 cm^2

15. 28 games

Chapter 8 Analysing Quadratic Functions
Preparing for the Chapter, page 396

4. a) $-3x - 13$ **b)** $-5x^2 - 17x$
c) $10x^2 + 10$ **d)** $2x^2 - 2x - 1$

5. a) $3x^2 - 15x$ **b)** $x^2 + x - 12$
c) $x^2 + 12x + 36$ **d)** $x^2 - 10x + 25$
e) $2x^2 + 4x + 2$ **f)** $3x^2 - 12x + 12$
g) $-2x^2 + 16x - 32$ **h)** $-4x^2 - 4x - 1$

6. a) 6.5 **b)** -6 **c)** 11.5 **d)** 2.25 **e)** -10.25 **f)** -1.25

7. a) -1 **b)** -16 **c)** 19 **d)** 6 **e)** 1 **f)** 5

8. a) $y = -5x + 3$ **b)** $y = \frac{1}{2}x + \frac{7}{4}$

c) $y = -\frac{7}{4}x - \frac{5}{4}$ **d)** $y = -\frac{1}{2}x + 3$

e) $y = \frac{3x}{2}$ **f)** $y = -\frac{1}{4}x + 2$

9. a) ii) 12   **iii)** −4

b) ii) 10   **iii)** None

c) ii) 0   **iii)** 0, 4

10. b) $y = -0.25x^2 - 2$

Practise Your Skills, page 405

1. a) 3   **b)** 6   **c)** 2   **d)** −7   **e)** −1   **f)** −7

2. a) $x - (-1)$ **b)** $x - (-2)$ **c)** $x - (-4)$

d) $x - (-7)$ **e)** $x - (-9)$ **f)** $x - (0)$

8.1 Exercises, page 406

6. a) $a = 1, p = -3, q = 0$ **b)** $a = 1, p = 0, q = 4$

c) $a = 1, p = 3, q = 0$ **d)** $a = 1, p = 0, q = -4$

e) $a = 5, p = 0, q = 0$ **f)** $a = -1, p = 2, q = 0$

g) $a = \frac{1}{4}, p = 0, q = 0$ **h)** $a = -3, p = 0, q = 0$

7. a) ii   **b)** iv   **c)** i   **d)** iii

8. a) i   **b)** iii   **c)** iv   **d)** ii

8.2 Exercises, page 413

1. a) i) Positive   **ii)** (−1, −2)

b) i) Negative   **ii)** (3, 4)

c) i) Negative   **ii)** (1, 3)

2. a) Maximum, 4, 1   **b)** Minimum, −4, 3

c) Minimum, −1, −2   **d)** Maximum, 2, −3

e) Minimum, −8, 6   **f)** Maximum, 3, 2

3. a) i) (5, 6)   **ii)** Up   **iii)** 6, Minimum

b) i) (−2, 4)   **ii)** Up   **iii)** 4, Minimum

c) i) (3, 1)   **ii)** Down   **iii)** 1, Maximum

d) i) (−5, 2)   **ii)** Down   **iii)** 2, Maximum

e) i) (0, 7)   **ii)** Down   **iii)** 7, Maximum

f) i) (−4, 0)   **ii)** Down   **iii)** 0, Maximum

4. a) $y = (x - 0)^2 + 2$ **b)** $y = -2(x - 0)^2 + 5$

c) $y = \frac{1}{2}(x - 4)^2 + 0$ **d)** $y = -(x + 3)^2 + 0$

5. a) Yes, −5   **b)** No   **c)** Yes, 3

6. a) i) 5   **ii)** Minimum   **iii)** 3

b) i) −3   **ii)** Minimum   **iii)** −1

c) i) 4   **ii)** Maximum   **iii)** 1

d) i) −6   **ii)** Maximum   **iii)** −2

e) i) −9   **ii)** Minimum   **iii)** 0

f) i) 7   **ii)** Maximum   **iii)** 0

7. Answers may vary.

a) i) $y = (x + 2)^2 + 4$ **ii)** $y = (x - 3)^2 - 7$

iii) $y = -(x - 5)^2 - 1$

8. a) i) 2, 2, positive   **ii)** 1

iii) $y = (x - 2)^2 + 2$

b) i) −2, −1, negative   **ii)** −2

iii) $y = -2(x + 2)^2 - 1$

c) i) 1, 1, positive   **ii)** 0.5

iii) $y = 0.5(x - 1)^2 + 1$

9. a) i) (3, −5)   **ii)** Up

b) i) (6, 13)   **ii)** Up

c) i) (−2, 10)   **ii)** Up

d) i) (−1, 8)   **ii)** Down

Practise Your Skills, page 422

1. a) $x^2 + 10x + 25$ **b)** $x^2 + 4x + 4$

c) $x^2 - 10x + 25$ **d)** $x^2 + 6x + 9$

e) $-x^2 + 12x - 36$ **f)** $4x^2 - 8x + 4$

g) $-3x^2 - 30x - 75$ **h)** $2x^2 - 12x + 18$

8.3 Exercises, page 422

1. a) $y = x^2 - 6x + 14$ **b)** $y = -x^2 - 2x + 3$

c) $y = 2x^2 - 12x + 18$ **d)** $y = -2x^2 - 4x$

3. b) i) −6, 2, (−2, −16)   **ii)** −8, 0.25, (−3.9, 68.1)

iii) 0.5, 5, (2.75, −10.125)   **iv)** −1, 2, (0.5, −13.5)

5. a) i   **b)** (3, 1)

6. a) Yes   **b)** No   **c)** No

8. a) −2, 3   **b)** −4, 2   **c)** −7, 0   **d)** 5

9. b) $x > 5, x < -5$ **c)** $y > 11$ **d)** $-25 < y < -16$

10. b) $1 < x < 9$ **c)** $x < 1, x > 9$ **d)** $2 < x < 8$

11. b) $-2 < x < 2$ **c)** $y < -9$ **d)** $7 < y < 16$

12. b) $x < 1, x > 3$ **c)** $y > 20$ **d)** $20 < y < -12$

13. a) i) 0.845, 3.155   **ii)** −4, Minimum

b) i) 1.177, 3.823   **d)** −3.5, Minimum

14. c) The line $y = 4x + 3$

15. d) i) (−3, −6)   **ii)** (3, −6)   **iii)** (−2, −1)   **iv)** (2, −1)

v) (−1, 2)   **vi)** (1, 2)   **vii)** (0, 3)

8.4 Exercises, page 429

1. a) 12 m²   **b)** 4 m   **d)** 12.25 m²   **e)** Square

2. b) 51.375 m   **c)** 4.1 s

3. b) 86.5 m   **c)** 3.06 s

4. b) 35 m   **c)** 8.85 m

5. b) 9 A   **c)** 40.5 W

6. a) 0.71 s   **b)** 0.41 s   **c)** 1.12 s

7. a) 255 m   **b)** 500 m   **c)** −625 m   **e)** 18 s

8. b) Approximately (15.32, 1173)

c) 12 s

9. 0.44 s

10. 95 cm

11. 76 cm

Testing Your Knowledge, page 436

2. a) $y = (x + 2)^2 + 3$ **b)** $y = 2(x - 1)^2$ **c)** $y = -x^2 - 4$

3. a) i) -8 **ii)** Maximum **iii)** -5
   b) i) -9 **ii)** Minimum **iii)** 2
   c) i) 7.5 **ii)** Maximum **iii)** 3
   d) i) 7 **ii)** Minimum **iii)** 0
   e) i) 12 **ii)** Minimum **iii)** 8
   f) i) 0 **ii)** Maximum **iii)** 7

4. a) i) $(0, 8)$ **ii)** $x = 0$ **iii)** $-4, 4; 8$ **iv)** 8
   b) i) $(-2, -3)$ **ii)** $x = -2$ **iii)** $-3, -1; 9$ **iv)** -3

5. a) $y = -x^2$ **b)** $y = \frac{1}{2}(x - 4)^2$
   c) $y = 2(x - 4)^2 - 1$ **d)** $y = -2(x + 4)^2 + 1$

6. a) i) $(-7, 0)$ **ii)** $-7; 147$
     iii) Up **iv)** 0
   b) i) $(3, 4)$ **ii)** $1.59, 4.41; -14$
     iii) Down **iv)** 4
   c) i) $\left(\frac{1}{2}, -\frac{3}{4}\right)$ **ii)** None; -1
     iii) Down **iv)** $-\frac{3}{4}$

7. a) Yes **b)** Yes **c)** Yes

8. a) i) $(3, -4)$ **ii)** $1, 5$
   b) i) $(2, -13)$ **ii)** $-0.55, 4.55$
   c) i) $(3, 7)$ **ii)** $1.47, 4.53$

9. a) $2 < x < 4$ **b)** $y > -13$ **c)** $-5 < y < 7$

10. b) 42.55 m **c)** 1 s and 5 s **d)** 45 m

11. b) 4.9 m **c)** 5.2 m **d)** 2.5 m **e)** 12.7 m

Cumulative Review, page 438

1. b) 125 min **c)** $80\ 000$ barrels

2. a) The n-intercept of the graph is not 0, so the relation is a partial variation.
   b) $n = 4.5a - 1$

3. a) $y = 7(x - 7) - 2$
   b) $y = 3(x + 2) - 5$ or $y = 3(x - 2) + 7$

4. a) -6 **b)** $\frac{2}{3}$ **c)** $\frac{1}{6}$ **d)** -2

6. a) $y = 2x + 6$ **b)** $y = \frac{5}{2}x - 5$ **c)** $y = -\frac{12}{5}x + 12$

7. a) $(7, -6)$ **b)** $(3, 11)$ **c)** $(2, 2)$
   d) $(2, 20)$ **e)** $(-4, 5)$ **f)** $(2, -7)$

8. a) $\$1.90$ **b)** $\$1.10$

9. $10\ 500$ V

10. 118 s

11. Answers may vary.
   a) 1 cm $= 1$ m **b)** 8.0 m by 7.0 m
   c) 11.5 m^2 **d)** 9.6 m^2

12. a) $x \doteq 138.5, y \doteq 216.7$ **b)** $x = 13.\overline{3}, y = 3.\overline{3}$
   c) $x = 5$

13. $\$44$

14. 2823 m

15. $10.8°$

16. a) Approximately 26 m **b)** Approximately 12.5 m

18. a)

x	y
−3	27
−2	12
−1	3
0	0
1	3
2	12
3	27

b)

x	y
−6	−18
−4	−8
−2	−2
0	0
2	−2
4	−8
6	−18

19. a) Vertex $(0, 0)$, opens up **b)** Vertex $(0, 0)$, opens up
   c) Vertex $(0, 0)$, opens down **d)** Vertex $(0, 3)$, opens up
   e) Vertex $(0, -5)$, opens up **f)** Vertex $(0, 4)$, opens down

20. a) Vertex $(1, -9)$, zeros -2 and 4
   b) Vertex $(4, -2)$, zeros 2 and 6
   c) Vertex $(2, -5)$, zeros 1 and 3

21. a) $2x + 6$ **b)** $5x + 35$ **c)** $x^2 + 4x$
   d) $-4x - 12$ **e)** $15x^2 + 6x$ **f)** $-10x^2 - 15x$

22. a) $3(x^2 + 8)$ **b)** $-5(x + 2)$ **c)** $7(x^2 + 3)$
   d) $x(x^2 + 2)$ **e)** $2x(x + 8)$ **f)** $-5x(2x - 3)$

23. a) $x^2 + x - 6$ **b)** $x^2 - 3x - 40$ **c)** $10x^2 + 13x - 3$
   d) $4x^2 + 12x + 9$ **e)** $24x^2 - 29x + 7$ **f)** $75 + 10x - x^2$
   g) $x^2 + 7x + 6$ **h)** $x + 1$ **i)** $x^2 + 19x + 8$

24. a) $(x + 1)(x - 10)$ **b)** $(x - 3)(x + 5)$
   c) $(x - 4)(x - 8)$ **d)** $(x + 1)(x + 13)$
   e) $(x - 7)(x + 6)$ **f)** $(x - 8)(x + 3)$
   g) $(x + 6)(x - 6)$ **h)** $(5x + 7)(5x - 7)$
   i) $(10 - 3x)(10 + 3x)$

25. a) $3, -8$ **b)** $-1, 9$ **c)** $\frac{7}{5}, -\frac{7}{5}$

26.

	Vertex	Direction of opening	Transformation
a)	$(3, 5)$	down	reflect in x-axis; translate right 3, up 5
b)	$(6, -20)$	up	expand vertically by a factor of 3; translate right 6, down 20
c)	$(-1, 2)$	up	compress vertically by a factor of one-half; translate left 1, up 2

27. a) IV **b)** III **c)** II **d)** I

28. a) II **b)** III **c)** IV **d)** I

29. b) 200 m^2 **c)** Length $= 20$ m, width $= 10$ m

Glossary

acute angle: an angle measuring less than 90°

acute triangle: a triangle with three acute angles

additive inverses: a number and its opposite; the sum of additive inverses is 0; for example, $+3 + (-3) = 0$

algebraic expression: a mathematical expression containing a variable: for example, $6x - 4$ is an algebraic expression

alternate angles: angles that are between two lines and are on opposite sides of a transversal that intersects the two lines

Angles 1 and 3 are alternate angles.
Angles 2 and 4 are alternate angles.

altitude: the perpendicular distance from the base of a figure to the opposite side or vertex

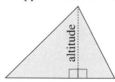

angle: the figure formed by two rays from the same endpoint

angle bisector: the line that divides an angle into two equal angles

approximation: a number close to the exact value; the symbol $\doteq$ means "is approximately equal to"

area: the number of square units needed to cover a region

array: an arrangement in rows and columns

average: a single number that represents a set of numbers; see *mean*, *median*, and *mode*

axis of symmetry: a line that divides a figure into two congruent parts

balance: the result when money is added to or subtracted from an original amount

bar notation: the use of a horizontal bar over a decimal digit to indicate that it repeats; for example, $1.\overline{3}$ means 1.333 333 …

base: the side of a polygon or the face of a solid from which the height is measured; the factor repeated in a power

bias: an emphasis on characteristics that are not typical of the entire population

binomial: a polynomial with two terms; for example, $3x - 8$

bisector: a line that divides a line segment into two equal parts

The dotted line is a bisector of AB.

broken-line graph: a graph that displays data by using points joined by line segments

Calculator-Based Ranger™ unit: a sonic motion detector that collects data and displays them on the calculator screen

capacity: the amount a container can hold

Cartesian plane: See *coordinate plane*

CBR™: See *Calculator-Based Ranger™*

centroid: the point where the three medians of a triangle intersect

circle: the set of points in a plane that are a given distance from a fixed point (the centre)

circumcentre: the point where the perpendicular bisectors of the sides of a triangle intersect

circumcircle: a circle drawn through all three vertices of a triangle; the centre of this circle is the circumcentre

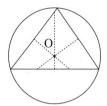

circumference: the distance around a circle, and sometimes the circle itself

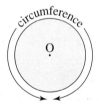

coefficient: the numerical factor of a term; for example, in the terms $3x$ and $3x^2$, the coefficient is 3

collinear points: points that lie on the same line

commission: a fee or payment given to a sales-person, usually a specified percent of the person's sales

common denominator: a number that is a multiple of each of the given denominators; for example, 12 is a common denominator for the fractions $\frac{1}{3}, \frac{5}{4}, \frac{7}{12}$

common factor: a monomial that is a factor of each of the given monomials; for example, $3x$ is a common factor of $15x$, $9x^2$, and $21xy$

commutative property: the property stating that two numbers can be added or multiplied in either order; for example, $6 + 8 = 8 + 6$ and $4 \times 7 = 7 \times 4$

complementary angles: two angles whose sum is 90°

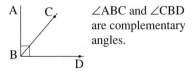

∠ABC and ∠CBD are complementary angles.

composite number: a natural number with three or more factors; for example, 8 is a composite number because its factors are 1, 2, 4, and 8

compound interest: see *interest*; if the interest due is added to the principal and thereafter earns interest, the interest earned is compound interest

cone: a solid formed by a region and all line segments joining points on the boundary of the region to a point not in the region

congruent: figures that have the same size and shape

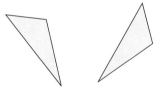

consecutive numbers: integers that come one after the other without any integers missing; for example, 34, 35, 36 are consecutive numbers, so are −2, −1, 0, and 1

constant term: a number

constant of proportionality: the constant k in a direct variation equation of the form $y = kx$; the slope of the graph of this equation

Consumer Price Index (CPI): a measure of the change in the cost of living, compared with a base year; for example, in 1999 the CPI was 110.5 compared to 1992, so you would pay $110.50 to purchase goods and services that cost $100 in 1992

coordinate axes: the x- and y-axes on a grid that represents a plane

coordinate plane: a two-dimensional surface on which a coordinate system has been set up

coordinates: the numbers in an ordered pair that locate a point in the plane

corresponding angles: angles that are on the same side of a transversal that intersects two lines and on the same side of each line

Angles 1 and 3 are corresponding angles.
Angles 2 and 4 are corresponding angles.
Angles 5 and 7 are corresponding angles.
Angles 6 and 8 are corresponding angles.

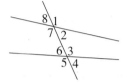

corresponding angles in similar triangles: two angles, one in each triangle, that are equal

cosine: for an acute $\angle A$ in a right triangle, the ratio of the length of the side adjacent to $\angle A$ to the length of the hypotenuse

cube: a solid with six congruent, square faces

cubic units: units that measure volume

cylinder: a solid with two parallel, congruent, circular bases

data: facts or information

database: an organized collection of facts or information, often stored on a computer

degree of a term: the sum of the exponents of the variables in that term
The degree of $3x^2y^3$ is 5.
The degree of $2ab^2$ is 3.
The degree of $-4x$ is 1.
The degree of a constant, such as 7, is 0.

degree of a polynomial: the degree of its highest-degree term
The degree of $x^3 + 2xy + y^2$ is 3.
The degree of $3x - 2y + 5$ is 1.

denominator: the term below the line in a fraction

density: the mass of a unit volume of a substance

diagonal: a line segment that joins two vertices of a polygon, but is not a side

diameter: the distance across a circle, measured through the centre; a line segment through the centre of the circle with its endpoints on the circle

difference: the quantity resulting from the subtraction of one quantity by another

difference of squares: a polynomial that can be expressed in the form $x^2 - y^2$; the product of two monomials that are the sum and difference of the same quantities, $(x + y)(x - y)$

digit: any of the symbols used to write numerals; for example, in the base-ten system the digits are 0, 1, 2, 3, 4, 5, 6, 7, 8, and 9

dilatation: a transformation in which the image is the same shape as the object, but is enlarged or reduced in size

direct variation: a relation that has an equation of the form $y = mx$

Distributive Law: the property stating that a product can be written as a sum or difference of two or more products; for example, for all real numbers a, b, and c: $a(b + c) = ab + ac$ and $a(b - c) = ab - ac$

equation: a mathematical statement that two expressions are equal

equidistant: the same distance

equilateral triangle: a triangle with three equal sides

evaluate: to substitute a value for each variable in an expression and simplify the result

even number: an integer that has 2 as a factor; for example, 2, 4, −6

event: any set of outcomes of an experiment

expanding: multiplying a polynomial by a polynomial

experiment: an operation, carried out under controlled conditions that is used to test or establish a hypothesis

exponent: a number, shown in a smaller size and raised, that tells how many times the number before it is used as a factor; for example, 2 is the exponent in 6^2

expression: a mathematical phrase made up of numbers and/or variables connected by operations

extrapolate: to estimate a value beyond known values

extremes: the highest and lowest values in a set of numbers

factor: to factor means to write as a product; to factor a given integer means to write it as a product of integers, the integers in the product are the factors of the given integer; to factor a polynomial with integer coefficients means to write it as a product of polynomials with integer coefficients

formula: a rule that is expressed as an equation

fraction: an indicated quotient of two quantities

frequency: the number of times a particular number occurs in a set of data

function: a rule that gives a single output number for every valid input number

greatest common factor: the greatest factor that two or more terms have in common; $4x^2$ is the greatest common factor of $8x^3 + 16x^2 - 64x^4$

grouping property of addition (and multiplication): when three or more terms are added (or multiplied), the operations can be performed in any order

hectare: a unit of area that is equal to $10\ 000\ m^2$

height: the distance from the top to the bottom

hexagon: a six-sided polygon

horizontal intercept: the horizontal coordinate of the point where the graph of a function or a relation intersects the horizontal axis

hypotenuse: the side that is opposite the right angle in a right triangle

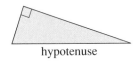

hypotenuse

hypothesis: something that seems likely to be true

identity for addition: a number that can be added to any number without changing the number; 0 is the identity for addition of real numbers

identity for multiplication: a number that can be multiplied by any number without changing the number; 1 is the identity for multiplication of real numbers

image: the figure that results from a transformation

incentre: the point where the three angle bisectors of a triangle intersect

incircle: a circle drawn inside a triangle, with its centre at the incentre and with the radius the shortest distance from the incentre to each side of the triangle

inequality: a statement that one quantity is greater than (or less than) another quantity

integers: the set of numbers… −3, −2, −1, 0, +1, +2, +3,…

interest: money that is paid for the use of money, usually according to a predetermined percent

interpolate: to estimate a value between two known values

intersecting lines: lines that meet or cross; lines that have one point in common

interval: a regular distance or space between values

inverse: see *additive inverses* and *multiplicative inverses*

irrational number: a number that cannot be written in the form $\frac{m}{n}$, where m and n are integers ($n \neq 0$)

isometric view: a representation of an object as it would appear in three dimensions

isosceles acute triangle: a triangle with two equal sides and all angles less than 90°

isosceles obtuse triangle: a triangle with two equal sides and one angle greater than 90°

isosceles right triangle: a triangle with two equal sides and a 90° angle

isosceles triangle: a triangle with at least two equal sides

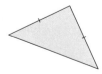

kite: a quadrilateral with two pairs of equal adjacent sides

lattice point: on a coordinate grid, a point at the intersection of two grid lines

legs: the sides of a right triangle that form the right angle

light-year: a unit for measuring astronomical distances; one light-year is the distance light travels in one year

like terms: terms that have the same variables; for example, $4x$ and $-3x$ are like terms

line of best fit: a line that passes as close as possible to a set of plotted points

line segment: the part of a line between two points on the line

line symmetry: a figure that maps onto itself when it is reflected in a line is said to have line symmetry; for example, line l is the line of symmetry for figure ABCD

linear equation: an equation in which the degree of the highest-degree term is 1; for example, $x = 6$, $y = 2x - 3$, and $4x + 2y - 5 = 0$ are linear equations

linear function: a function whose defining equation can be written in the form $y = mx + b$, where m and b are constants

linear system: two or more linear equations in the same variables

mass: the amount of matter in an object

mean: the sum of a set of numbers divided by the number of numbers in the set

median: the middle number when data are arranged in numerical order

median of a triangle: a line from one vertex to the midpoint of the opposite side

midpoint: the point that divides a line segment into two equal parts

mode: the number that occurs most often in a set of numbers

monomial: a polynomial with one term; for example, 14 and $5x^2$ are monomials

multiple: the product of a given number and a natural number; for example, some multiples of 8 are 8, 16, 24,…

multiplicative inverses: a number and its reciprocal; the product of multiplicative inverses is 1; for example, $3 \times \frac{1}{3} = 1$

natural numbers: the set of numbers 1, 2, 3, 4, 5,…

negative number: a number less than 0

non-linear relation: a relation that cannot be represented by a straight-line graph

numeracy: the ability to read, understand, and use numbers

numerator: the term above the line in a fraction

obtuse angle: an angle greater than 90° and less than 180°

obtuse triangle: a triangle with one angle greater than 90°

octagon: an eight-sided polygon

odd number: an integer that does not have 2 as a factor; for example, 1, 3, −7

operation: a mathematical process or action such as addition, subtraction, multiplication, or division

opposite angles: the equal angles that are formed by two intersecting lines

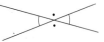

opposites: two numbers whose sum is zero; each number is the opposite of the other

optimization: making a system or object as efficient as possible

order of operations: the rules that are followed when simplifying or evaluating an expression

ordered pair: a pair of numbers, written as (x, y), that represents a point on a coordinate grid

order property of addition (and multiplication): two terms that are added (or multiplied) can be added (or multiplied) in any order

orthocentre: the point where the altitudes of a triangle intersect

outcome: a possible result of an experiment or a possible answer to a survey question

parabola: the name given to the shape of the graph of a quadratic function

parallel lines: lines in the same plane that do not intersect

parallelogram: a quadrilateral with both pairs of opposite sides parallel

partial variation: a relation that has an equation of the form $y = mx + b$

pentagon: a five-sided polygon

per capita: for each person

percent: the number of parts per 100; the numerator of a fraction with denominator 100

perfect square: a number that is the square of a whole number; a polynomial that is the square of another polynomial

perimeter: the distance around a closed figure

perpendicular: intersecting at right angles

perpendicular bisector: the line that is perpendicular to a line segment and divides it in two equal parts

The dotted line is the perpendicular bisector of AB.

pi (π): the ratio of the circumference of a circle to its diameter; π ≐ 3.1416

piecewise linear function: a function whose graph consists of points that lie on line segments, but not on a single line

plane geometry: the study of two-dimensional figures; that is, figures drawn or visualized on a plane

point of intersection: a point that lies on two or more figures

point slope form: the equation of a line in the form $y = m(x - p) + q$, where (p, q) are the coordinates of a point on the line and m is its slope

polygon: a closed figure that consists of line segments; for example, triangles and quadrilaterals are polygons

polynomial: a mathematical expression with one or more terms, in which the exponents are whole numbers and the coefficients are real numbers

population: the set of all things or people being considered

positive number: a number greater than 0

power: an expression of the form a^n, where a is called the base and n is called the exponent; it represents a product of equal factors; for example, $4 \times 4 \times 4$ can be expressed as 4^3

prime number: a natural number with exactly two factors, itself and 1; for example, 3, 5, 7, 11, 29, 31, and 43

prism: a solid that has two congruent and parallel faces (the *bases*), and other faces that are parallelograms

product: the quantity resulting from the multiplication of two or more quantities

proportion: an equation of the form $\frac{a}{b} = \frac{c}{d}$

pyramid: a solid that has one face that is a polygon (the *base*), and other faces that are triangles with a common vertex

Pythagorean Theorem: for any right triangle, the area of the square on the hypotenuse is equal to the sum of the areas of the squares on the other two sides

quadrant: one of the four regions into which coordinate axes divide a plane

quadratic equation: an equation of the form $ax^2 + bx + c = 0$, where $a \neq 0$; for example, $x^2 + 5x + 6 = 0$ is a quadratic equation

quadratic function: a function with defining equation $y = ax^2 + bx + c$, where a, b, and c are constants, and $a \neq 0$

quadrilateral: a four-sided polygon

quotient: the quantity resulting from the division of one quantity by another

radical: the indicated root of a number; for example, $\sqrt{400}$

radical sign: the symbol $\sqrt{}$ that denotes the positive square root of a number

radius (plural, radii): the distance from the centre of a circle to any point on the circumference, or a line segment joining the centre of a circle to any point on the circumference

random numbers: a list of digits in a given range such that each digit has an equal chance of occurring

random sample: a sampling in which all members of the population have an equal chance of being selected

range: the difference between the highest and lowest values (the *extremes*) in a set of data

rate: a certain quantity or amount of one thing considered in relation to a unit of another thing

ratio: the quotient of two numbers or quantities

rational number: a number that can be written in the form $\frac{m}{n}$, where m and n are integers ($n \neq 0$)

ray: a figure formed by a point P on a line and all the points on the line on one side of P

real numbers: the set of rational numbers and the set of irrational numbers; that is, all numbers that can be expressed as decimals

reciprocals: two numbers whose product is 1; for example, $\frac{3}{4}$ and $\frac{4}{3}$ are reciprocals, 2 and $\frac{1}{2}$ are reciprocals

rectangle: a quadrilateral that has four right angles

rectangular prism: a prism that has rectangular faces

rectangular pyramid: a pyramid with a rectangular base

reflection: a transformation that maps every point P onto an image point P' such that P and P' are equidistant from line l, and line PP' is perpendicular to line l

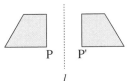

reflex angle: an angle between 180° and 360°

regular polygon: a polygon that has all sides equal and all angles equal

relation: a connection between a pair of quantities, often expressed in words, as a table of values, a graph or an equation

relationship: see *relation*

rhombus: a quadrilateral with four equal sides

right angle: a 90° angle

right circular cone: a cone in which a line segment from the centre of the circular base to the vertex is perpendicular to the base

right triangle: a triangle that has one right angle

rise: the vertical distance between two points

root of an equation: a value of the variable that satisfies the equation

rotation: a transformation in which the points of a figure are turned about a fixed point

rotational symmetry: a figure that maps onto itself in less than one full turn is said to have rotational symmetry; for example, a square has rotational symmetry about its centre O

run: the horizontal distance between two points

sample/sampling: a representative portion of a population

scale: the ratio of the distance between two points on a map, model, or diagram to the distance between the actual locations; the numbers on the axes of a graph

scale factor: the ratio of corresponding lengths on two similar figures

scalene triangle: a triangle with no two sides equal

scatter plot: a graph of data that is a set of points

scientific notation: a way of expressing numbers as the product of a number greater than -10 and less than -1 or greater than 1 and less than 10, and a power of 10; for example, 4700 is written as 4.7×10^3

semicircle: half a circle

similar figures: figures with the same shape, but not necessarily the same size

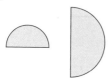

sine: for an acute $\angle A$ in a right triangle, the ratio of the length of the side opposite $\angle A$ to the length of the hypotenuse

slope: describes the steepness of a line or line segment; the ratio of the rise of a line or line segment to its run

slope y-intercept form: the equation of a line in the form $y = mx + b$, where m is the slope and b is the y-intercept

spreadsheet: a computer-generated arrangement of data in rows and columns, where a change in one value results in appropriate calculated changes in the other values

square: a rectangle with four equal sides

square of a number: the product of a number multiplied by itself; for example, 25 is the square of 5

square root: a number which, when multiplied by itself, results in a given number; for example, 5 and -5 are the square roots of 25

statistics: the branch of mathematics that deals with the collection, organization, and interpretation of data

straight angle: an angle measuring 180°

straightedge: a strip of wood, metal, or plastic with a straight edge, but no markings

sum: the quantity resulting from the addition of two or more quantities

supplementary angles: two angles whose sum is 180°

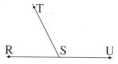

$\angle RST$ and $\angle TSU$ are supplementary angles.

symmetrical: possessing symmetry; see *line symmetry* and *rotational symmetry*

tangent: for an acute $\angle A$ in a right triangle, the ratio of the length of the side opposite $\angle A$ to the length of the side adjacent to $\angle A$

term: of a fraction is the numerator or the denominator of the fraction; when an expression is written as the sum of several quantities, each quantity is called a term of the expression

tetrahedron: a solid with four triangular faces

three-dimensional: having length, width, and depth or height

trajectory: the curved path of an object moving through space

transformation: a mapping of the points of a figure that results in a change in position, shape, size, or appearance of the figure; for example, translations, rotations, reflections, and dilatations are transformations

translation: a transformation in which the points of a figure are moved in a straight line to another position in the same plane

transversal: a line intersecting two or more lines

trapezoid: a quadrilateral that has only one pair of parallel sides

triangle: a three-sided polygon

trinomial: a polynomial with three terms; for example, $3x^2 + 6x + 9$

two-dimensional: having length and width, but no thickness, height, or depth

unit fraction: a fraction that has a numerator of 1

unit price: the price of one item, or the price for a particular mass or volume of an item

unit rate: the quantity associated with a single unit of another quantity; for example, 6 m in 1 s is a unit rate

unlike terms: terms that have different variables, or the same variable but different exponents; for example, $3x$, $-4y$ and $3x^2$, $-3x$

variable: a letter or symbol representing a quantity that can vary

vertex (plural, vertices): the corner of a figure or a solid

vertex of a parabola: the point where the axis of symmetry of a parabola intersects the parabola

vertical intercept: the vertical coordinate of the point where the graph of a function or a relation intersects the vertical axis

volume: the amount of space occupied by an object

whole numbers: the set of numbers 0, 1, 2, 3,…

x-axis: the horizontal number line on a coordinate grid

x-intercept: the x-coordinate of the point where the graph of a function or a relation intersects the x-axis

y-axis: the vertical number line on a coordinate grid

y-intercept: the y-coordinate of the point where the graph of a function or a relation intersects the y-axis

zero of a function: the horizontal coordinate of the point where the graph of a function intersects the horizontal axis drawn through the origin

Zero Principle: the sum of opposites is zero

Index

problem solving
    using algebra, 115, 123,
        138–140, 143, 144, 145,
        146–147, 149, 150, 151
    using graphing calculator,
        140–142, 143–144, 145,
        147
solving by
    elimination, 131–134, 148
    substitution, 124–128, 148
solving with
    graphing calculators,
        116–119, 140–142, 148
    graphs, 115–116
    tables, 122–123

M

Maps, 193, 198, 203, 204, 213,
    215

Mathematical model
    definition, 2

Maximum value or minimum
    value of a quadratic function,
    334, 340, 409–411, 435
    determining with technology,
    338–339

Measures of volume, in cubic
    units and capacity units,
    33–35

Misleading diagrams, 99, 187,
    204, 205, 209–210

Mondrian, Piet, 48, 49

Monomial, multiplying a
    polynomial, 350–351, 369

Motion detector, collection
    and analysis of data, 36–39,
    321–323

Multiplying
    in arithmetic compared to
        expanding in algebra, 355
    of polynomial by monomial,
        350–351, 369
    of two binomials, 362–365,
        369, 370, 391

Murden, Teri, 203

N

Naismith, James, 192

Non-linear function, 15, 16, 46
    definition, 16
    see also Quadratic functions,
        Parabolas

Non-proportional relationships,
    164, 166, 167

O

O'Neal, Shaquille, 212

Output number, 15, 53

P

Parabolas, 302
    congruence of, 398–405
    constructing, 303, 304–305,
        311, 444–446
    estimating best fit, 315–316,
        324, 397
    properties, 304, 340
    shape, 304
    step pattern, 305–308, 340
    vertex, 303, 304, 332, 340,
        409–411
    see also Quadratic functions

Partial variation, 56–58, 60,
    107, 164
    definition, 58

Pashby, Dr. Tom, 26

Payette, Julie, 169, 172

Pelletier, Annie, 426–428

Percent
    expressing as a decimal, 163
    expressing as a fraction, 163
    problems involving, 160, 162,
        173, 174, 189–191

Perimeter of rectangles, 5, 7, 14

Piecewise linear functions,
    75–80, 85–89, 93–100
    definition, 76, 107
    extrapolating, 79
    graphing with technology,
        85–89, 97–98

graphing without using
    technology, 75–79, 93, 94,
    97
interpolating, 77
situations involving, 75–84,
    85, 90–92, 93
tables of values, 75, 77, 78, 85
variations of given conditions,
    75, 80, 90, 91

Plotting points and lines using
    The Geometer's Sketchpad, 43

Point of intersection, finding,
    see Linear systems

Point slope form of linear
    equation, 62–65, 107

Polynomial expressions
    expanding, 350–351, 355, 357,
        369–371, 419
    factoring, *see* Factoring
    simplifying, 362–365,
        369–371, 419
    subtracting, 370

Projectile motion, 298, 330–331

Projects
    curve stitching designs,
        444–450
    indoor skiing, 156–158
    medicine doses, 292–296

Proportional reasoning, 167,
    168, 169, 176–181
    inconsistent representations,
        172, 185
    misleading diagrams and
        graphs, 99, 187, 204,
        205, 209–210
    problems involving, 168–169,
        171–175, 180–181,
        183–188, 189–191, 205

Proportional relationships,
    53–54, 164, 165, 167, 168,
    179, 180, 181, 183–188,
    189–191, 192–195
    definition, 211
    problem solving, 168–175,
        179–188

Pythagorean Theorem, 220,
    253, 254, 255, 268, 270, 272,
    274, 275, 286

PHOTO CREDITS AND ACKNOWLEDGMENTS

The publisher wishes to thank the following sources for photographs, illustrations, articles, and other materials used in this book. Care has been taken to determine and locate ownership of copyright material used in the text. We will gladly receive information enabling us to rectify any errors or omissions in credits.

PHOTOS

Cover Bryan Peterson/FPG; **Inside Front Page** Bryan Peterson/FPG; **p. iv** Steve Tomlinson/Tomlinson Photography; **p. v** Joe McBride/Tony Stone Images; **p. vi** C. Cheadle/First Light; **p. viii** T. Stewart/First Light; **p. ix** Dave Starrett; **p. x** Ian Crysler; **p. xi** Pronk&Associates; **p. 2–3** (background), Ian Crysler; **p. 3** (centre), Ian Crysler; **p. 4** (top right), Ian Crysler; **p. 8** Artbase Inc.; **p. 10** Billy Hustace/Tony Stone Images; **p. 12** Artbase Inc.; **p. 13** Walter Hodges/Tony Stone Images; **p. 19** Ian Crysler; **p. 22** Grant Black/First Light; **p. 23** Artbase Inc.; **p. 24** Dave Starrett; **p. 26** (top left), D. Stoecklein/First Light; **p. 26** (centre right), Dr. Tom Pashby Sports Safety Fund; **p. 27** Artbase Inc.; **p. 37** Ian Crysler; **p. 40** Dave Starrett; **p. 45** Ian Crysler; **p. 46** Dick Hemingway; **p. 48** Alan Marsh/First Light; **p. 49** Giraudon/Art Resource, New York; **p. 50** (bottom right), Paul A. Souders; **p. 50** (top right), Douglas Peebles/CORBIS; **p. 50–51** (background), Rick Doyle/CORBIS; **p. 51** (centre left), David Leah/Allsport USA; **p. 51** (top right), Steve Tomlinson/Tomlinson Photography; **p. 51** (bottom right), Paul A. Souders/CORBIS; **p. 52** E.B. Graphics/Tony Stone Images; **p. 54** Artbase Inc.; **p. 59** 95 MUGSHOTS/First Light; **p. 60** Artbase Inc.; **p. 61** Ian Crysler; **p. 68** (centre right), Artbase Inc.; **p. 68** (bottom right), Artbase Inc.; **p. 69** Artbase Inc.; **p. 72** (centre right), Artbase Inc.; **p. 72** (bottom right), Artbase Inc.; **p. 73** Artbase Inc.; **p. 74** (top right), Gary Withey/Bruce Coleman; **p. 74** (centre right), Artbase Inc.; **p. 75** Artbase Inc.; **p. 76** Howard Davies/CORBIS; **p. 78** Artbase Inc.; **p. 81** Artbase Inc.; **p. 82** Steve Myers/International Stock; **p. 84** Artbase Inc.; **p. 85** Artbase Inc.; **p. 86** Bruce Ayres/Tony Stone Images; **p. 90** Dick Hemingway; **p. 92** Artbase Inc.; **p. 93** Jamie Squire/Allsport USA; **p. 97** Steve Tomlinson/Tomlinson Photography; **p. 101** Ian Crysler; **p. 108** Dick Hemingway; **p. 110** Douglas Peebles/CORBIS; **p. 112** (bottom), Joe McBride/Tony Stone Images; **p. 113** (centre), Sanders/The Stock Market/First Light; **p. 113** (top left), Jim Sugar Photography/CORBIS; **p. 114** Ian Crysler; **p. 115** (bottom left), Artbase Inc.; **p. 121** Artbase Inc.; **p. 123** Alan Marsh/First Light; **p. 126** Ian Crysler; **p. 129** The Purcell Team/CORBIS; **p. 136** CP Picture Archive(Frank Gunn); **p. 138** Neal Preston/CORBIS; **p. 139** Artbase Inc.; **p. 140** Dave Starrett; **p. 143** Artbase Inc.; **p. 144** (centre right), Artbase Inc.; **p. 144** (bottom right), Dave Starrett; **p. 145** Artbase Inc.; **p. 147** Joe McBride/Tony Stone Images; **p. 149** Ron Watts/First Light; **p. 150** Artbase Inc.; **p. 151** Gunter Marx/CORBIS; **p. 152** Artbase Inc.; **p. 154** Artbase Inc.; **p. 155** Artbase Inc.; **p. 156** (top left), John Kelly/Image Bank; **p. 156** (bottom left), J. A. Kraulis; **p. 157** (top right), John P. Kelly/Image Bank; **p. 158** (top left), John Kelly/Image Bank; **p. 158** (centre left), Dave Starrett; **p. 162** (top right), Ian Crysler; **p. 163** The Toronto Star/K. Faught; **p. 164** Michael Ventura/International Stock; **p. 168** Dave Starrett; **p. 169** AFP/CORBIS; **p. 170** Michael Herman; **p. 171** (top right), Westlight (c) W. Cody/First Light; **p. 171** (centre right), Artbase Inc.; **p. 171** (bottom right), Artbase Inc.; **p. 173** (top right), Dave Starrett; **p. 173** (centre right), Dave Starrett; **p. 173** (bottom right), Dave Starrett; **p. 175** (centre right), Artbase Inc.; **p. 177** Bill Brooks/Masterfile; **p. 179** UPI/CORBIS-BETTMANN; **p. 183** Artbase Inc.; **p. 184** (top right), Reuters Newmedia Inc./CORBIS; **p. 184** (bottom right), Jonathan Daniel/Allsport USA; **p. 186** Mike Hutmacher/The Wichita Eagle; **p. 188** Copyright, Vix Cox/Peter Arnold, Inc.; **p. 192** (centre right), Artbase Inc.; **p. 197** (centre left), Tery Murphy/Animals Animals; **p. 197** (centre right), Nuridsany et Perennou/Photo Researchers; **p. 197** (bottom), Dave Starrett; **p. 202** (top left), Artbase Inc.; **p. 202** (top right), WorldSat International Inc., 2000/www.worldsat.ca All Rights Reserved.; **p. 202** (centre left), Artbase Inc.; **p. 202** (centre right), S.P.L./Photo Researchers; **p. 202** (bottom left), Andrew Syred/S.P.L./Photo Researchers; **p. 202** (bottom right), R.E. Litchfield/S.P.L./Photo Researchers; **p. 203** (centre left), Bob Jordan/Associated Press AP; **p. 203** (bottom right), Artbase Inc.; **p. 205** (top right), Fred Lum/The Globe and Mail; **p. 206** Dave Starrett; **p. 208** (top right), David Kent Madison/Bruce Coleman Inc.; **p. 208** (centre right), Otto Greule Jr./All-Sport USA; **p. 208** (bottom right), Glen Short/Bruce Coleman Inc.; **p. 209** (top right), Jim Wiley/Ivy Images; **p. 209** (centre right), Jeremy Horner/CORBIS; **p. 210** (centre right), Nathan Bilow/Allsport USA; **p. 211** (centre), Artbase Inc.; **p. 212** (centre left), D. Cooper/The Toronto Star; **p. 212** (bottom left), SuperStock; **p. 213** (bottom), Trans Canada Trail Foundation; **p. 215** (top right), Lester V. Bergman/CORBIS; **p. 217** (centre left), David Schaarf/Peter Arnold, Inc.; **p. 217** (centre right), Joe Bryksa/Winnipeg Free Press; **p. 218–219** (centre right), Artbase Inc.; **p. 218-219** (background), Artbase Inc.; **p. 218-219** (top right), Warren Morgan/CORBIS; **p. 219** (bottom right), Artbase Inc.; **p. 219** (bottom centre), Artbase Inc.; **p. 220** Bettman/CORBIS; **p. 226** (centre left), Ian Crysler; **p. 231** Michael Herman; **p. 237** (centre right), Steve MacEachern; **p. 238** (centre left), C. Cheadle/First Light; **p. 246** (centre right), Dave Starrett; **p. 251** Ian Crysler; **p. 257** Dave Starrett; **p. 274** (centre right), Dave Starrett; **p. 274** (bottom right), Dave Starrett; **p. 276** (centre right), Dave Starrett; **p. 276** (bottom right), Dave Starrett; **p. 277** (top left), Dave Starrett; **p. 278** (bottom), Artbase Inc.; **p. 278** (bottom right), Artbase Inc.; **p. 288** (centre), Artbase Inc.; **p. 288** (centre right), Artbase Inc.; **p. 289** Artbase Inc.; **p. 292** Artbase Inc.; **p. 293** Gary Gladstone/Image Bank; **p. 294** Artbase Inc.; **p. 296** (top left), Artbase Inc.; **p. 296** (top right), Dave Starrett; **p. 298–299** Dave Starrett; **p. 300** Dave Starrett; **p. 301** Bob Seaquist/International Human Powered Vehicle Association; **p. 302** (centre), Albert Normandin/Masterfile; **p. 302** (bottom), Artbase Inc.; **p. 303** Dave Starrett; **p. 305** Ian Crysler; **p. 311** (top right), Ian Crysler; **p. 311** (centre right), Ian Crysler; **p. 314** Ian Crysler;

p. 315 (centre right), Dave Starrett; **p. 318** Dave Starrett; **p. 319** Dave Starrett; **p. 320** WL, Huber/First Light; **p. 321** Ian Crysler; **p. 325** Charlie Palek/Tom Stack & Associates; **p. 328** Artbase Inc.; **p. 329** Artbase Inc.; **p. 330** Pronk&Associates; **p. 337** Wally McNamee/CORBIS; **p. 341** Ian Crysler; **p. 343** Al Harvey/The Slide Farm; **p. 344** T. Stewart/First Light; **p. 345** Ian Crysler; **p. 346** (bottom right), Artbase Inc.; **p. 346-347** (background), Artbase Inc.; **p. 347** (background), Artbase Inc.; **p. 347** (bottom), Canadian Press CP; **p. 348** Ian Crysler; **p. 360** (top right), Canadian Press CP; **p. 360** (background), Artbase Inc.; **p. 362** Ian Crysler; **p. 364** Dave Starrett; **p. 369** Dave Starrett; **p. 387** T. Stewart/First Light; **p. 393**, Artbase Inc.; **p. 394** (centre right), Jon Ferrey/Allsport USA; **p. 395** (top left), Zack Gold/CORBIS; **p. 395** (bottom right), Artbase Inc.; **p. 396** (top right), Dave Starrett; **p. 396** (centre left), Johan Elzenger/Tony Stone Images; **p. 398** Dave Starrett; **p. 404** Dave Starrett; **p. 406** Dave Starrett; **p. 415** Dave Starrett; **p. 425** Artbase Inc.; **p. 426** Amy Sancetta/AP Wide World Photos Inc./Canapress; **p. 429** (bottom right), First Light; **p. 429** (top right), Artbase Inc.; **p. 430** D. P. Hershkowitz/Bruce Coleman; **p. 431** SuperStock; **p. 433** Ezra O. Shaw/Allsport USA; **p. 434** Artbase Inc.; **p. 437** Duomo/CORBIS; **p. 438** Artbase Inc.; **p. 440** Artbase Inc.; **p. 441** Ian Crysler; **p. 443** Artbase Inc.; **p. 444** Dave Starrett; **p. 446** (top left), Dave Starrett; **p. 446** (centre right), Dave Starrett; **p. 447** Dave Starrett; **p. 448** Dave Starrett; **p. 449** Dave Starrett; **p. 450** (top left), Dave Starrett; **p. 450** (centre right), Dave Starrett.

ILLUSTRATIONS

Brian Hughes **p. 196** (bottom), **p. 199**, **p. 200**, **p. 216** (centre), **p. 282** (top left)

Dave McKay **p. 282** (top left)

David Bathurst **p. 146** (top right), **p. 146** (centre)

Greg Stevenson **p. 160–161** (centre right)

Ian Phillips **p. 253**, **p. 267**

Lee Dunnette **p. 194**

Jane Whitney **p. 95**, **p. 96**, **p. 187**, **p. 198** (bottom), **p. 203** (centre right), **p. 240** (centre left), **p. 261** (top), **p. 275** (top right), **p. 277** (centre right), **p. 287** (top left), **p. 287** (centre right), **p. 287** (bottom right), **p. 432** (top right)

Jun Park **p. 197** (top), **p. 255**, **p. 260**, **p. 264**, **p. 271** (centre right), **p. 275** (centre right), **p. 439** (bottom), **p. 298–299**

Margo Davies Leclair **p. 11**, **p. 47** (top right), **p. 56**, **p. 109**, **p. 115** (top), **p. 204** (top), **p. 204** (bottom), **p. 205** (centre right)

Martha Newbigging **p. 193** (centre left)

Michael Herman **p. 14**, **p. 20**, **p. 25**, **p. 33**, **p. 53** (centre), **p. 60** (bottom right), **p. 165** (top right), **p. 226** (top right), **p. 231** (top right), **p. 249**, **p. 270** (bottom right), **p. 275** (bottom right), **p. 383**

Pronk&Associates **p. 4** (centre), **p. 6** (centre), **p. 106**, **p. 215** (bottom), **p. 290**, **p. 335**, **p. 400**

Steve Attoe **p. viii** (bottom right), **p. 32**, **p. 100**, **p. 134**, **p. 137**, **p. 175** (top), **p. 185**, **p. 205** (bottom right), **p. 212** (top right), **p. 308**, **p. 390**, **p. 411**

Steve MacEachern **p. 237** (centre right), **p. 242**, **p. 244**, **p. 245**, **p. 256**, **p. 269**, **p. 270** (centre right), **p. 270** (top right)